Advances in Meat Alternatives

Principles, Engineering, and Safety Challenges

nova
science publishers

www.novapublishers.com

Food Science and Technology Series

Xanthan Gum: Science, Technology, and Applications
Hedayat Hosseini, Amin Abbasi, PhD, Amin Paydareh, Negin Kamali, Tina Jamshidi, Yasaman Kordmaslak, Maryam Ebrahimi and Anahita Ghorbani Tajani, PhD
ISBN: 979-8-89530-628-4
eBook ISBN: 979-8-89530-651-2

Process Engineering in the Obtention and Preservation of Food Bioactive Ingredients
Pedro Aguilar-Zárate, PhD, Romeo Rojas, PhD, Mariela R. Michel, PhD and Guillermo Cristian G. Martínez-Ávila, PhD (Editors)
ISBN: 979-8-89113-985-5
eBook ISBN: 979-8-89530-043-5

The Cassava Crop: Cultivation, Potential Uses, and Food Security
Pedro Manuel Villa, PhD, Rajesh Jha, PhD, Ima Mulyama Zainuddin, PhD and Alexis Manuel Faneite Noguera (Editors)
ISBN: 979-8-89113-844-5
eBook ISBN: 979-8-89113-931-2

Food Byproducts: Valorization Through Nutraceutical Production
Sandra Teresita Martín-del-Campo, PhD, Anaberta Cardador-Martínez, PhD and Jessica del Pilar Ramírez-Anaya, PhD (Editors)
ISBN: 979-8-88697-612-0
eBook ISBN: 979-8-88697-876-6

Bioprospecting of Natural Compounds in Food, Pharmaceutical and Biomedical Science
Siddhartha Pati, PhD, Tanmay Sarkar, PhD and Dibyait Lahiri, PhD (Editors)
ISBN: 979-8-88697-361-7
eBook ISBN: 979-8-88697-431-7

Proceedings of BIOSPECTRUM: The International Conference on Biotechnology and Biological Sciences: Biotechnological Intervention Towards Enhancing Food Value
Sanket Joshi, PhD, Susmita Mukherjee and Moupriya Nag, PhD (Editors)
ISBN: 978-1-68507-985-7
eBook ISBN: 979-8-88697-069-2

More information about this series can be found at
https://novapublishers.com/product-category/series/food-science-and-technology-series/

Amin Abbasi
Amin Paydareh
Irandokht Vakili
Melika Dashab
Mahdis Nadian
Fahimeh Ebrahimi
Hedayat Hosseini

Advances in Meat Alternatives

Principles, Engineering, and Safety Challenges

nova science publishers
www.novapublishers.com

DOI: https://doi.org/10.52305/MLWF4236

Library of Congress Cataloging-in-Publication Data

ISBN: 979-8-89530-714-4 (Hardcover)
ISBN: 979-8-89530-801-1 (eBook)

Published by Nova Science Publishers, Inc. † New York

Endorsements

"*Advances in Meat Alternatives: Principles, Engineering, and Safety Challenges* is a must-read for anyone invested in the future of food production. The authors' thorough examination of cultured meat, mycoprotein, and plant-based meat provides a clear view of the complex scientific, environmental, and ethical issues we face in the food industry today. This book is a critical contribution to the growing conversation about how we can build a more sustainable and humane food system."

— Dr. Anahita Ghorbani Tajani, PhD,
University of Wyoming, Laramie, USA

"This book offers a timely and comprehensive exploration of the technologies behind alternative proteins. The authors expertly bridge the gap between cutting-edge research and real-world application, making this an indispensable guide for anyone navigating the challenges of scaling cultured and plant-based meat. It's a critical resource for those shaping the future of sustainable food."

— Dr. Kiana Ghafourian, MD,
Shahid Beheshti University of Medical Sciences, Tehran, Iran

"*Advances in Meat Alternatives*" is a game-changing resource that highlights both the promise and the challenges of transitioning to alternative protein sources. The depth of analysis, coupled with the authors' interdisciplinary approach, makes this a key reference for industry professionals, researchers, and policymakers who are working toward a more sustainable and ethical food system.

— Dr. Hossein Samadi Kafil, PhD,
Head of Applied Pharmaceutical Research Center
Tabriz University of Medical Sciences, Tabriz, Iran

Dedication

This book is dedicated to the pioneers of sustainable food systems and the relentless innovators who have dedicated their lives to transforming the future of food production. To the scientists, researchers, and engineers who are pioneering the way for alternative proteins, your vision, and commitment are reshaping the world for future generations.

It is also dedicated to those who continue to seek a balance between food security, animal welfare, and environmental sustainability. May your work inspire others to embrace the possibility of a more ethical, healthy, and sustainable food future.

Lastly, this book is dedicated to our families, whose unwavering support, patience, and understanding have made this journey possible.

Contents

Foreword xix
Sahar Sabahi

Preface xxi

Acknowledgments xxiii

Part 1. Cultured Meat 1

Chapter 1 **The Process and Drivers for Cellular Meat Production** 3
Abstract.. 3
Introduction 3
Definition and Process of Cellular Meat Production 4
Source of Cells.. 5
Growth Media for Proliferation, Merging and Maturation 8
Biomaterial/Scaffold, Electrostimulation and Mechanical Stimulation.................................... 9
Scaling Up and Industrial Production.................... 10
Harvesting and Manufacture of Cellular Meat Products 12
Animal Welfare 14
Regulation, Labelling, and Hazard and Critical Control Points 15
Conclusion.. 17
References 17

Chapter 2 **Structural and Technological Characteristics of Scaffold Manufacturing for Cultured Meat**.............21
Abstract.. 21
Introduction 21

Current Scenario 24
Understanding of Compositional, Cellular,
and Molecular Communications 25
Major Components for Cultured Meat Production 28
Utilization of Cells *29*
Biomaterial Scaffolds and Potential
Manufacturing Technologies *32*
Solvent-Casting/Porogen Leaching *33*
Physical Crosslinking *34*
Chemical Crosslinking *35*
Decellularization *36*
Self-Assembly *38*
Film Stacking *39*
Micromolding Technology *39*
Electrospinning and Other
Fiber-Production Strategies *39*
Additive Manufacturing Methods *40*
Laser Assisted Methods *41*
Extrusion-Based Printing *42*
Combined Technologies *43*
Media and Bioreactors *43*
Challenges and Opportunities 45
Conclusion 46
References 47

Chapter 3 **The Critical Role of Stem Cells in the Production of Cultured Meat** 53
Abstract 53
Introduction 53
Stem Cells for Muscle Tissue 55
Muscle Development *55*
Pluripotent Stem Cells *55*
Muscle Stem Cells *56*
Mesenchymal Stem/Stromal Cells *57*
Genetic Modification *58*
Fat Culture for Cultivated Meat Production 59
Fat Development *59*
Stromal Vascular Fraction (SVF) *60*
Dedifferentiated FAT (DFAT) *61*
Scaffold and Tissue Engineering 61

Production Cost of Cultivated Meat 63
Safety of Cell-Based Food Materials.................................65
Perspectives 66
Conclusion..67
References 68

Chapter 4 **Packaging Requirements and Labeling Design of Cultured Meat**..73
Abstract...73
Introduction 73
Processing and Packaging Requirements of Cultured Meat..76
Maintaining Quality and Shelf Life of Cultured Meat Products...78
Proposed Packaging of the Cultured Meat *79*
Labeling Design to Increase Intention to Purchase Cultured Meat Products..83
Policy and Regulations-Based Labeling Design *84*
Nutrition-Based Labeling *85*
Sensory Quality-Based Labeling *86*
Health Claim-Based Labeling *86*
Environmental-Based Labeling *87*
Conclusion..88
References 88

Chapter 5 **Risk Assessment of Cultured Meat** 95
Abstract...95
Introduction 95
Key Technologies and Possible Risks that Could Be Introduced During Cultured Meat Production.............98
Cell Collection.. *98*
Cell Culture *100*
Biomass Accumulation *102*
Food Processing.. *103*
Risk Assessment Strategies in Cultured Meat Production 104
The Premise of Safety-Standardized Production of Cultured Meat................................ *104*

Risk Assessment Strategies in Cultured Meat Production 104
Routine Biochemical Detection *104*
Allergenicity Testing *106*
Toxicity Testing *106*
Kinetics Testing *109*
Regulations or Legislations of Cultured Meat 110
Conclusion and Future Perspectives 112
References 113

Part 2. Mycoprotein 117

Chapter 1 **Fungi in the Food Chain** 119
Abstract 119
Introduction 119
Fungal Biomass as Food 120
Mycoprotein *120*
Beverages, Baked Goods, and Novel Food Products *121*
Fungi-Based Food Manufacturing 122
Control of Food Properties *122*
Fungi-Based Food Products on the Market *125*
Environmental Impact of Fungi-Based Foods *126*
Summary and Research Needs 126
Conclusion 127
References 128

Chapter 2 **Mycoprotein: A Promising Source of Meat Alternative** 133
Abstract 133
Introduction 133
Application and Formulation of Mycoproteins in Meat Products 135
Environmental Effects 136
Carbon Footprints *136*
Water Footprints *136*
Land Use *137*
Reactive Nitrogen Species (RNS) *137*
Economic Aspects and Market Affordability 138

Health and Safety 139
Effects on Total Cholesterol, Low-Density Lipoprotein (LDL) Cholesterol and High-Density Lipoprotein (HDL) Cholesterol 139
Effects on Satiety 139
Effects on Glycemic Responses 139
Foodborne Pathogens 140
Allergenic Effects 141
Mycotoxin Analysis 141
Waste Use 142
Ethics and Halal Concerns 143
Conclusion and Future Prospects 143
Conclusion 145
References 145

Chapter 3 The Critical Role of Mycoprotein in Supporting Muscle and Metabolic Health 151
Abstract 151
Introduction 151
Mycoprotein 152
Mycoprotein, Dietary Fiber and Cardio-Metabolic Health 153
Mycoprotein and Skeletal Muscle Adaptation 156
Conclusions and Future Directions 158
Key Points 159
References 159

Chapter 4 The Main Manufacturing Methods of Mycoprotein 167
Abstract 167
History of Mycoprotein 167
Methods of Production 168
Solid-State Fermentation (SSF) 169
Submerged Fermentation (SmF) 171
Surface Culture Method (SCM) 172
Nutritional Status 173
Safety Assessment 174
Future Prospects 174
Conclusion 175
References 176

Part 3. Plant Based Meat 179

Chapter 1 **Promising and Sustainable Meat Analogues** 181
Abstract 181
Introduction 181
Meat Analogue 184
Key Ingredients 184
Soy Protein *186*
Mushroom *187*
Wheat Gluten *189*
Egg Albumen *189*
Carbohydrates and Gum *190*
Flavorings and Miscellaneous Ingredients *191*
Major Constraints and Future Challenges 193
Conclusion 194
References 195

Chapter 2 **Plant-Based Meat Alternatives** 197
Abstract 197
Introduction 197
Formulation and Processing of Plant-Based Meat Alternatives 200
Texturization of Protein Enriched Fractions and Formulations *200*
Other Considerations *202*
Recommendations for Future Food-Science Related Research 202
Ingredients *203*
Processing *204*
Encapsulated Ingredients *204*
Consumer Acceptance/Regulations *204*
Safety *205*
Other Considerations *206*
Conclusion 206
References 207

Chapter 3 **Main Factors Affecting the Functionality of Plant-Based Proteins** 213
Abstract 213
Introduction 213
Factors Influencing Plant-Based Proteins 215

Plant Genetics ... 215
Supply Chain Challenges ... 217
Sustainability ... 218
Plant-Based Protein Manufacture ... 219
Nutritional Value of Plant-Based Proteins ... 220
Physical and Physiological Functionality ... 222
Consumer Acceptance ... 223
Climate Change ... 225
Safety and Allergenicity ... 226
Formulating with Plant-Based Protein ... 227
Conclusion ... 228
References ... 229

Chapter 4 **Digestibility and Bioavailability of Plant-Based Proteins** ... 235
Abstract ... 235
Digestibility and Bioavailability ... 235
Different Sources of Plant-Proteins in Meat Analogues ... 237
Cereals ... 238
Legumes ... 240
Oilseeds ... 242
Nuts ... 243
Chia Seeds ... 243
Effect of Main Development Processes on Digestibility and Bioavailability of Meat Analogues ... 244
Green Processing Methods ... 245
Enzymatic Treatment ... 246
Ultrasound and Microwave Treatments ... 247
High-Pressure Process (HPP) ... 250
Electric Field ... 251
Conclusions and Future Perspective ... 253
References ... 255

Chapter 5 **Synthetic Biology Prospects for Plant-Based Meat Production** ... 263
Abstract ... 263
Introduction ... 263
Current Challenges of Plant-Based Meat ... 264
Processing Technologies of Plant-Based Meat ... 266

Structuring Process of Plant-Based Meat 266
Texture Simulation of Plant-Based Meat............... 267
Flavor Improvement of Plant-Based Meat 268
Nutritional Enhancement of Plant-Based Meat 270
Synthetic Biology Approaches for Developing Plant-Based Meat 271
Conclusion...272
References 273

Chapter 6 **The Prospects of 3D Printing in Meat Materials** 277
Abstract..277
Introduction 277
3D Printing of Traditional Meat Materials 279
Challenges of 3D Printing in Meat Materials 280
Poor Printability.. 280
Meat-Related 3D Printers Need Further Development.. 287
Improvement of Nutritional and Sensory Properties.. 289
Opportunities of 3D Printing in Meat Applications.........290
3D Printing of Cultured Meat Products 290
Exploitation of Meat By-Product Ink..................... 291
New Technology Combined with Modeling Methods to Improve Printing Accuracy................ 292
Conclusion..294
References 295

Chapter 7 **Novel Meat Analogue Foodstuffs** 305
Abstract..305
Nutrition of Meat Analogues..305
Beef Burger Products 306
Beef Meatball Products 307
Pork Ham Products 308
Chicken Nugget Products 308
Summary of Nutritional Composition 309
Conclusion..309
References 310

Chapter 8 **Microalgae: A Promising Accessible Ingredient to Produce Meat Analogues** 313
Abstract....................313
Introduction 313
Microalgae as an Alternative Protein Resource....................315
Protein Content and Productivity....................316
Amino Acid Composition....................316
Bioavailability....................317
Physicochemical Properties of Microalgal Protein....................317
Nutraceutical Compounds from Microalgae....................318
Safety and Allergenicity of Microalgae....................320
Approved Microalgae for Foods....................320
Industrial Cultivation of Microalgae 321
Cultivation System and Cultivation Mode....................322
Food Processing Technologies 324
Texturing Techniques....................324
Flavor Adjustment Techniques....................327
Fat Imitation Techniques....................328
The Potential Technological Routes to Produce Microalgae-Based Meat Analogues....................329
Bottlenecks....................329
Conclusion....................336
References 337

About the Authors 345

Index 349

Foreword

The global food landscape is at a pivotal crossroads. As the demand for protein continues to rise, we face an urgent need to reconsider how we produce and consume food. The environmental, ethical, and health challenges posed by conventional meat production have fueled innovation and sparked an exciting new era in food technology. The emergence of meat alternatives, including cultured meat, mycoprotein, and plant-based meat, has the potential to revolutionize the food industry, offering solutions to some of the world's most pressing challenges. This book, *Advances in Meat Alternatives: Principles, Engineering, and Safety Challenges*, provides an essential roadmap to understanding the science and technologies behind these groundbreaking innovations. With contributions from leading experts in the field, it delves into the complexities of cellular agriculture, tissue engineering, and alternative proteins. The work explores not only the scientific and technical aspects of these emerging technologies but also addresses the social, ethical, and regulatory dimensions that are key to their success and widespread acceptance. As we stand on the brink of a major shift in food production, this book serves as an invaluable resource for anyone interested in the future of food. It will inspire researchers, industry professionals, policymakers, and educators to engage in the critical conversations surrounding sustainable food systems. Together, we have the opportunity to shape a future that provides nourishment for all while protecting the planet for future generations. The journey to a more sustainable and ethical food system will not be easy, but with continued research, innovation, and collaboration, the possibilities are endless. This book offers a glimpse into the potential of alternative proteins and the promising future they hold.

Dr. Sahar Sabahi, PhD,
Ahvaz University of Medical Sciences, Ahvaz, Iran

Preface

The global food industry is at a pivotal moment, facing increasing demands for sustainable, ethical, and health-conscious alternatives to traditional meat production. The quest to balance food security, environmental sustainability, and consumer acceptance has catalyzed significant advancements in alternative protein sources. This book, *Advances in Meat Alternatives: Principles, Engineering, and Safety Challenges*, aims to provide a comprehensive exploration of cutting-edge research and development in this transformative field. Spanning cultured meat, mycoprotein, and plant-based alternatives, it offers insights into the principles, methodologies, and challenges associated with these innovations.

Overview of the Content

The book is divided into three major sections, each dedicated to a crucial domain of meat alternatives:

1. Cultured Meat: The first section delves into the science and engineering behind cultured meat, also known as lab-grown or cell-based meat. Chapters in this section explore topics such as the cellular processes involved in meat production, the development of scaffolds and bioreactors, and the role of stem cells. Additionally, it addresses critical considerations such as cost-effectiveness, regulatory frameworks, and consumer acceptance. The section provides an in-depth analysis of the environmental and ethical drivers, as well as the safety and labeling challenges unique to cultured meat.
2. Mycoprotein: The second section focuses on mycoprotein, a promising protein source derived from fungi. It examines its nutritional and functional properties, environmental benefits, and potential applications in meat analogues. Key topics include the

manufacturing methods of mycoprotein, its impact on metabolic and muscle health, and its role in addressing global protein needs. Special emphasis is placed on the health benefits and safety assessments of mycoprotein-based products, as well as their market potential.

3. Plant-Based Meat: The third section investigates plant-based meat alternatives, emphasizing their formulation, processing, and market dynamics. Topics range from the functionality of plant proteins to the use of innovative technologies like 3D printing and synthetic biology. This section also discusses the environmental footprints of plant-based products, regulatory considerations, and strategies to enhance consumer perception and acceptance.

This book is intended to serve as a valuable resource for researchers, industry professionals, policymakers, and educators who are invested in the future of sustainable food systems. We encourage readers to engage critically with the content, drawing on the evidence and perspectives presented to inspire further innovation and collaboration in this rapidly evolving field.

Your feedback is invaluable to us. If you have comments, or insights regarding this book, please do not hesitate to reach out. Together, we can advance the dialogue and drive progress toward a more sustainable and inclusive global food system.

We extend our deepest gratitude to all contributors, reviewers, and supporters who made this publication possible. Let this book be a stepping stone toward achieving a transformative impact in the realm of meat alternatives.

Dr. Amin Abbasi (aminabasi.tbz.med.ac@gmail.com)
Mr. Amin Paydareh (amin.paidare80@gmail.com)
Ms. Mahdis Nadian (mahdisnaadian@gmail.com)
Ms. Melika Dashab (melikadashab1400@gmail.com)
Ms. Irandokht Vakili (irandokht_vakili@yahoo.com)
Dr. Fahimeh Ebrahimi (fahimeh.ebrahimi1373@gmail.com)
Professor Hedayat Hosseini (hedayat@sbmu.ac.ir) (Corresponding author)

Acknowledgments

We would like to express our deepest gratitude to everyone who contributed to the completion of this book. Without their support, expertise, and commitment, this work would not have been possible.

First and foremost, we thank our colleagues, researchers, and industry experts who shared their valuable knowledge and insights. Their contributions have been instrumental in shaping the content of this book and advancing the dialogue surrounding alternative proteins. Special thanks to the scientists and engineers at the forefront of cultured meat, mycoprotein, and plant-based meat technologies, whose dedication continues to drive innovation in this critical field.

We are immensely grateful to our families for their patience, understanding, and unwavering encouragement throughout this journey. Their support has been a constant source of strength and motivation.

Additionally, we acknowledge the publisher, Nova Science Publishers, for their expertise and support in bringing this book to fruition. We also extend our thanks to the editorial team, whose diligence and attention to detail helped shape this work into its final form.

Lastly, we would like to thank all the readers, researchers, and professionals who will engage with this book. Your curiosity and commitment to a more sustainable food system are what drive this field forward.

Part 1. Cultured Meat

Chapter 1

The Process and Drivers for Cellular Meat Production

Abstract

Cell-based meat, also known as lab-grown or synthetic meat, has gained attention due to its potential benefits for the environment, animal welfare, and food security. However, challenges such as establishing cost-effective growth media, scaling production, regulatory issues, and replicating the texture and flavor of traditional meat persist. Consumer acceptance is uncertain, but specific market segments may embrace the product. The high costs and slow commercialization process mean widespread availability is still far off. Meanwhile, the success of plant-based alternatives like the Impossible™ Burger could overshadow cellular meat development, especially as they more readily mimic the attributes of conventional meat.

Keywords: cellular meat, cultured meat, bioreactors, sustainability

Introduction

The provision of safe, affordable protein is critical to human nutrition and food security (Waughray, 2018). Protein, particularly from animal-sources, has been shown to play a key role in brain development and function in all humans, particularly in the elderly and in children (Mann, 2018). Although many vegetarian protein sources are available as alternatives, meat continues to be very attractive to most consumers (Van Der Weele and Tramper 2014; Piazza et al., 2015).

As a global community, the drivers for the agriculture and processing industry are focused on food security. For these reasons, there has been serious consideration of, and advances in, the production of some nutrients from cellular agriculture. At a consumer level, the proponents of cellular meat have focused their communication strategy around the societies' present concerns around meat consumption, including environmental issues, animal welfare, safety and health (Hocquette, 2016). The increase in urbanization of human beings and associated loss of connection to the farm, and how meat is produced, has also led to some consumers' lack of understanding and acceptance of meat production on farms.

Definition and Process of Cellular Meat Production

The definition of cellular meat production is meat made from stem cells, which tries to mimic traditional meat (Hocquette, 2015). It is essentially a 'substitute for meat made from animals. The common terms used for this type of meat production include lab meat, cultured meat, *in vitro* meat, artificial meat, synthetic meat and for those keen to promote the industry, 'clean meat.' Nomenclature for cell-based meat is widely debated, and for an excellent discussion on the various terms and their history and socio-political connotations, see Stephens et al., (2019). Producers, investors and advocates of cell-based meat production prefer to use the term 'clean meat' in order to promote consumer acceptance as it has positive connotations (Bryant and Barnett, 2019). Conversely, those arguing against the term 'clean' meat point out that the term implies that conventional meat production is 'dirty.' The term cell-based meat has been selected for use throughout this chapter due to the linkage with cellular agriculture, although it is recognized that company marketing and brands will choose different nomenclature, which may be dictated by regulators.

The development of cellular meat production from cells involves a number of biological principles regarding how muscle cells are made *in vivo*. Replicating these processes at an industrial scale presents a number of technical challenges (Hocquette, 2016). Hence the development of novel meat and protein products is utilizing ground-breaking technologies designed to meet the issues facing the conventional meat industry (Bonny et al., 2017).

Muscle comprises muscle cells (myocytes), fat cells (adipocytes), connective tissue cells (fibroblasts) and vascular tissue (endothelial cells). Extensive research has been conducted on conditions for allowing muscle, fat and connective tissue cell culture *in vitro*, but the aims of these studies have been very diverse. Due to the complexity of conditions involved for culture of each individual cell type, successful and continued co-culture of these cells has not yet been achieved. Different cell types, particularly stem cells, can be used as a source of material to generate cellular meat, and the following section focuses on muscle cell production. The general stages of development of a mature muscle fiber, from a stem cell, including the growth factors required are shown in Figure 1. The stages of production of meat cells, including the critical decision points, are shown in Figure 2.

Source of Cells

In order to manufacture cellular meat, cells with a large capacity for multiplication must be identified and sourced. Ideally these cells have a self-renewing capacity and can infinitely continue to divide, wherein lies the first challenge. Myogenesis starts after the formation of the embryo, continues through the life of the fetus and is mostly completed at birth (Orzechowski, 2015). At birth, all muscle cells (fibers) in the muscle tissue are fully formed and mature, with the main changes during growth and maturation of the animal being changes in muscle cell width (hypertrophy) and some changes in muscle fiber type. Muscle fibers are multinucleate, being a result of fusion of myoblasts which are uninucleate. In animal skeletal muscle tissue, myosatellite stem cells lie just outside the muscle cell, between the sarcolemma and the basement membrane and are usually in a quiescent state (non-dividing) (Datar and Betti, 2010). When stimulated, for example, when muscle cell damage occurs, these satellite cells can divide and provide new nuclei. These new nuclei, once formed, then are transferred from extracellular, to intracellular within the muscle cell. The majority of myogenesis occurs during the development of the embryo and fetus, but myogenesis can also occur through recruitment of myosatellite cells for muscle regeneration after injury or as an adaptation to workload (Orzechowski, 2015). As an organism ages, the regenerative potential of the myosatellite population rapidly decreases, hence the preference for harvesting myosatellite cells from neonate animals (Datar and Betti, 2010).

A stem cell has a unique ability to develop into many different cell types in the body, and all stem cells can self-renew (make copies of themselves through division) and differentiate (develop into more specialized cells). Stem cells vary widely in their capabilities and the circumstances under which they can perform certain functions. Different types of stem cells, depending on their origin, have differing longevity and persistence in terms of the number of replications *in vivo* (Kadim et al., 2015). Longevity and derivation are crucial factors in muscle cell culture, as a short-lived source of stem cells requires constant replenishment with new tissue from a living animal. The four types of stem cells relevant to this article are embryonic, tissue-specific, induced pluripotent, and myosatellite.

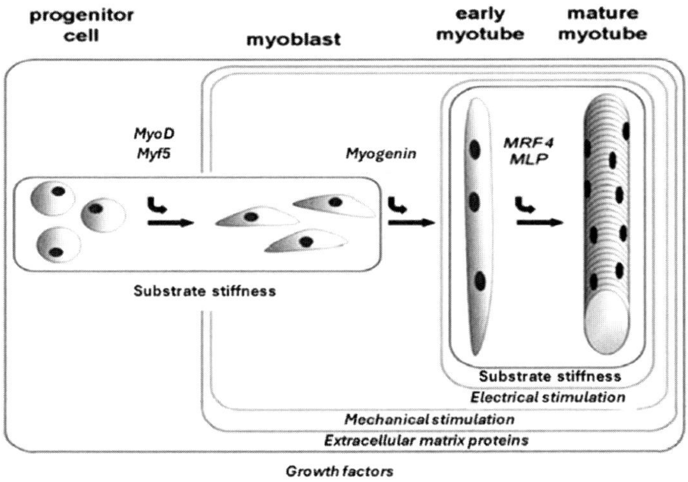

Figure 1. Factors affecting muscle cell proliferation, differentiation and maturation. Substrate stiffness supports myosatellite cell proliferation and myotube maturation, while mechanical stimulation aids myoblast alignment and myotube development. Extracellular matrix proteins and growth factors play crucial roles in both differentiation and maturation stages. Key myogenic regulatory growth factors include MyoD (myoblast determination protein 1), Myf5 (myogenic factor 5), myogenin (MyoG, myogenic factor 4), MRF4 (myogenic factor 6 or herculin), and MLP (muscle LIM protein), each contributing to specific stages of muscle cell development. These factors work in tandem to regulate the complex process of muscle formation and maturation.

Embryonic stem cells are obtained from the inner cell mass of the blastocyst, which forms 3 to 5 days after an egg is fertilized by sperm. Embryonic stem cells are pluripotent (capable of giving rise to every cell type in a fully developed body), have long-term persistence, and may have an unlimited number of divisions (Kadim et al., 2015). Tissue-specific stem cells, also known as adult stem cells, can generate cell types specific to the tissue from which they are derived. Their divisions are typically limited to 50 to 60 cycles (see Kadim et al., (2015) for more detailed description). Induced pluripotent stem cells are cells that have been engineered in the laboratory (*in vitro*) to behave 'like' embryonic stem cells. These have variable, and often unknown, persistence and number of replications. Muscle tissue removed by biopsy from a living mammal can be used to derive 'myosatelite stem cells,' which can only differentiate into a muscle cell. For myosatelite stem cells derived from adult animals, the number of divisions can be as low as 20 *in vitro* (Mouly et al., 2005).

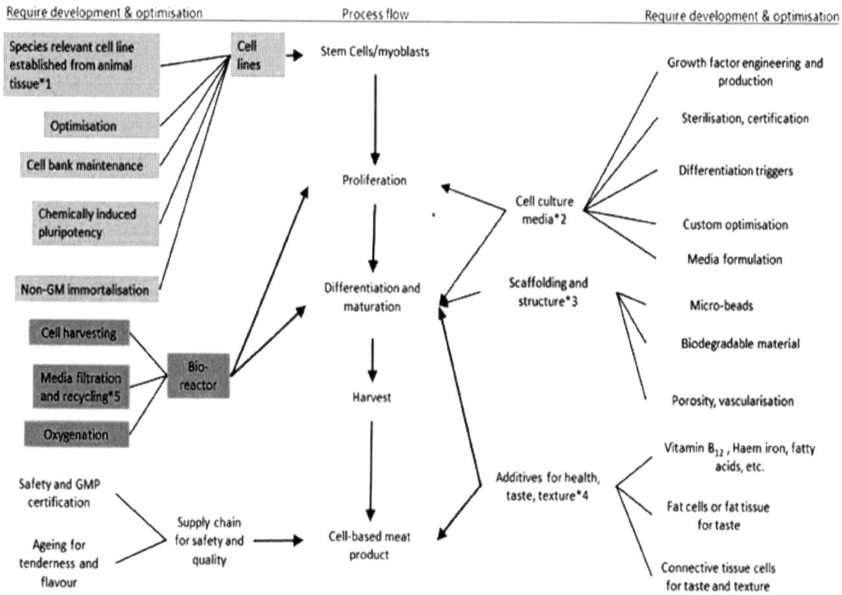

Figure 2. outlines the stages of cellular meat production, highlighting process flow (center), areas for research and optimization (left and right), and key decision points (*). Critical decisions for both small- and large-scale production include: (1) selecting tissue sources for cell derivation, (2) developing growth media for cell differentiation, proliferation, and maturation, (3) designing scaffolds for cell growth, (4) determining the timing for adding nutrients or fat tissue to mimic sensory and nutritional attributes, and (5) deciding on water and nutrient recycling for growth media. These factors are crucial for refining and scaling cellular meat production processes.

The most promising cell type for cellular meat production is the myosatellite cell, as these are the primary adult stem cell for muscle, although the isolation, culturing and maintenance of a high-quality population of myosatellite cells *in vivo* is challenging (Post, 2012). Embryonic stem cells would be an ideal starting source for cellular meat production due to the pluripotent nature, but the proliferation and differentiation can be difficult (Kadim et al., 2015). Mosa Meat in the Netherlands (https://www.mosameat.com/) harvests muscle tissue from a living cow with a biopsy, deriving adult myosatellite stem cells, BioFood Systems (https://www.biofood-systems.com/) wants to produce beef products using bovine embryonic stem cells.

Growth Media for Proliferation, Merging and Maturation

Growth of cells in culture requires nutrients, similar to cells in living tissue. In the absence of a blood supply to provide nutrients and remove waste, cells are immersed in a culture medium that supplies the essential nutrients and growth factors. The growth factors required for proliferation, differentiation, and maturation are typically provided by adding 10% to 20% growth media (Stephens et al., 2018). While the nutrients needed, such as carbohydrates, lipids, amino acids, and vitamins, are relatively straightforward to predict and calculate, different growth factors (and hormones) are required at different stages, as shown in Figure 2. This represents the second series of challenges and decision points in cellular meat production: determining how to supply the appropriate growth media for each stage (decision point 2 in Figure 2).

The culture media needs to be formulated to ensure high rates of cell growth with not only the appropriate level of nutrients, but also the appropriate myogenic regulatory growth factors (myoblast determination protein 1 (MyoD), myogenic factor 5 (Myf5), myogenin factor 4 (MYOG), myogenic factor 6 (MRF4) and muscle LIM protein (MLP); see Figure 1) and hormones (insulin, thyroid hormone, growth hormone). A description of the requirements for growth factors and hormones at different stages for muscle cells from different species is given in Burton et al., (2000). All of these are necessary to culture the cells and to allow them to proliferate and differentiate (Bonny et al., 2017). It should be noted that proliferation (increase in number of cells) is possible only at the satellite cell stage and the mononucleate myoblasts stage. Myotubes (multinucleate fused myoblasts) and myofibers (multinucleate mature muscle cells) do not proliferate and hence the importance of extensive and rapid proliferation at the stem cell and myoblast stages (Datar and Betti, 2010). Traditionally stem cells are cultivated in a medium containing nutrients and fetal, or calf, bovine serum, and the precise composition of this serum is not defined (Burton et al., 2000). Attempts have been made to synthesize serum-free growth media from mushrooms, algae and plants, but there has been limited success with these synthetic growth media (Datar and Betti, 2010; Stephens et al., 2018). *In vitro* cell culture is usually conducted in an aseptic environment, due to risk of contamination which can result in bacterial contamination and cell death (Sanders, 2012; Phelan and May, 2015). Cellular meat production requires a preservative, such as sodium benzoate, added to the growth media, to protect the growing cells from yeast and fungus. Sodium benzoate is a common preservative added to processed meat products (Hoang and Vu, 2016). Furthermore, antibiotics are often added to the growth media of cells in long-term culture to prevent any infection from bacteria (Burton et al., 2000, Renzini et al., 2018, Stephens et al., 2018). Patents for industrial production of cell-based meat state that the process will be achieved without antibiotics (or hormones) (Van Eelen, 2007). It remains to be seen whether antibiotics, or antibacterials, will be routinely or occasionally required during muscle cell culture.

Biomaterial/Scaffold, Electrostimulation and Mechanical Stimulation

Similar to the need for biochemical stimulation by growth factors and hormones described earlier, biomechanical, biophysical and electrical stimulation are also necessary for proliferation, differentiation, maturation and fully functioning skeletal muscle cells.

Application of a biomechanical stimulus to satellite cells can induce the formation of muscle precursor cells (myoblasts, see Figure 2) (Langelaan et al., 2010). Biophysical stimuli are also crucial in the maturation process towards functional muscle cells with a high level of functional sarcomeres (Kosnik et al., 2003). In addition, neuronal activity is required for the development of mature muscle fibers (cells), and this can be simulated through the application of electrical stimulation (Figure 1) (Langelaan et al., 2010).

Mechanical stretch is another biophysical stimulus which can be applied and appears to be crucial in myogenesis, proliferation and differentiation. Mechanical stretch applies to various stages of the muscle cell process and facilitates; alignment of myotubes, fusion of myotubes to form muscle fibers, hypertrophy of both myotubes and muscle fibers, proliferation of myoblasts and activation of satellite cells (Vandenburgh and Karlisch, 1989, Tatsumi et al., 2001).

At the stage of differentiation from stem cells into myotubes, the cells must be attached to a mesh scaffold, such as a collagen meshwork, or to a carrier like microcarrier beads, which can also be made from collagen (Bhat et al., 2015). The myotubes can then fuse into myofibers, provided the appropriate cues are present in the growth media and environment. *In vivo*, the cells attach to the insoluble network of extracellular matrix proteins through integrin receptors located in the sarcolemma, allowing the transmission of applied force to the cytoskeleton (Langelaan et al., 2010). Therefore, *in vitro*, this process needs to be replicated to enable biomechanical and biophysical stimulation. The choice of material for the scaffold is crucial in cellular meat production, as indicated by decision point 3 in Figure 2. The scaffold-based method is limited to producing a cell layer only 100 to 200 μm thick in static culture due to the absence of blood supply and the diffusion limits of nutrients and growth factors in the growth medium (Datar and Betti, 2010).

Scaling Up and Industrial Production

A major challenge in cell-based meat production is to produce the nutrients, hormones and growth factors in large quantities and compatible with human food (Hocquette, 2016). All these compounds will need to be produced by the chemical industry, and the waste and pollution generated will need to be included in the environmental analysis of the effects of cellular meat production.

Identification of a self-replicating linc of satellite cells is important to the process, in order to remove the reliance on animals. As myosatellite stem cells derived from a biopsy are presently the most effective source for cellular meat production, this remains a challenge for the scaling-up process. Availability of cost-effective growth medium for cell proliferation and growth is a critical component of the scale-up process. Bovine fetal calf serum is the preferred growth media, due to its effectiveness, but is prohibitively expensive (Stephens et al., 2018), and it is derived from animals.

For differentiation and, in particular, growth, a continuous supply of nutrients, growth factors, oxygen, and the removal of waste products is required in the culture medium. This is typically achieved at large scale by bioreactors with continuous agitation and flow, allowing the perfusion of required gases and chemicals into and out of the cells. The lack of a vascular system in current muscle cell culture systems, which impedes both nutrient supply and waste removal, currently limits the thickness of muscle cell tissue to 1 to 2 cells thick, or 100 to 200 μm.

Although cellular meat production is generally considered to be most feasible at an industrial scale, some companies (e.g., FM Technologies) are proposing two possible model systems where cellular meat production could occur at a small business or even individual consumer scale. FM Technologies calls this a 'distributive approach to sustainable manufacturing' (https://www.future-meat.com/). It proposes that localized production could be achieved through several models. The first model could be where farmers maintain a population of livestock for periodically deriving a source of myosatellite cells through biopsy. In this model, for example, animals could be kept in backyards, or on animal-friendly urban farms, and serve as living donors (Van Der Weele and Tramper 2014). Alternatively, the model proposed by FM Technologies involves centralized stores of stem cells, which are periodically shipped to local communities. These communities would have small-scale bioreactors to supply small villages or regions, with technological expertise provided from a central point. This distributive approach partially addresses criticism that cellular meat production could be controlled by multinational corporations. Additionally, if production were centralized or limited to specific countries, the environmental costs of transportation could potentially offset any environmental benefits (Mattick et al., 2015a). These models at this stage are purely theoretical and remain to be tested or proven, and the proposed approach may struggle to be cost-effective, sustainable or efficient without involvement of large corporations and multinationals.

Recycling of growth media can significantly reduce the water and nutrient footprint of industrial-scale cellular meat production, as shown in Figure 2 at decision point 5. The management of metabolic waste through disposal, recycling, or upgrading must be considered when designing and implementing a hazard and critical control points (HACCP) procedure (see the 'Regulation, labelling and hazard and critical control points' section) (Stephens et al., 2018). Cell-based meat factories will need to consider constructing on-site treatment or recycling systems to limit emitted pollution, which may be required by state and federal regulations (Mattick et al., 2015a). If recycling is implemented, the emissions from excess nutrients in the waste water could be similar to, or lower than, a poultry operation (Mattick et al., 2015a). Recycling of culture medium can be enabled through the replenishment of utilized nutrients such as glucose, glutamine and other amino acids and carbohydrates and also through removal of waste products such as lactate and ammonia (Moritz et al., 2015). It is also possible that growth factors and cytokines produced by muscle cells can be reused in subsequent culture to stimulate cell growth. It is proposed that recycling would involve cells staying in the reactor and in-line recycling of media, with removal of waste products through chromatographic purification (Moritz et al., 2015). In addition, it is likely that micro-carrier beads can be recycled. The inclusion of waste water recycling in the lifecycle analysis to determine the impact on environment, emissions and water use is an important consideration.

Harvesting and Manufacture of Cellular Meat Products

Obviously the products of muscle cell culture, using collagen scaffold or micro-beads for structure, do not have the same structure as skeletal muscle tissue from animals, and can only be used for processed meat products (Datar and Betti, 2010). In addition, if the scaffold or carriers are edible, they can be included in the cellular meat mix and harvested from the bioreactor without separating cells from scaffold/carrier (Stephens et al., 2018). But this is not always desirable, as the matrix may need to be reused, the matrix may contribute an unacceptable texture, flavor or functionality to the final product or the matrix may be non-edible. In this case, removal of the cells from the matrix can be achieved through chemical means or through environmental manipulation of pH, temperature and so on (van der Weele et al., 2019). Any chemicals used obviously need to pass food safety tests and achieve regulatory approval (see the 'Regulation, labelling and hazard and critical control points' section).

The flavor of meat is derived from more than 750 compounds. Flavor consists of volatiles developed during cooking that contribute to the aroma or odor, which are detected in the retro-nasal cavity, and the taste, which is sensed on the tongue by taste receptors (Watkins et al., 2013; Frank et al., 2016). Meat from whole muscle contains a mixture of fat cells, fatty acids in both membranes and fat cells, fibroblasts, the protein collagen, muscle and endothelial cells, and compounds such as haem-iron, creatine, carnitine, glutamate, and other compounds that uniquely and collectively contribute to the overall flavor experience. The flavor of meat develops during cooking, with certain fatty acids and amino acids prevalent in meat (e.g., linoleic acid and inosine monophosphate) playing a significant role in its unique flavor profile. Whether cell-based meat products will be able to replicate the more than 750 compounds associated with the flavor of whole muscle remains an unanswered question and a significant challenge.

The texture of meat, mouthfeel and breakdown during oral processing contribute to the overall sensory experience. The texture, particularly hardness, of meat changes during cooking, in response to temperature, as proteins denature at different temperatures, causing shrinkage in the muscle at a macro- and micro- level and changes in the gel matrix binding proteins together (Purslow et al., 2016). Adipocytes are the main fat deposits in muscle tissue and are a rich source of aroma, taste and juiciness in the consumer's mouth (Orzechowski, 2015). Stroma, or connective tissue, comprises cells called fibroblasts, which extrude collagen fibrils to form the extracellular matrix of muscle. Both stroma and muscle cell protein integrity have important contributions not only to texture and tenderness, but also to particle breakdown and size in the mouth, and thus are pivotal in the consumer's sensory experience. Furthermore, upon prolonged heating in appropriate conditions, collagen breaks down to gelatine, which has a unique and desirable flavor, as well as contains important amino acids for human muscle function and health (e.g., glutamate) (Arihara, 2006; Toldra et al., 2012).

The tenderness and flavor of whole muscle develop during the ageing or maturation process postmortem. During ageing, proteases such as calpains degrade proteins in the microstructure, particularly the myofibrils (Koohmaraie et al., 1991). Meat ageing occurs in an anaerobic environment over a period of days to weeks, typically by packaging the meat in a sealed bag, removing all air through vacuum application, and storing it at temperatures between −1 and 1°C. Since cell-based meat products currently rely on the formulation of a burger or patty using only cells, the product does not require ageing to improve tenderness. For instance, beef patties are often made from cow beef, which is inherently tough, but once ground or minced, the toughness of the whole muscle becomes irrelevant.

Assuming future success in producing whole muscle products from cells, attention to ageing for optimal tenderness and flavor will be necessary. Therefore, a product made from purely muscle cells, such as cell-based meat, will have a different texture and mouthfeel compared to a product made from whole muscle derived from the post-mortem carcass of a living animal. This presents a challenge in the manufacturing of cell-based meat products and requires the addition of binders, such as carrageenan, collagen powder, xanthan gum, or mannitol (Post, 2012), to simulate the texture of whole muscle.

The production of cell-based meat will be classed as a 'manufactured' meat, as it will involve the addition of ingredients for health, functionality, texture and flavor. This will likely push meat from animals into the premium end of the market, and cell-based meat products could supply the bulk, cheap end of the market if conventional meat products become more expensive and the palatability and versatility of manufactured meats improve (Bonny et al., 2017). The generation of a whole muscle, complete with blood supply, connective tissue, fat cells, muscle cells and associated structure, has neither been achieved in human biology and medicine nor in cellular meat production. Hence, as pointed out by Hocquette (2016), the process of producing meat products from muscle cells should be called cellular muscle production, as it contains only muscle cells. The cell-based production of a whole steak is technologically far more complex and requires technological break-throughs in providing a vascular supply through a thick layer of muscle tissue as well as determining the right growth media for co-culture of muscle, fat, connective tissue and endothelial cells.

Animal Welfare

Ethical issues are increasingly important in our food choices, and the consequence has been the development of societal concerns. According to Cornish et al., (2016) 'the production of food from animals poses many ethical challenges. These societal concerns are varied and encompass the right to kill animals and to prevent any suffering of animals (Hocquette, 2016). Implicit in the prevention of suffering, often advocated by vegans and vegetarians, is the mistaken belief that all animals raised on farms experience suffering. Similarly, some consumers use anthropomorphic considerations and assume that animals in pens, intensive livestock systems, or feedlots undergo suffering. These views are predominantly held by consumers in industrialized, wealthy, Western cultures, while consumers in less industrialized, lower-income economies typically do not share these perspectives (Cornish et al., 2016). If cell-based meat production requires regular or irregular biopsy samples from animals, it remains unclear whether this would alleviate the ethical concerns held by some consumers regarding animal production.

A major limitation in cell-based meat production is the need for a supply of bovine serum for the culture medium. About 50 l of bovine serum is required to make one burger (Mark Post cited on https://www.wired.com.uk), and this volume of serum requires blood from 91 to 333 fetuses (3-month-old fetus has 150 ml of blood, near full-term fetus has 550 ml of blood) (Jochems et al., 2002). Hence until animal-free serum is available for muscle cell culture at industrial scale, the production of cell-based meat products will not meet consumers' demands for animal welfare. Consideration of animal welfare in traditional animal production, and in cell-based meat production, is likely the most polarizing and contentious of the consumer concerns.

Regulation, Labelling, and Hazard and Critical Control Points

Before it can reach the shelves, cell-grown meat will be required to go through regulatory hurdles and paperwork. There are presently 44 companies listed on the Australian clean meat's website (https://www.cleanmeats.com.au), and of these, 19 are in the US. For the cellular meat industry in the US and other countries to succeed, companies need assurance that their product will be responsibly regulated, in the case of the US, by the United States Department of Agriculture (USDA) and the Food and Drug Administration (FDA). Although literature on the regulation of cultured meat has existed since 2013-2014 in both the US and the European Union (EU), it identified that the existing regulations at the time were inadequate to address cell-based meat production without significant development (Stephens et al., 2018). In the US, if the production is defined as explant (expanding existing muscle tissue), it would require an FDA New Animal Drug application, whereas if scaffold-based, it should follow FDA food additive regulations. In the EU, cell-based meat production would fall under novel food regulation via the European Food Safety Authority (EFSA) risk assessment. However, EFSA's novel food regulation excludes genetically modified foods, which may not apply to cell-based meat (Stephens et al., 2018).

In November 2018, the USDA and FDA announced an agreement for a joint regulatory framework for the production of cell-cultured food products derived from livestock and poultry. The agreement outlines the point of transfer as follows. 'FDA oversees cell collection, cell banks, and cell growth and differentiation. A transition from FDA to USDA oversight will occur during the cell harvest stage. The United States Department of Agriculture will then oversee the production and labelling of food products derived from the cells of livestock and poultry' (Piper, 2019; Rollins and Rumley, 2019). This agreement was formalized on 7 March 2019 in a memorandum signed by principals from USDA and FDA laying out the delegation of responsibility (Piper, 2019; Rollins and Rumley, 2019); however the details concerning inspection and the labelling process are still to be worked out by the respective agencies.

In some states in the US (e.g., Missouri), legislation has been introduced and, in some cases, passed, prohibiting cell-based meat companies from using the label 'meat.' However, the United States Department of Agriculture's (USDA) labelling authority supersedes that of individual states, meaning the USDA's future ruling on what cell-based meat can be called will, in theory, be final. Cultured meat has already faced resistance from the US meat industry and is likely to encounter a labelling dispute once it reaches the market. Consequently, lawsuits and legal battles will likely continue in court, regardless of the labelling ultimately approved by the USDA (Piper, 2019).

In Europe, a Brussels agriculture committee has approved a ban on producers of vegetarian food using nomenclature usually deployed to describe meat (Boffey, 2019). Instead proposed that terms such as 'veggie discs' rather than 'veggie hamburger' be used. The proposed terms will now be voted on by the full parliament after May's European elections, before being put to member states and the European Commission. After the vote in the European Parliament on revisions to a food-labelling regulation, it is likely that any introduction of cellular meat products will face similar restrictions in labeling, particularly as Europe tends to be more stringent in banning foods associated with increased chemical use during production, which is likely to be the case with cell-based meat. However, the potential for cell-based meat to offer benefits for animal welfare may influence European regulators to adopt a more lenient stance, given that Europe is generally strongly pro-animal welfare.

In order to develop regulatory guidelines, the procedures for production must be clearly defined. At this early stage, these procedures are not yet established; therefore, regulations will need to evolve alongside the development of documented processes, and multiple pathways for production are likely to emerge. The regulations currently being developed in the USA and EU will be closely monitored, as they may provide guiding principles for regulators in other countries where a cell-based meat production industry is emerging.

Regulatory guidelines, HACCP plans, and auditing will need to address several steps and aspects, including governance at local, state, and federal levels. The initial collection of samples will be governed by animal welfare acts. A HACCP plan is designed as a preventative food safety system in which every step in the manufacturing, storage, and distribution of a food product is scientifically analyzed for microbiological, physical, and chemical hazards. Therefore, for the production and processing of cell-based meat, a HACCP plan will be required to specify auditing procedures for the identification of potential pathogens, physical contamination risks, and the safety of added chemicals. This plan will also include methods to prevent contamination at each HACCP step, which are typically overseen by state or federal government agencies, depending on whether the product is for local consumption or export. Auditing procedures for cell-based meat production will also likely need to incorporate monitoring and quality assurance functions at each stage, testing of genetic stability, and management of metabolic waste through disposal or recycling (Stephens et al., 2018).

Conclusion

This chapter has provided a comprehensive overview of the process and drivers behind cellular meat production. As we explored, cellular meat offers a promising solution to many of the global challenges associated with traditional meat production, including environmental impact, animal welfare, and food security. However, significant technical and economic hurdles remain in making cellular meat a viable and scalable alternative. Key challenges such as the development of cost-effective growth media, efficient scaling of production, and regulatory frameworks continue to require innovative solutions.

Despite these obstacles, the progress made in the field of cellular agriculture is encouraging. With continued advancements in biotechnology, materials science, and regulatory frameworks, the future of cellular meat production looks increasingly promising. As consumer demand for more sustainable and ethically produced food continues to grow, cellular meat could play a critical role in meeting global protein needs while minimizing the environmental and ethical concerns associated with conventional livestock farming.

In conclusion, while cellular meat technology is still in its early stages, its potential to transform the global food system is vast. Ongoing research, innovation, and collaboration across disciplines will be essential in overcoming the challenges ahead, ensuring that cellular meat can be produced safely, sustainably, and economically for consumers worldwide.

References

Arihara, K. 2006. Strategies for designing novel functional meat products. *Meat Science* 74, 219–229.

Bhat, Z. F., Kumar S and Fayaz H 2015. *In vitro* meat production: challenges and benefits over conventional meat production. *Journal of Integrative Agriculture* 14, 241–248.

Bhat, Z. F., Kumar S and Fayaz H 2015. *In vitro* meat production: challenges and benefits over conventional meat production. *Journal of Integrative Agriculture* 14, 241–248.

Boffey, D. 2019. 'Veggie discs' to replace veggie burgers in EU crackdown on food labels.

Bonny, S. P. F., Gardner GE, Pethick DW and Hocquette J-F 2017. Artificial meat and the future of the meat industry. *Animal Production Science* 57, 2216–2223.

Bryant, C. J. and Barnett JC 2019. What's in a name? Consumer perceptions of *in vitro* meat under different names. *Appetite* 137, 104–113.

Burton, N. M., Vierck Jf, Krabbenhoft L, Bryne K and Dodson MV 2000. Methods for animal satellite cell culture under a variety of conditions. *Methods in Cell Science* 22, 51–61.

Cederberg, C., Henriksson M and Berglund M 2013. An LCA researcher's wish list – data and emission models needed to improve LCA studies of animal production. *Animal* 7, 212–219.

Cliceri, D., Spinelli S, Dinnella C, Prescott J and Monteleone E 2018. The influence of psychological traits, beliefs and taste responsiveness on implicit attitudes toward plant- and animal-based dishes among vegetarians, flexitarians and omnivores. *Food Quality and Preference* 68, 276–291.

Cornish, A., Raubenheimer D and McGreevy P 2016. What we know about the public's level of concern for farm animal welfare in food production in developed countries. *Animals* 6, 74.

Crosson, P., Shalloo L, O'Brien D, Lanigan GJ, Foley PA, Boland TM and Kenny DA 2011. A review of whole farm systems models of greenhouse gas emissions from beef and dairy cattle production systems. *Animal Feed Science and Technology* 166–167, 29–45.

Daly, H. E. 1990. Toward some operational principles of sustainable development. *Ecological Economics* 2, 1–6.

Datar, I. and Betti M 2010. Possibilities for an *in vitro* meat production system. *Innovative Food Science & Emerging Technologies* 11, 13–22.

Hopkins, P. D. 2015. Cultured meat in western media: the disproportionate coverage of vegetarian reactions, demographic realities, and implications for cultured meat marketing. *Journal of Integrative Agriculture* 14, 264–272.

Jochems, C. E. A., van der Valk JBF, Stafleu FR and Baumans V 2002. The use of fetal bovine serum : ethical or scientific problem? *Alternatives to Laboratory Animals* 30, 219–227.

Jones, A. D., Hoey L, Blesh J, Miller L, Green A and Shapiro LF 2016. A systematic review of the measurement of sustainable diets. *Advances in Nutrition* 7, 641–664.

Kadim, I. T., Mahgoub O, Baqir S, Faye B and Purchas R 2015. Cultured meat from muscle stem cells: a review of challenges and prospects. *Journal of Integrative Agriculture* 14, 222–233.

Koohmaraie, M., Whipple G, Kretchmar DH, Crouse JD and Mersmann HJ 1991. Postmortem proteolysis in longissimus muscle from beef, lamb, and pork carcasses. *Journal of Animal Science* 69, 617–624.

Kosnik, P. E., Dennis RG and Vandenburgh H 2003. Tissue engineering skeletal muscle. In *Functional tissue engineering* (eds. F Guilak, DL Butler and SA Goldstein), pp. 377–392. Springer-Verlag, New York, NY, USA.

Langelaan, M. L. P., Boonen KJM, Polak RB, Baaijens FPT, Post MJ and van der Schaft DWJ 2010. Meet the new meat: tissue engineered skeletal muscle. *Trends in Food Science & Technology* 21, 59–66.

Lynch, J. and Pierrehumbert R 2019. Climate impacts of cultured meat and beef cattle. *Frontiers in Sustainable Food Systems* 3, doi: 10.3389/fsufs.2019.00005.

Mattick, C. S., Wetmore JM and Allenby BR 2015b. An anticipatory social assessment of factory-grown meat. *IEEE Technology and Society Mag.*

McAuliffe, G. A., Takahashi T and Lee MRF 2018. Framework for life cycle assessment of livestock production systems to account for the nutritional quality of final products. *Food and Energy Security* 7, e00143.

Moritz, M. S. M., Verbruggen SEL and Post MJ 2015. Alternatives for large-scale production of cultured beef: a review. *Journal of Integrative Agriculture* 14, 208–216.

Mouly, V., Aamiri A, Bigot A, Cooper RN, Di Donna S, Furling D, Gidaro T, Jacquemin V, Mamchaoui K, Negroni E, Périé S, Renault V, Silva-Barbosa SD and Butler-Browne GS 2005. The mitotic clock in skeletal muscle regeneration, disease and cell mediated gene therapy. *Acta Physiologica Scandinavica* 184, 3–15.

Nelson, M. E., Hamm MW, Hu FB, Abrams SA and Griffin TS 2016. Alignment of healthy dietary patterns and environmental sustainability: a systematic review. *Advances in Nutrition* 7, 1005–1025.

Orzechowski, A. 2015. Artificial meat? Feasible approach based on the experience from cell culture studies. *Journal of Integrative Agriculture* 14, 217–221.

Phelan, K. and May KM 2015. Basic techniques in mammalian cell tissue culture. *Current Protocols in Cell Biology* 66, 1.1.1–1.1.22.

Piazza, J., Ruby MB, Loughnan S, Luong M, Kulik J, Watkins HM and Seigerman M 2015. Rationalizing meat consumption. The 4Ns. *Appetite* 91, 114–128.

Renzini, A., Benedetti A, Bouché M, Silvestroni L, Adamo S and Moresi V 2018. Culture conditions influence satellite cell activation and survival of single myofibers. *European Journal of Translational Myology* 28.

Ritchie, H. and Roser M 2017. *Meat and seafood production & consumption*. Retrieved on 18 July 2019 from https://ourworldindata.org/meat-and-seafoodproduction-consumption.

Rollins, B. and Rumley R 2019. *The Regulation of 'Cell-Cultured Meat.'*

Sala, S., Anton A, McLaren SJ, Notarnicola B, Saouter E and Sonesson U 2017. In quest of reducing the environmental impacts of food production and consumption. *Journal of Cleaner Production* 140, 387–39.

Sanders, E. R. 2012. Aseptic laboratory techniques: plating methods, *JoVE,* e3064. Retrieved on 18 July 2019 from https://www.jove.com/video/3064.

Smetana, S., Mathys A, Knoch A and Heinz V 2015. Meat alternatives: life cycle assessment of most known meat substitutes. *International Journal of Life Cycle Assessment* 20, 1254.

Soussana, J. F., Tallec T and Blanfort V 2010. Mitigating the greenhouse gas balance of ruminant production systems through carbon sequestration in grasslands. *Animal* 4, 334–350.

Stephens, N., Di Silvio L, Dunsford I, Ellis M, Glencross A and Sexton A 2018. Bringing cultured meat to market: technical, socio-political, and regulatory challenges in cellular agriculture. *Trends in Food Science & Technology* 78, 155–166.

Stephens, N., Sexton AE and Driessen C 2019. Making sense of making meat: key moments in the first 20 years of tissue engineering muscle to make food. *Frontiers in Sustainable Food Systems* 3, doi: 10.3389/fsufs.2019.00045.

Toldra, F., Aristoy MC, Mora L and Reig M 2012. Innovations in value-addition of edible meat by-products. *Meat Science* 92, 290–296.

Tuomisto, H. L. and Teixeira de Mattos MJ 2011. Environmental impacts of cultured meat production. *Environmental Science & Technology* 45, 6117–6123.

Van Der Weele, C. and Tramper J 2014. Cultured meat : every village its own factory. *Trends in Biotechnology* 32, 294–296.

Van Eelen, W. F. 2007. *Industrial production of meat using cell culture methods.* US Patent 7,270,829 B2.

Wiedemann, S., McGahan E, Murphy C and Yan M 2016a. Resource use and environmental impacts from beef production in eastern Australia investigated using life cycle assessment. *Animal Production Science* 56, 882–894.

Wiedemann, S. G., McGahan EJ and Murphy CM 2018. Environmental impacts and resource use from Australian pork production determined using life cycle assessment. 2. Energy, water and land occupation. *Animal Production Science* 58, 1153–1163.

Wiedemann, S. G., Yan MJ and Murphy CM 2016b. Resource use and environmental impacts from Australian export lamb production: a life cycle assessment. *Animal Production Science* 56, 1070–1080.

Chapter 2

Structural and Technological Characteristics of Scaffold Manufacturing for Cultured Meat

Abstract

In vitro cultured meat is an emerging research field that uses tissue engineering to address global food demands. Lab-grown meat production offers a solution to environmental, public health, and animal welfare concerns. While significant progress has been made in cultivating artificial meat, many challenges hinder large-scale commercialization. This chapter focuses on the production of edible scaffolds for cultured meat. It begins with an overview of artificial meat cultivation and its current market status. The composition, cellular, and molecular interactions in muscle tissue essential for scaling production were discussed. It also shows key components such as cells, biomaterial scaffolds, media, and bioreactors used in cultured meat production, followed by recent advancements. Finally, the chapter discusses existing challenges, opportunities, and future research directions to advance large-scale cultured meat production.

Keywords: scaffold manufacturing, cultured meat, biomaterial engineering, tissue engineering

Introduction

The global population is continuously growing, which in turn increases the demand for a more consistent food supply, influenced by regional, cultural, and personal preferences. Among protein-based food sources, the rising global demand for meat has had significant health and environmental impacts on human life, in addition to the ethical concerns surrounding animal cruelty, including suffering on farms and during slaughter. Therefore, a more efficient and safer alternative to conventional animal meat is highly desired, one that can meet the global meat protein demand without reducing consumption or relying on direct sourcing from slaughtered animals.

In response, cellular agriculture, or the cultivation of artificial meat, has emerged as a promising alternative technology, gaining increasing interest from both the industry and the public (Bryant and Barnett, 2020). Cellular agriculture can be categorized into two major areas: (1) Fermentation-based cellular agriculture and (2) Tissue engineering-based cellular agriculture. Meat culture, as a subset of cellular agriculture, is expected to contribute positively to human health, with minimal contaminants and antibiotics during cultured meat production (Djisalov et al., 2021). Thus, to address the growing demand for meat, *in vitro* cultivation of meat, also known as cultured meat, has gained substantial interest. Artificial meat offers a transformative approach for developing alternative proteins while prioritizing human nutrition, food safety, livestock-related environmental issues, animal welfare, and other relevant factors (Post et al., 2020; Rubio, Xiang, and Kaplan, 2020).

General research efforts in developing potential cultured meat are shown in Figure 1. The technology for meat cultivation aims at producing meat via *in vitro* or *ex vivo* culturing of animal cells, in which muscle stem cells (MSCs) are expanded in an undifferentiated state using extracellular matrix (ECM), inflammatory factors, and physicochemical considerations (Guan et al., 2022). By leveraging established principles and research in *in vitro* cell culturing, stem cell biology, bioprocess technology, materials science, tissue engineering, animal sciences, and other engineering fields, this approach shows considerable potential for advancing cultured meat production (Reiss, Robertson, and Suzuki, 2021; Guan et al., 2022).

However, one of the major challenges in the large-scale commercial production of cultured meat is the development of efficient scaffolds, the choice of cell lines, the use of serum-free media, and the design of bioreactors for upscaling production (Datar and Betti, 2010). Scaffolds act as support matrices that facilitate cellular activities such as adhesion, proliferation, differentiation, and regulation, while also promoting the transport of oxygen and essential nutrients into thicker tissues (Gil et al., 2010; Chen, Zhou, et al., 2020). Therefore, in cultured meat production, the selection of scaffold biomaterials is crucial and should meet the criteria of being edible, digestible, and acceptable. Additionally, the scaffolds should provide specific taste, texture, nutritional value, thermal stability, and be available on a large scale at a low cost (Post et al., 2020). Furthermore, it is important to emphasize that the conventional approach to meat production has a significant adverse impact on the environment due to the resources it consumes, such as land, energy, and water, as well as its environmental footprint, including deforestation, greenhouse gas emissions, and biodiversity loss. Therefore, the meat production process needs to be more sustainable, less harmful, and more animal-friendly (Steinfeld et al., 2006). The initiation of cultured meat production requires cells that can proliferate over time. In general, well-characterized cell lines are used for meat production, as they can proliferate continuously over extended periods. Moreover, access to cell lines is another major hurdle in the up scalability of meat production at commercial levels (Shaikh et al., 2021). Further, cells specific for cultured meat production has conventionally been proliferated on fetal bovine serum (FBS), an extremely expensive, unsustainable, and inconsistent component, which is fundamentally adversative to the aims an object of cultured meat (Stout et al., 2022). Moreover, to minimize the production cost of cultured meat, it is essential to implement more automated and efficient processes with minimal reliance on manual labor. This necessitates the development of advanced bioreactors tailored specifically to the final form of meat production.

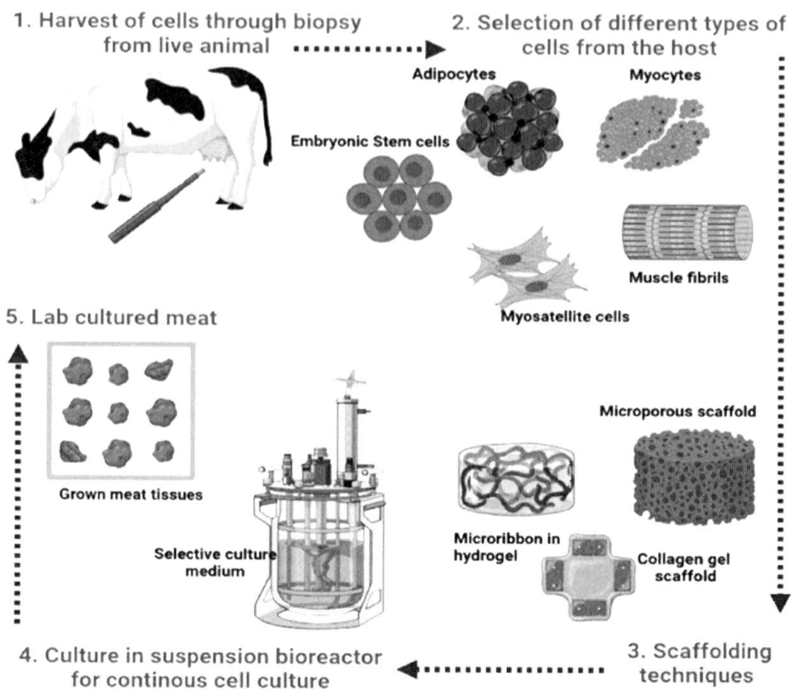

Figure 1. provides a schematic representation of the steps involved in producing cultivated meat, including sourcing cells, isolating and expanding the selected cells in a bioreactor, and culturing the cells on scaffolds to create a suitable cultured meat product.

Recently, several specific reviews have been published on artificial meat. The review by Post et al., focuses on the scientific, social, and regulatory challenges (Post et al., 2020). Chen et al., presented the bioprocess technologies related to the manufacturing of artificial meat, including various factors and challenges (Chen, Guttieres, et al., 2022). Additionally, a brief overview of scaffolds for meat culture was discussed in another review (Seah et al., 2022). Zhang et al., summarized the development status of various meat alternatives and the associated technological challenges (2022), while the technical requirements for meat culturing were reviewed by Ramani et al., (2021). Most of these reviews briefly addressed the scientific challenges, limitations, and technological requirements for artificial meat production. However, a detailed review focusing on innovations in the manufacturing of edible scaffolds for meat culturing is still needed.

Current Scenario

The current animal-based food products are being questioned in terms of social, economic sustainability, and environmental issues. Therefore, an efficient transition from traditional agriculture (livestock) to more sustainable cellular agriculture through the application of cell-culture techniques has gained favor for the production of cultured meat. Additionally, a key benefit of using cell-culture techniques is the potential to enhance the nutritional profile of cultured meat through precisely regulated bioprocessing approaches. This advancement was highlighted in 2013 with the production of the first lab-grown beef hamburger, which generated significant media and investor interest and led to a remarkable increase in the number of cultured meat companies. Since then, various startup companies have emerged in the cultured meat sector, developing a wide range of products, including seafood, duck meat, chicken meat, pork, and beef (Choudhury, Tseng, and Swartz 2020; Hong et al., 2021). However, there are still various challenges such as the growth media, meat-like characteristics, scalable production, safety and regulatory framework, and consumer perception for becoming cultured meat as commercial or consumer product.

Dietary protein supplies of nearly 9.8 billion populations worldwide need to be satisfied in a healthful and feasible manner by 2050 (Kamalapuram, Handral, and Choudhury 2021). In this regard, the business models of cultured meat are growing with funding opportunities and efforts for commercialization. Currently, several multinational companies and startups from Asia, Mediterranean, Pacific, South America, North America, European Union, and United Kingdom are involved in cultured meat production and established research infrastructure (Choudhury, Tseng, and Swartz 2020; Miller 2020). Out of the 32 major cultured meat companies, at least 25% focus on beef, 22% on cultured poultry (e.g., chicken, duck), and 19% each on pork and cultured seafood (e.g., shrimp, fish). In addition, two companies expressed interest in investigating mouse meat as a substitute for pet food, while another two companies showed interest in kangaroo and horse meat. Between 2015 and 2020, publicly disclosed capital investments reached approximately 320 million USD. Recently, various companies and startups have shown significant interest in cultured meat production across different meat categories, utilizing both business-to-consumer (B2C) and business-to-business (B2B) models. Furthermore, some government funding agencies have started supporting the growth of the cultured meat industry (Choudhury, Tseng, and Swartz, 2020). As a result, a wide range of opportunities is available at various stages of the cultured meat development process for stakeholders, including entrepreneurs, investors, and strategic partners, all of whom have significant potential to capitalize on the global shift occurring in meat production (Morach et al., 2021). Owing to it, various public-private collaborations among universities, companies, and nonprofit establishments are fast-tracking the thrust to set-up funding initiatives and research frameworks. Cost-effective and balanced cultured meat products can be developed with a focus on food acceptability, sustainability, and the security aspects of the meat industry, supported by upcoming research initiatives from both public and private sectors, along with substantial funding opportunities (Choudhury, Tseng, and Swartz 2020; Kamalapuram, Handral, and Choudhury 2021).

Understanding of Compositional, Cellular, and Molecular Communications

The quality of meat is expressed by the nutritional composition and consumer perceptions such as visual appearance, smell, rigidity, juiciness, flavor or taste, and tenderness (Maltin et al., 2003; Muniz et al., 2022). Therefore, the understanding of compositional, cellular, and molecular communications in skeletal muscles is very important for designing and developing cultured meat (*in vitro*) with natural meat-like characteristics for better prospects and acceptance of the consumers. Meat is the autopsy aspect of a complex biological muscle tissue (shaped by greatly specialized contractile fibers bound together in a complicated connective tissue-network that form tendons and adhesions connecting directly or indirectly to bones). Comprehensively, the composition of meat consists of approximately 75% water, 19% protein, 3.5% soluble non-protein substances (including inorganic components), and 2.5% fat. Muscle proteins can be categorized into three groups: (1) sarcoplasmic proteins (soluble in water or dilute salt solutions), (2) myofibrillar proteins (insoluble in water but soluble in concentrated salt solutions), and (3) stromal proteins (2%, proteins of connective tissue and other structural components, insoluble in concentrated salt solutions, especially at low temperatures). These proteins play various roles in meat processing. Myofibrillar proteins consist of long fibrillar networks arranged into repeated sections (i.e., sarcomeres) that contract by sliding the thick (myosin) and thin (actin) filaments laterally against each other. Additionally, sarcoplasmic proteins comprise various types of proteins, including heme pigment proteins (myoglobin) and enzymes. The solubility of both myofibrillar and sarcoplasmic proteins can serve as important indicators of muscle quality, and consequently, meat quality (e.g., color and water-holding capacity). Furthermore, the intramuscular connective tissue (stromal protein) is a complicated-network of extracellular proteins that upholds muscle structure and conveys contraction forces into tendon and bones (López-Bote 2017; Vieira et al., 2018). A schematic illustration of the assembly of skeletal muscle is provided for better understanding of compositional and sensory sites.

As shown in Figure 2, the myofibrillar matrix provides the physical space where the majority of endogenous water is trapped through capillary forces, hydrogen bonding, and ion-dipole interactions. The fibrous protein complexes, along with collagen fibrils (perimysium and endomysium), contribute to the unique textural impression when cooked meat is eaten. In contrast, in minimally processed meat (e.g., muscle chunks or coarse-ground meat), the chewiness of cooked meat is primarily due to the aggregation of muscle fibers and bundles. Overall, muscle proteins perform a crucial function in creating a type of viscoelastic textural features in manufactured and processed muscle foods. The rigidity, tenderness, juiciness, and mouthfeel of meat and meat-analogues are the terms utilized to describe the textural characteristics. The organoleptic texture formed in processed meat significantly depends on the gelling and emulsifying capability of myofibrillar proteins, including myosin and actomyosin (Asghar et al., 1985; Sha and Xiong 2020).

In the *in vivo* system, ECM of skeletal muscle (Figure 3) upholds tissue elasticity, creates adhesion sites for cells, offers a 3D environment, and controls biological progressions. As shown in Figure 3A, the extracellular matrix (ECM) is a complex 3D environment composed of various molecules that provide structural support, facilitate cellular interactions, and mediate signaling responses to injuries. It is also responsible for maintaining the architecture of skeletal muscles. ECM adheres to, networks with, and safeguards the muscle cells, enables biochemical signaling, supplies structural support that are vital during myogenesis (Ju Lee et al., 2018), including several ECM proteins contributing to cell-matrix interactions as well as matrix assembly (Gillies and Lieber 2011; Ahmad et al., 2021). In addition to biological functions, proteins mainly collagen and glycosaminoglycans in ECM influence the texture of the tissue and overall meat quality. Furthermore, ECM is also crucial to support and regulate the developmental progressions of myotubes during an early stage of myogenic differentiation (DiMaio 2019; Zhang et al., 2021) through the interaction of ECM proteins and regulation of the functions and the MSCs phenotypes (Ahmad et al., 2021) (Figure 3B).

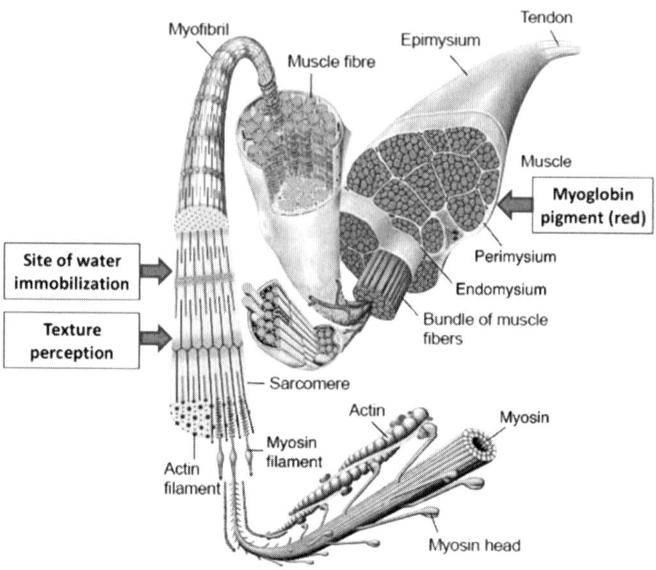

Figure 2. presents a schematic illustration of skeletal muscle assembly, highlighting the differentiated sites that correspond to the sensory characteristics of meat.

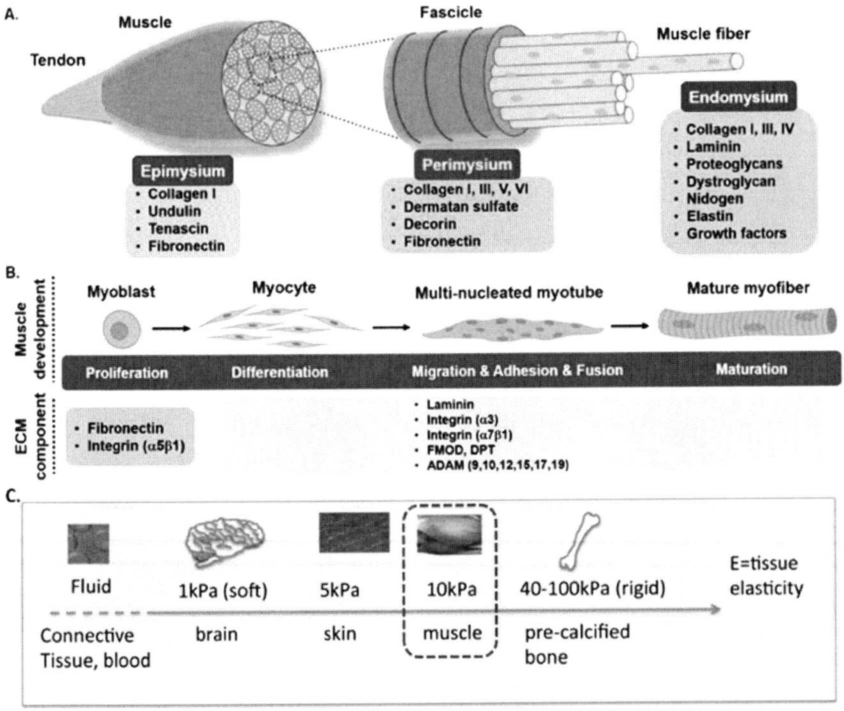

Figure 3. illustrates the ECM of skeletal muscle: (A) a three-layered organization of linked ECM modules, (B) a schematic of ECM-associated protein events during different phases of myogenesis, and (C) a diagram showing the elasticity of various tissues, with a range of stiffness represented by the elastic modulus.

However, the ability of artificial constructs to mimic the native extracellular matrix (ECM)-like characteristics, such as those found in tissue-engineered scaffolds, remains a significant challenge. In this context, the mechanical properties (e.g., stiffness) of the ECM are also crucial factors in achieving meat-like properties (Figure 3C).

Major Components for Cultured Meat Production

Muscle is a complex tissue that requires a suitable strategy for cultivation in *in vitro* meat products. There are three major components for cultured meat production: cells, scaffolds, and media. Broadly, the cultivation of cells for artificial meat can be divided into four steps: (1) isolation or acquisition of cells and initial cell proliferation, (2) large-scale cell expansion, (3) differentiation of cells or muscle/fat tissue formation (maturation) on 3D biomaterial scaffolds, and (4) processing into food products, either unstructured (e.g., burgers) or structured (e.g., "whole-cut" like steaks).

Utilization of Cells

For the purpose of artificial meat synthesis, the breakthrough of stem cells or progenitor cells has provided a better opportunity for tissue engineering applications and thereby for *in vitro* meat production (*in vitro* myogenesis and adipogenesis). The process of cultured meat production involves isolating stem cells or progenitor cells from a biopsy or living donor animals, followed by the multiplication of these isolated cells (*ex vivo*) and the application of tissue engineering techniques. Industrial-scale production of cultured meat heavily depends on the advancement of cell proliferation and expansion processes. Herein, for this process, two types of stem and progenitor cell types are of high interest: (1) pluripotent stem cells (embryonic stem cells (ESCs) and induced pluripotent stem cells (iPSCs)), and (2) adult stem cells (mesenchymal stem cells (MSCs), adipose-derived stem cells (ADSCs), fibro-adipogenic progenitors (FAPs), including resident muscle stem cells (rMuSCs)/muscle satellite cells (MuSCs), also known as myosatellites and myoblasts as progenitors (Knežić et al., 2022). Moreover, these cells can differentiate into myocytes (muscle cells), adipocytes (fat cells), fibroblasts, or chondrocytes. For example, multipotent muscle-derived satellite cells play a critical role in developing and regenerating skeletal muscle tissue and are considered the most promising cell type for initiating cultured meat production due to their easy and efficient myogenic differentiation (Yin, Price, and Rudnicki, 2013). Embryonic stem cells (ESCs), originating from the inner cell mass of embryos at the blastocyst phase, are suitable for cellular agriculture due to their ability to proliferate easily and differentiate into any cell type (pluripotency). However, the isolation of ESCs is relatively expensive due to the use of specialized media and precise conditions required for their maturation into desired tissue types (Gandolfi et al., 2012; Bhat et al., 2020), and they raise ethical concerns related to embryo destruction.

Further, in addition to MuSCs, PSCs also show immense potential for exploitation. For example, iPSCs, produced by reprogramming somatic cells into an embryonic-like pluripotent state (inducing pluripotency - related genes by encoding species-dependent transcription factors), show great *in vitro* proliferation capability, and exhibit an indefinite ability of renewal as well as the differentiation into any cell types existing in the meat, unlike MuSCs (Takahashi and Yamanaka 2006; Yu et al., 2007; Reiss, Robertson, and Suzuki 2021). Furthermore, MSCs are easy to isolate and highly suitable for muscle development, with the ability to differentiate into adipocytes (Huang et al., 2012), myocytes (Ramírez-Espinosa et al., 2016; Okamura et al., 2018), endothelial cells, and fibroblasts (Bosnakovski et al., 2005). During muscle development, most MSCs participate in myogenic differentiation, forming myofibers and enhancing the population of MuSCs. Another important component in cellular agriculture is intramuscular fat (mesenchymal FAPs), which exhibit high adipogenic potential and contribute to the "marbling" of meat, playing a crucial role in enhancing its palatability, flavor, and texture (Guan et al., 2017; Li et al., 2020; Dohmen et al., 2022), in addition to other fat tissues. However, pro-adipogenic FAP subgroups are of the highest importance for cellular agriculture objectives. In another example, differentiated fat (DFAT) cells have received interest for cultivated fat production and these cells are fibroblast-like multipotent and proliferative cells derived from mature lipid-holding adipocytes. Owing to their prominent proliferative ability through long-standing culture and high adipogenic prospects, porcine DFAT cells are having good potential for large-scale cultured fat production (Knežić et al., 2022).

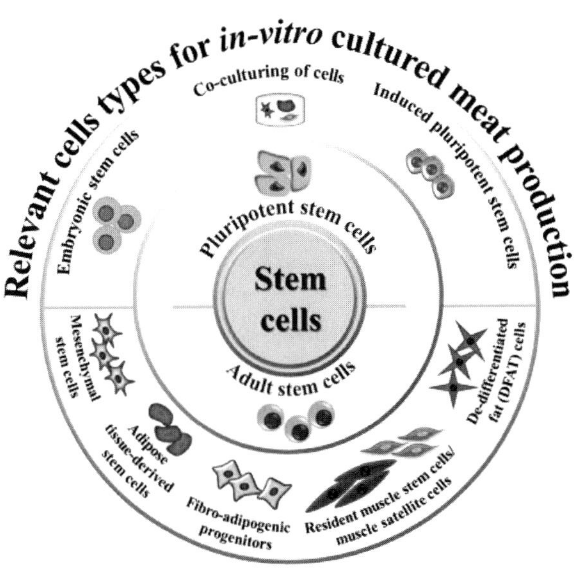

Figure 4. Illustrates the relevant cell types involved in cellular agriculture, such as those used in *in vitro* cultured meat production.

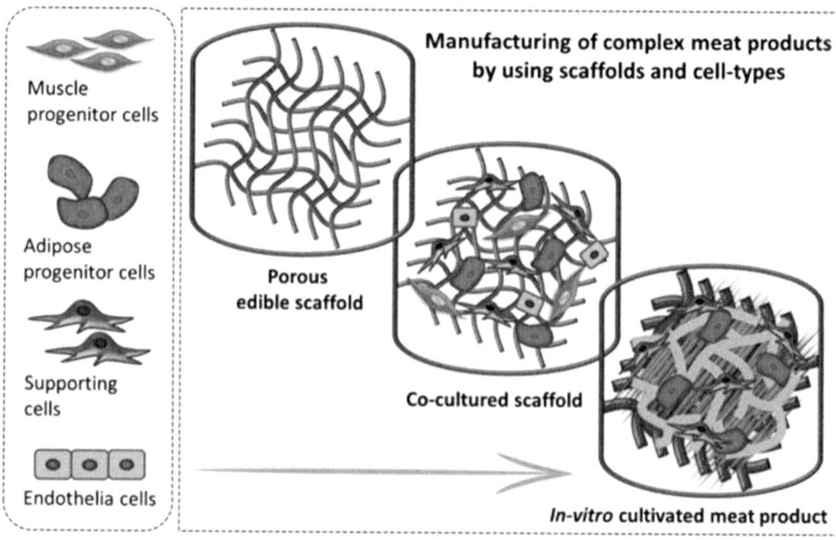

Figure 5. This figure provides a schematic illustration of the manufacturing process for complex meat products, utilizing muscle, fat, connective tissue, and vascular cells on scaffolds.

A schematic representation of relevant cell types is provided in Figure 4. Moreover, *in vitro* co-culturing method is promising and feasibly mimics *in vivo* microenvironment by observing interactions among the cells and between various cell lines (e.g., myogenic and adipogenic). Therefore, the co-culturing of myocytes and adipocytes cells on scaffolding biomaterials is imperative and essential to improve the quality of cultured meat in terms of texture, juiciness, flavor, tenderness, nutritional, and functional characteristics by enhancing intramuscular fat. These methods can be categorized into two main groups: indirect and direct methods. In indirect methods, cells are physically separated and communicate only through secreted molecules, whereas direct methods allow cell-to-cell interactions between various cell types. However, co-culturing methods present particular challenges, as cells secrete different metabolites that can influence signaling pathways (i.e., cellular molecular mechanisms) involved in cell proliferation and differentiation. For example, the co-culturing of myoblasts and adipocytes enhances adipogenic gene expression in myogenic cells (Choi et al., 2013; Knežić et al., 2022), while the co-culturing of porcine pre-adipocytes and muscle stem cells (MuSCs) promotes cell growth and proliferation but inhibits differentiation (Yan et al., 2013). Therefore, the co-culturing of myoblasts and adipocytes is challenging and should be performed under precise conditions that are necessary to optimize the proliferation and differentiation of both cell types (Knežić et al., 2022). Additionally, the efficient supply of oxygen and nutrients, along with the removal of metabolites, is essential for cell survival and the formation of skeletal tissue. It is well-established that parallel myogenesis, adipogenesis, and vasculogenesis are crucial for skeletal tissue formation, enabling the creation of functional 3D muscle tissue with intramuscular fat. However, the vasculogenesis process is complex and time-consuming, making it less suitable for cultured meat production, where blood vessels do not significantly impact the taste and texture of the meat. Instead, the vascularization process can be replaced with a perfusion system to support cellular survival (Pajčin et al., 2022).

Biomaterial Scaffolds and Potential Manufacturing Technologies

In tissue engineering technology, biomaterial-based 3D scaffold as a supportive and temporary structure helps in providing 3D environment for cellular consequences (e.g., cell adhesion, growth, differentiation, and tissue formation) under possible native-like physiological environment. Therefore, the porous scaffold is a critical component or template not only in cell culturing techniques but also in cellular agriculture, playing a key role in the development of cultured or lab-grown meat. The scaffold provides both physical and biochemical cues that enable cells to attach, proliferate, and differentiate into essential cell types, facilitating spatial heterogeneity in the final product. This may result in a structure that closely resembles the native meat assembly and texture (Specht et al., 2018).

The porous scaffold-network permits the flow of oxygen and nutrient and removal of waste products for maintaining metabolic functions of the cells by balancing the morphology, structure, and chemistry. The scaffold must be palatable, safe to eat (cooked or un-cooked), economical, and readily available at large-scale production (Post et al., 2020). However, the main challenge is to manufacture a suitable scaffolding-network for cultivating meat-like product with appropriate characteristics, like composition, microstructure, appearance, taste or flavor, specific texture, color, and content of fat and protein (Datar and Betti 2010). The biomaterials for cultured meat production should be edible, digestible, animal-free, stable, and acceptable with nutritional qualities (Post et al., 2020). In this regard, biopolymers are preferred and promising choices for manufacturing scaffolds used in the cultivation of meat-like products. However, the selection of biomaterials, whether alone or in combination, depends on their ability to target specific meat tissues, such as collagen, gelatin, agarose, alginate, xanthan gum, as well as plant proteins or decellularized plant materials (Post et al., 2020; Ng and Kurisawa, 2021). Biomaterials derived from animals should be avoided (Post et al., 2020). Moreover, it is speculated that careful manipulation of the biophysical characteristics of cells on different scaffolds could regulate these properties by mimicking systematic vascularization outcomes. Due to the complex nature of tissues, a single biomaterial is insufficient to replicate the native characteristics of meat. Therefore, exploring new approaches to engineer composite or hybrid scaffolds, while minimizing costs, is critical for cultivating meat-like edible products (i.e., cultured meat). The tissue-engineered scaffolds should also promote vascularization to recruit cells, allow their habitation, and support the development of the tissue matrix. Proper structural design, along with mechanical, biological, and surface characteristics of the scaffolds, is crucial for facilitating cellular activities. Therefore, various manufacturing technologies have been employed, either individually or in combination, to achieve and optimize the desired properties. A schematic illustration of the manufacturing process for complex meat products from muscle, fat, connective tissue, and vascular cells using scaffolds is shown in Figure 5.

Due to the complex and multicellular nature of meat culturing, the scaffolds to be used for artificial meat production should offer the ability for cell attachment, proliferation, differentiation, and thereby tissue formation. Furthermore, the consumer view-point is also a major concern toward artificial meat because of various cultures and demographics (Zhang et al., 2021). Moreover, it is crucial to achieve the appropriate or optimized architecture, mechanical performance, surface characteristics, and biological properties through the use of processing techniques, either individually or in combination, to influence cellular fate (Seah et al., 2022). In addition to the selected conventional fabrication methods, various advanced fabrication technologies for scaffold production in artificial meat production are discussed in detail in the following sections.

Solvent-Casting/Porogen Leaching

Solvent-casting method is a traditional method for scaffold designing and are often used in the fabrication of films (Wahid et al., 2018). This method comes with the advantage of being very economical, less time duration for preparation, and applicability for variation in reaction conditions. This method involves the complete dissolution of polymers in a solvent, ensuring uniformity, which is a primary requirement for an ideal scaffold (Choi, Zhang, and Xia, 2010). The solvent casting process is highly influenced by the choice of solvent and processing conditions, which in turn affect the surface structure of the polymer (Ghosal et al., 2018). Properties such as surface heterogeneity, swelling behavior, deformation rates, and reorientation of surface crystal segments are significantly impacted by both the choice of solvent and the processing conditions, further influencing the applicability of the designed scaffold (Koide et al., 2013).

The principle of solvent casting involves the dissolution of the polymer in the solvent, and then, transfer of the solution into a predefined three-dimensional (3D) mold. The technique utilizes porogens, which are dispersed into the molded structure. Once the structure has set, the solvent is evaporated and the porogens are dissolved, resulting in the formation of pores. Highly porous scaffolds with a wide range of pore sizes have been fabricated using the solvent-casting method, which relies on the use of different pore-sized porogens (Hulko et al., 2019). Although this method has been extensively studied, control over the pore shape and inter-pore connectivity in the fabricated scaffolds remains a challenge. Another drawback of this technique is the mixing process of the polymer and porogens. Since the porogens are completely enveloped by the polymer solution, the complete removal of salt particles with water becomes a tedious task. As a result, the porosity of the scaffolds obtained using this technique is limited to 90%, with a wide range of pore sizes (5 to 600 μm) and irregular microstructures (Page, Flood, and Reynaud 2013). In addition, the organic solvents utilized may prevent the cells seeding onto these scaffolds owing to their lethal properties. Due to these associated disadvantages, only a few successful preclinical trials have been demonstrated (bone marrow and hard bone regeneration) using scaffolds synthesized by solvent casting method (Sola et al., 2019). The process of solvent casting is often supplemented with crosslinking agents taking either physical or chemical routes for their utility.

Physical Crosslinking

Physically crosslinked scaffolds are formed because of physical interactions that may comprise ionic interactions, temperature triggered, and dehydrothermal crosslinking (DHT) techniques (Chan et al., 2015; Nonoyama et al., 2020). Significant attention is being given to the design of physically crosslinked scaffolds for tissue engineering applications, as the crosslinking process typically occurs under mild conditions without the use of chemical agents. In the case of ionic interactions, the crosslinking agent forms ionic bridges with the polymer backbone, whereas (Russo, Malinconico, and Santagata 2007), temperature-triggered crosslinking mainly relies on the thermal behavior of the crosslinking agent. In the case of DHT, the polymer is exposed to high temperatures for a specific period under strict vacuum conditions to remove water, thereby forming crosslinks (Chen, Guo, et al., 2020). In recent years, extensive research has focused on crosslinking hydrogels through physical interactions for their application in tissue engineering platforms. The process of physical crosslinking via ionic interactions has also been explored for the synthesis of artificial meat. This process was explored by Orellana et al., to develop a new scaffold (microstructured film) via micromolding technology to grow muscle fibers (2020). The edible films representing artificial meat were made using the solvent casting method with non-mammalian biopolymers that were crosslinked using calcium chloride solution. To prepare hydrogels for soft tissue, Mahapatra et al., prepared hydrogels via dual networks crosslinking using positively charged quaternary polyethylene imine and micelles formed by Pluronic F127 diacrylate. The hydrogels exhibited superior strength and exceptional stretchability. This work could be further explored for the artificial meat industry by utilizing the designed material to grow muscle cells and adipocytes. Additionally, further investigation is needed to explore the use of DHT-based scaffold materials for artificial meat culturing, as these scaffolds may offer the necessary mechanical properties and structure for meat cultivation. Evidence suggests that DHT treatment enhances the structural and mechanical properties of collagen films, indicating its potential applicability in this context (Chen, Zhou, et al., 2020).

Chemical Crosslinking

Till date, numerous chemical crosslinking methods have been reported that tends to form covalent bonds with the polymer chains during scaffold synthesis. Chemical crosslinking agents include small molecules, photo-induced crosslinking, and enzymatic-induced crosslinking (Xue et al., 2022). Compared to physical crosslinking methods, chemical crosslinking tends to provide greater stability to the designed scaffold along with improved mechanical properties. Over time, small crosslinking molecules such as genipin, dopamine, glutaraldehyde, and tannic acid have been extensively studied as effective crosslinking agents for scaffold synthesis (Oryan et al., 2018). However, glutaraldehyde has limited applications in tissue engineering platforms due to its toxic nature. In addition, inspired by the adhesion process of mussel organisms, catecholic compounds such as dopamine and caffeic acid, which contain catechol functional groups, have been frequently explored in the fields of tissue engineering and regenerative medicine (Li et al., 2016). Moreover, photo-triggered hydrogels have emerged as a promising class of biomaterials. The gelling properties of these materials depend on the response of the crosslinkers to light with specific wavelengths (Piluso et al., 2020). Conventional chemical crosslinking methods (photo crosslinking, covalent crosslinking) result in scaffold formation with enhanced stiffness and rapid gelation times. However, the cytotoxicity associated with chemical crosslinkers remains a challenge. To address this issue, enzyme crosslinkers have received significant attention, as they not only improve material properties but also aim to eliminate the harmful effects of chemicals. This strategy can be utilized in synthesizing artificial meat as enzymes will also be helpful in maintaining the microenvironment required for artificial scaffolds to mimic meat.

Decellularization

Native ECM can serve as a potential scaffold material, preserving its ultrastructure and maintaining the most similar biophysical and biochemical characteristics (i.e., 3D microenvironment) of the native tissue. This native ECM can be obtained by the decellularization process through the removal of the cells and nucleic acids from a plant or fungal or animal tissues. Therefore, this decellularization could be a promising method for culturing artificial meat for commercial translation.

Decellularization of a tissue is typically carried out by using physical, chemical, and enzymatic methods. In case of physical methods for decellularization, the most common technique is impulsive freezing or freeze-thawing. Freezing causes the formation of intracellular ice crystals within tissues, which disrupt cellular membranes and trigger cell lysis (Burk et al., 2014). Freeze-thawing is a simple and straightforward technique for performing uniform decellularization. The key requirement of this technique is precise temperature control, as temperature fluctuations directly affect the size of the ice crystals formed, thereby minimizing unnecessary damage to the extracellular matrix (ECM) (Jackson, Windler, and Simon 1990). The process of physical decellularization cannot be self-sustaining; however, it is an effective method when combined with enzymatic and chemical treatments to aid in cell lysis and the proper removal of cell debris. Detergents are commonly used to disrupt membrane integrity and disturb the structural arrangement of cell membranes. Among these, Triton-X-100 is frequently employed (Woods and Gratzer, 2005). Triton-X-100 is particularly useful in tissues where the major matrix components are proteins, as it primarily targets lipid–lipid and lipid–protein interactions, with minimal effect on protein–protein interactions. In addition to Triton-X-100, Sodium dodecyl sulfate (SDS) is another detergent that has been extensively studied for its role in decellularization (Mendibil et al., 2020). SDS is known to solubilize both external and nuclear membranes, while also denaturing proteins, significantly altering the native structure of the entire matrix. This makes SDS a potent decellularizing agent, demonstrating effective results even in short treatment periods (Chen et al., 2004).

There has been significant use of enzymes as a promising agent for decellularization. Enzymes offer the advantage of eliminating cell waste and other unwanted components of the ECM during the process of decellularization. However, the removal of enzymes after the completion of the process possesses serious limitations to the utility of enzymatic treatments (Keane, Swinehart, and Badylak 2015). This in turn extends the duration of the complete process as further steps are to be inculcated to remove enzymes (use of detergents or nonenzymatic agents). Among various enzymes, trypsin is commonly used for decellularization. It is typically utilized in combination with EDTA, which disrupts cell-matrix integrity, and often targets the C-side bonds in amino acids such as arginine and lysine (Rieder et al., 2004). Prolonged exposure to trypsin-EDTA treatment significantly alters the structural integrity of tissues, resulting in acute mechanical weakness. In addition to trypsin, pepsin is another enzyme of considerable importance in the decellularization process. Like trypsin, pepsin targets bonds between amino acids and can damage the extracellular matrix (ECM) with extended exposure (Poon et al., 2013). Up recently, Yuksel et al., developed multilayered decellularized amniotic membranes that could be used as biomimetic 3D scaffold systems for meat production. This scaffold system with mechanical durability and flexibility facilitates the attachment and survival of the cells (Yüksel et al., 2022).

Self-Assembly

The field of tissue engineering offers the possibility of recapitulating tissues and organs that require replacement. In this context, self-assembly has been identified as a promising tissue-engineering technique (Hu and Athanasiou, 2006). The self-assembly process mimics developmental events, leading to the generation of functional tissue with characteristics similar to those of native tissue. Several mechanisms have been proposed for the process of self-assembly. The first mechanism is known as differential adhesion hypothesis that reports cell-cell binding behavior and minimization of free energy as the driving force for self-assembly (Raghothaman et al., 2014). This intercellular interaction occurs due to the adhesion of proteins on the cell surface, leading to the formation of a cell mass that behaves similarly to a liquid, reducing its surface tension, known as tissue surface tension. This surface tension dictates the cell's movement. Cells with higher surface tension tend to move toward the center to enhance intercellular adhesion, which in turn minimizes surface tension. Similarly, in the process of self-assembly, a nonadherent substrate drives a consistent cell population with intercellular adhesion, resulting in the minimization of free energy (Ofek et al., 2008).

The second proposed mechanism for self-assembly is differential interfacial tension, which is also linked to differential adhesion. In the case of differential adhesion/differential interfacial tension, the minimization of free energy governs cellular behavior, with cell movement driven by forces generated by the cytoskeleton at the cellular membrane. Furthermore, cells with similar interfacial tensions tend to aggregate, while cells with different interfacial tensions tend to remain apart (Manning et al., 2010; Lee, An, & Chua, 2017).

In tissue engineering, the self-assembly of peptides has been extensively studied by several research groups (Maude, Ingham, & Aggeli, 2013; Chen & Zou, 2019; Gelain et al., 2021). Recently, Kang et al., demonstrated the *in vitro* construction of engineered steak-like tissue assembled from three types of bovine cell fibers (i.e., steak-like meat) using tendon-gel integrated bioprinting. This approach mimics actual meat, which consists of an aligned assembly of fibers connected to tendons for contraction and relaxation (Kang et al., 2021). Recapitulating tissue is a complex process involving a cascade of events and steps. However, mimicking native tissue provides valuable insight into the structural and physiological architecture of the tissue under investigation.

Film Stacking

The cell sheet and cell sheet stacking techniques demonstrate great potential in tissue engineering and regenerative medicine. For example, a biomimetic 3D lamellar corneal tissue construct has been developed by seeding human corneal fibroblasts (hCFs) onto optically transparent surface-patterned thin silk films, which are then stacked onto porous silk using the film stacking method (Koo et al., 2019).

Micromolding Technology

Micromolding is a promising technology that can be utilized as a robust device to produce small-dimensional molded microstructures with a high aspect ratio, meeting the requirements for the precise production of microstructures at a large scale (Weber et al., 1996). In this context, edible microstructured films are also considered an effective approach for growing functional muscle fibers for *in vitro* meat production. In this context, a microstructured mold with parallel microchannels has been fabricated with acrylic material by using laser cutting machine and focused to potentially grow muscle fiber (diameter: 70–90 μm). In this study, edible films composed of salmon gelatin (having RGD-sequences), calcium alginate (as crosslinking agent), agarose (as gelling agent), and glycerol (as plasticizer) were fabricated by using cold-casting (micromolding) (Orellana et al., 2020).

Electrospinning and Other Fiber-Production Strategies

Electrospinning has been in the use for several decades to fabricate ECM-mimicking fibrous scaffolds (Karuppuswamy et al., 2014; Jiang et al., 2015). Electrospinning can be defined as a spinning technique that utilizes electrostatic forces to fabricate fibrous scaffolds from biocompatible polymers. A basic setup of electrospinning consists of four main parts namely, a glass syringe that contains the polymer solution, a metallic needle, power supply, and a metallic collector. A typical electrospinning process begins with the introduction of electric charges into the selected polymer solution using a magnetic needle. Due to the electric charges, the polymer solution becomes unstable, causing ultrafine nanofibers to emerge, which are then collected at the collector. The simple synthetic route and equipment setup make electrospinning a versatile method for fabricating fibrous scaffolds from various biocompatible polymers. In the field of tissue engineering, the electrospinning process has been extensively studied (Mo et al., 2019; Nemati et al., 2019; Rahmati et al., 2021). Precise control over the setup parameters (electric field, flow rate, distance between the needle and collector, and diameter of the metallic needle) and polymers parameter (viscosity, surface tension, concentration) could help in achieving the desired set of nanofibers with adequate properties. This technique could be further explored for artificial meat cultivation. Fibrous scaffolds made from biocompatible polymers using electrospinning may provide the necessary environment for culturing muscle cells and adipocytes. The fibrous structure of the scaffold will also help maintain the proper texture, closely mimicking real meat.

Another example of fiber-production technologies is an immersion rotatory jet-spinning (i.e., wet spinning) that applies centrifugal force to extrude polymer solution into a liquid precipitation bath. This method can produce fibers of varying diameters at a higher rate compared to electrospinning techniques. Additionally, a wide range of polymeric materials can be utilized in this method for fiber production. For instance, edible gelatin has been applied to produce fibers with certain textural characteristics resembling meat (MacQueen et al., 2019). Furthermore, various technologies, such as mixed or multilayer electrospinning (Kidoaki, Kwon, and Matsuda, 2005; Ju et al., 2010), coaxial electrospinning (Ji et al., 2010; Rim, Shin, and Shin, 2013), emulsion electrospinning (Nikmaram et al., 2017), and microfluidic spinning (Cheng et al., 2017), have been developed to produce electrospun scaffolds with specified constructs.

Additive Manufacturing Methods

In recent years, additive manufacturing methods (i.e., 3D printing) have been a continuous progressing field with their applicability in the field of tissue engineering (Seol, Kang, and Cho 2012; Serpooshan et al., 2017), medicine (Serrano, Terres, and Lalats 2018), and material sciences (Lee et al., 2017) engineering. 3D printing generates complex structures through layer-by-layer assembly and can utilize a variety of biomaterials. These characteristics have positioned 3D printing as a powerful fabrication technique for tissue engineering applications. It offers the ability to design a range of scaffolds with tissue-specific features (Do et al., 2015). In 3D bioprinting, a top-down approach is typically used, where cells are externally seeded onto the designed scaffolds. However, when fabricating human tissues, a bottom-up approach is employed (Yi, Lee, and Cho 2017). 3D printing necessitates precise control over the parameters and is further categorized under different methods namely; stereolithography, laser-assisted printing, extrusion-based, and inkjet printing. In lieu of the theme of the work, the two important methods, that is, laser-assisted printing and extrusion-based printing are explained in detail.

Laser Assisted Methods

The most common and conventional technique under the laser-assisted methods is laser machining. This technique represents a thermal energy-based noncontact-type advanced machining processes (Chryssolouris, Stavropoulos, and Salonitis 2015). The laser machining process is further accompanied by controlled switchboards to monitor the real-time progress of the fabrication technique. Further advancements in laser-assisted methods have led to the development of more sophisticated and effective laser-assisted fabrication techniques. Recently, Nawroth et al., reported a novel method that applies ultraviolet laser (UVL)-activated photosensitizers and engravers to create patterns on hydrogels. This technique offers a shorter fabrication time and a higher patterning volume compared to conventional methods (Nawroth et al., 2018). This novel method could be significant for tissue engineering applications, particularly in the field of culturing artificial meat. Additionally, another laser-assisted fabrication technique, known as Laser-induced forward transfer (LIFT) bioprinting, has emerged. This technique involves three layers, the first layer: an energy-absorbing layer on top; the second layer: a donor layer below the first layer; the third layer: the bioink layer, at the bottom. This technique involves the vaporization of the donor layer when exposed to a laser beam at specified locations resulting in the formation of high-pressure bubbles at the interface. This pressure results in the transfer of bioink onto the collection stage, resulting in the formation of a 3D model (Kattamis et al., 2007). LIFT offers the advantage of imparting high cell viability ratio together with readiness in using highly viscous materials (Mandrycky et al., 2016). Laser-assisted methods could be explored for designing patterned and layered scaffolds that could further be useful for meat culturing. The uniformity of the pattern will play a crucial role in designing molds and scaffolds for artificial meat culturing. During the cell attachment process, uniform layers will contribute to the proper texture of the cultured meat. Moreover, scalability and high cost of the apparatus might be an issue, and therefore, has less possibility for promising utilization in cultured meat production.

Extrusion-Based Printing

Extrusion-based 3D printing (EBP) dates back from the year 2002 and includes two major components, the first component is the extruder assembly and the second component is the positioning system. An EBP comprises of a heating element that can melt the material in use and expel it using a mechanical device. EBP-based 3D printing is more often used for fabricating biological scaffolds. In the case of EBP-based bioprinting the cells and the biomaterial are simultaneously and directly extruding on to substrate (Das et al., 2015). The EBP process is particularly useful when designing complex structures, as it involves multiple printing heads through which a high number of cells or bioinks can be introduced (Jung, Lee, and Cho, 2016). The EBP system offers the advantage of being able to print highly viscous bioinks in a continuous mode. In bioprinting, highly viscous materials are preferred as they prevent the collapse of the printed scaffold, thereby maintaining its integrity. EBP system could have its utility while designing scaffolds for artificial meat culturing. The ability to precisely design the construct will definitely help in culturing muscle cells and adipocytes, thus, ensuring proper texture and maintained integrity of the artificial synthesized meat. Therefore, adequate advances in extrusion-based bioprinting have shown great potential in cultured meat production industry. Owing to the various advantages of extrusion-based additive manufacturing, it can be used to replicate *in vivo* growth conditions and construct a 3D culture system *in vitro*. For instance, Li et al., fabricated porcine skeletal muscle tissue for cultured meat production using a 3D bioprinting technique, in which hydrogel bioinks were developed and characterized. The printed 3D network structures (4- and 6-layered constructs) demonstrated the formation of compact muscle fibers organized into multinucleated myotubes (Li et al., 2021).

Coaxial Nozzle-Assisted 3D Printing Process

Coaxial nozzle-assisted 3D printing process, the extension of extrusion-based printing process, has been considered a new strategy for alternative meat production (texturization) and effective for inserting a texture of meat analogs via artificial fiber insertion based on hydrocolloids. In a study, a 3D food printer was equipped with a coaxial nozzle for this purpose. In this study, a hydrocolloid-based muscle fiber simulation material was developed using various formulations of iota carrageenan (iCG), alginate (SA), and glucomannan (GM). Only the CG/GM (1.5/1.5) and CG/GM (2.5/1.5) formulations demonstrated elastic strength (strong and stable gel) comparable to that of actual meat (e.g., beef), due to the formation of the GM gel. Additionally, the CG/GM (2.5/1.5) formulation exhibited the highest hardness, closely resembling that of beef (Ko et al., 2021).

Combined Technologies

In a study, a patient-specific hepatic cell sheet manufactured from human chemically derived hepatic progenitor cells (hCdHs) on a multiscale fibrous scaffold by using the combination of electrospinning and 3D printing process. In this approach, a film or electrospun fibrous sheets were cultured with cells, and after 3 days of culturing, these sheets were stacked according to a designed formation. Before stacking, a collagen gel solution was coated on the cell-cultured sheets and cured at body temperature (37°C) with 5% CO2 to achieve a structurally strong stacking assembly. Furthermore, a single layer was formed with only hCdHs, a double layer was formed with hCdHs and human umbilical vein endothelial cells (HUVECs), and a triple layer was formed with HUVECs, hCdHs, and HUVECs (Kim et al., 2021). In another study, cell-seeded stacked porous silk film stacking was performed. In this case, the porous silk films were first soaked in P7 immortalized human corneal keratocytes (hCF) media for 30 minutes and cultured for 24 hours. Then, these cell-seeded porous films were stacked into 12-well plates, with a total of seven films stacked per assembly in each well. These stacks were pressed together near the film edges (Lawrence et al., 2009). A combination of approaches provides a platform where the synergistic effect of individual approaches could be explored for different applications.

Media and Bioreactors

Cells require culture media for their proliferation and differentiation with Fetal bovine serum being the most commonly utilized growth supplement. However, there are issues of inconsistency in components, contamination, and ethical concerns associated with animal serum as well as major cost contribution to the media (Gottipamula et al., 2013; Chen et al., 2022).

Therefore, it is crucial to utilize serum-free media for economical and feasible cultured meat production, particularly in large-scale operations. While several serum-free media formulations have already been developed or commercially available for tissue engineering and cell therapies, these preparations need to be of higher quality to support effective cellular activities in high-volume production at a lower cost (Gottipamula et al., 2013; Karnieli et al., 2017; Chen et al., 2022).

Approaches to developing serum-free media include the inclusion of ingredients from plants (e.g., wheat peptone) or microbial fermentations (e.g., yeast extract), wherein wheat peptone could be a potential substitute for FBS or even a complex media comprised of yeast extract/peptone/dextrose (YPD) and medicinal feed additive can supply similar nutrients as compared to Essential 8™ (O'Neill et al., 2021). The use of plant- or fungi-derived ingredients may provide effective and affordable substitutes for growth factors in media, offering lower costs, easy integration, and the potential to enhance the flavor of the final meat product. However, these ingredients may introduce variations in the final product, including impurities that could be cytotoxic. Additionally, genetically modified plants, microbes, or fungi serve as favorable expression hosts for producing recombinant growth factors, which are a primary cost driver in media preparations (Kuo et al., 2020; Chen et al., 2022). To manufacture a wide range of inexpensive growth factors with improved stability and effectiveness, protein engineering approaches show promise and should be further explored to reduce overall media costs (Barnes, Uruakpa, and Udenigwe 2015; Chen et al., 2022). Further, the media recycling technologies can reduce the cost of production by maintaining a nutrient footprint for large-scale cultured meat production; however, they are still in their infancy and require more developments. Therefore, continuous development is highly desired to execute cost-effective recycling technologies for media in industrial bioreactors for cultured meat production. Moreover, there are several preparations and compounds to be considered for developing media. High throughput screening technologies are required to develop cell-specific animal-free media (Burton et al., 2000; Guan et al., 2021; Stout et al., 2022).

For industrial-scale production of cultured meat, a sequence of bioreactors is desirable to offer a regulated environment for different parameters such as pH, temperature, dissolved oxygen, and carbon dioxide. Preferably, bioreactors should be designed to minimize shear stress and provide a standardized microenvironment to promote large-scale cell growth at an exponential rate, while minimizing energy use and consumable supplies

(Allan, De Bank, & Ellis, 2019). Additionally, well-designed bioreactors should be automated with real-time monitoring to ensure the continuous supply of nutrients and the removal of waste byproducts.

Challenges and Opportunities

Cultured meat production is an alternative approach toward resource-efficient and cost-effective artificial meat compared to orthodox animal meat. However, its full commercialization at industrial scale is still not achieved due to various technological, environmental, and economical challenges, including the societal perception (Fernandes et al., 2022; Hubalek, Post, and Moutsatsou 2022). It is critical to understand first, the composition, structure, cellular, and molecular communications across muscle tissue to develop a proficient edible scaffold for cultured meat manufacturing with sensorial, nutritional, and textural attributes. The development of sustainable meat alternatives faces various fundamental challenges, such as effective tissue engineering, adjusted cultured settings, safe and cost-effective serum-free culture media, large-scale bioreactors for scaled-up production, food safety as well as the issues concerning social and ethical restrictions (Orzechowski 2015; Tiberius, Borning, and Seeler 2019; Tuomisto 2019; Zhang et al., 2020). Though plant-based growth factors and serum-free media are constantly being explored they can impart possible variation in the ultimate meat product, including the impurities that could be cytotoxic in nature.

The technologies for artificial cultured meat have continuously been developed and upgraded. Here, appropriate cultivation of artificial meat could be achieved by using tissue engineering of biomaterials through various potential novel manufacturing technologies such as film stacking and solvent-casting (with or without the application of crosslinking). Among them, extrusion, shear cell, and spinning are the main technologies. However, apart from fabrication, the nutritional, health, sensory, and safety of artificially cultured meat are still major challenges.

Among various challenges, the fabrication of edible scaffolds with adequate cell responsiveness, without the use of harmful chemical agents, remains a significant obstacle in cultured meat production. To address these issues, innovative manufacturing technologies, such as additive manufacturing, have emerged (Levi et al., 2022). These technologies are more effective compared to traditional cellular agricultural methods, with 3D bioprinting offering meat analogues that meet better nutritional and sensory standards for the development of tailored animal protein-based products (Balasubramanian et al., 2021; Bhat et al., 2021). Additionally, the integration of nanotechnology-based sensors with the bioprocessing of cultured meat facilitates the efficient evaluation of quality throughout the food supply chain. The lab-cultured meat is the need of the hour, and with its high market demand, ethical involvements, sentimental values, and expectancy, it is anticipated that the coming few years will witness a shift in the era from farm meat to lab-culture meat.

Conclusion

This chapter has provided an in-depth exploration of the structural and technological characteristics of scaffold manufacturing for cultured meat. Scaffolds play a critical role in mimicking the structure and texture of traditional meat, providing essential support for the proliferation, differentiation, and maturation of cells. The advances in scaffold technologies, including various biomaterials, 3D printing methods, and electrospinning techniques, represent exciting breakthroughs in the field of cellular agriculture. However, challenges related to scalability, cost-effectiveness, and the biological compatibility of scaffolds with cultured cells remain significant barriers to large-scale production.

Despite these challenges, the continuous evolution of scaffold materials and manufacturing techniques holds the potential to significantly improve the quality and viability of cultured meat products. As technology advances, it is anticipated that more efficient and sustainable scaffolds will emerge, facilitating the growth of thicker, more complex tissues that closely resemble traditional meat in terms of texture, flavor, and nutritional composition.

In conclusion, the development of effective scaffolds is pivotal to the future success of cultured meat production. Continued research into novel scaffold materials, manufacturing methods, and bioreactor technologies will be essential in overcoming current limitations and ensuring the scalability of cellular meat. By addressing these challenges, we are one step closer to realizing the full potential of cultured meat as a sustainable and ethical alternative to conventional animal agriculture.

References

Ahmad, K., J. H. Lim, E. J. Lee, H. J. Chun, S. Ali S, S. S. Ahmad, S. Shaikh, and I. Choi. 2021. Extracellular matrix and the production of cultured meat. *Foods* 10 (12):3116.

Andreassen, R. C., S. B. Rønning, N. T. Solberg, K. G. Grønlien, K. A. Kristoffersen, V. Høst, S. O. Kolset, and M. E. Pedersen. 2022. Production of food-grade microcarriers based on by-products from the food industry to facilitate the expansion of bovine skeletal muscle satellite cells for cultured meat production. *Biomaterials* 286:121602. doi: 10.1016/j.biomaterials.2022.121602.

Asghar, A., K. Samejima, T. Yasui, and R. L. Henrickson. 1985. Functionality of muscle proteins in gelation mechanisms of structured meat products. *Critical Reviews in Food Science and Nutrition* 22 (1):27–106. doi: 10.1080/10408398509527408.

Barnes, M. J., F. Uruakpa, C, and Udenigwe, C. 2015. Influence of cowpea (Vigna unguiculata) peptides on insulin resistance. *Journal of Nutritional Health & Food Science* 3:1–3. doi: 10.15226/jnhfs.2015.00144

Ben-Arye, T., and S. Levenberg. 2019. Tissue engineering for clean meat production. *Frontiers in Sustainable Food Systems* 3:46. doi: 10.3389/fsufs.2019.00046

Ben-Arye, T., Y. Shandalov, S. Ben-Shaul, S. Landau, Y. Zagury, I. Ianovici, N. Lavon, and S. Levenberg. 2020. Textured soy protein scaffolds enable the generation of three-dimensional bovine skeletal muscle tissue for cell-based meat. *Nature Food* 1 (4):210–20. doi: 10.1038/s43016-020-0046-5

Bosnakovski, D., M. Mizuno, G. Kim, S. Takagi, M. Okumura, and T. Fujinaga. 2005. Isolation and multilineage differentiation of bovine bone marrow mesenchymal stem cells. *Cell and Tissue Research* 319 (2):243–53. doi: 10.1007/s00441-004-1012-5.

Bryant, C., and J. Barnett. 2020. Consumer acceptance of cultured meat: An updated review (2020). *Applied Sciences* 10 (15):5201. doi: 10.3390/app10155201.

Burk, J., I. Erbe, D. Berner, J. Kacza, C. Kasper, B. Pfeiffer, K. Winter, and W. Brehm. 2014. Freeze-thaw cycles enhance decellularization of large tendons. Tissue Engineering. Part C, *Methods 20* (4):276–84. doi: 10.1089/ten.TEC.2012.0760.

Chan, W. P., F. C. Kung, Y. L. Kuo, M. C. Yang, and W. F. Lai. 2015. Alginate/poly(γ-glutamic acid) base biocompatible gel for bone tissue engineering. *BioMed Research International* 2015:185841. doi: 10.1155/2015/185841.

Chen, X., L. Zhou, H. Xu, M. Yamamoto, M. Shinoda, M. Kishimoto, T. Tanaka, and H. Yamane. 2020. Effect of the application of a dehydrothermal treatment on the structure and the mechanical properties of collagen film. *Materials* 13 (2):377. doi: 10.3390/ma13020377.

Cheng, J., Y. Jun, J. Qin, and S.-H. Lee. 2017. Electrospinning versus microfluidic spinning of functional fibers for biomedical applications. *Biomaterials* 114:121–43. doi: 10.1016/j.biomaterials.2016.10.040

Choi, S. H., K. Y. Chung, B. J. Johnson, G. W. Go, K. H. Kim, C. W. Choi, and S. B. Smith. 2013. Co-culture of bovine muscle satellite cells with preadipocytes increases PPARγ and C/EBPβ gene expression in differentiated myoblasts and increases GPR43 gene expression in adipocytes. *The Journal of Nutritional Biochemistry* 24 (3):539–43. doi: 10.1016/j.jnutbio.2012.01.015.

Choudhury, D., T. W. Tseng, and E. Swartz. 2020. The business of cultured meat. *Trends in Biotechnology* 38 (6):573–7. doi: 10.1016/j. tibtech.2020.02.012.
Choudhury, D., T. W. Tseng, and E. Swartz. 2020. The business of cultured meat. *Trends in Biotechnology* 38 (6):573–7. doi: 10.1016/j. tibtech.2020.02.012.
Cook, W., and R. Rose. 1934. Solubility of gluten. *Nature* 134 (3384):380–1. doi: 10.1038/134380c0.

Das, S., F. Pati, Y. J. Choi, G. Rijal, J. H. Shim, S. W. Kim, A. R. Ray, D. W. Cho, and S. Ghosh. 2015. Bioprintable, cell-laden silk fibroin-gelatin hydrogel supporting multilineage differentiation of stem cells for fabrication of three-dimensional tissue constructs. *Acta Biomaterialia* 11:233–46. doi: 10.1016/j.actbio.2014.09.023.

Datar, I., and M. Betti. 2010. Possibilities for an *in vitro* meat production system. Innovative *Food Science & Emerging Technologies* 11 (1):13–22. doi: 10.1016/j.ifset.2009.10.007.

DiMaio, T. 2019. *This scientist is developing new cell lines for slaughter-free meat.* Washington, DC: The Good Food Institute.

Djisalov, M., T. Knežić, I. Podunavac, K. Živojević, V. Radonic, N. Ž. Knežević, I. Bobrinetskiy, and I. Gadjanski. 2021. Cultivating multidisciplinarity: Manufacturing and sensing challenges in cultured meat production. *Biology* 10 (3):204. doi: 10.3390/biology10030204.

Dohmen, R. G. J., S. Hubalek, J. Melke, T. Messmer, F. Cantoni, A. Mei, R. Hueber, R. Mitic, D. Remmers, P. Moutsatsou, et al., 2022. Muscle-derived fibro-adipogenic progenitor cells for production of cultured bovine adipose tissue. *NPJ Science of Food* 6 (1):6–12. doi: 10.1038/s41538-021-00122-2.

Elli, L., E. Dolfini, and M. T. Bardella. 2003. Gliadin cytotoxicity and *in vitro* cell cultures. *Toxicology Letters* 146 (1):1–8. doi: 10.1016/j. toxlet.2003.09.004.

Fernandes, A. M., O. de Souza Teixeira, A. L. Fantinel, J. P. P. Revillion, and ÂR L. de Souza. 2022. Technological prospecting: The case of cultured meat. *Future Foods* 6:100156. doi: 10.1016/j.fufo.2022.100156.

Gandolfi, F., G. Pennarossa, S. Maffei, and T. Brevini. 2012. Why is it so difficult to derive pluripotent stem cells in domestic ungulates? *Reproduction in Domestic Animals* 47:11–7. doi: 10.1111/j.1439-0531.2012.02106.x.

Ghosal, K., A. Chandra, P. G, S. S, S. Roy, C. Agatemor, S. Thomas, and I. Provaznik. 2018. Electrospinning over solvent casting: Tuning of mechanical properties of membranes. *Scientific Reports* 8 (1):5058. doi: 10.1038/s41598-018-23378-3.

Gil, E. S., S. H. Park, J. Marchant, F. Omenetto, and D. L. Kaplan. 2010. Response of human corneal fibroblasts on silk film surface patterns. *Macromolecular Bioscience* 10 (6):664–73. doi: 10.1002/mabi.200900452.

Ianovici, I., Y. Zagury, I. Redenski, N. Lavon, and S. Levenberg. 2022. 3D-printable plant protein-enriched scaffolds for cultivated meat development. *Biomaterials* 284:121487. doi: 10.1016/j.biomaterials.2022.121487.

Jackson, D. W., G. E. Windler, and T. M. Simon. 1990. Intraarticular reaction associated with the use of freeze-dried, ethylene oxide-sterilized bone-patella tendon-bone allografts in the reconstruction of the anterior cruciate ligament. *The American Journal of Sports Medicine* 18 (1):1–10.

Jaques, A., E. Sánchez, N. Orellana, J. Enrione, and C. A. Acevedo. 2021. Modelling the growth of in-vitro meat on microstructured edible films. *Journal of Food Engineering* 307:110662. doi: 10.1016/j. jfoodeng.2021.110662.

Jones, J. D., A. S. Rebello, and G. R. Gaudette. 2021. Decellularized spinach: An edible scaffold for laboratory-grown meat. *Food Bioscience* 41:100986. doi: 10.1016/j.fbio.2021.100986.

Jung, J. W., J.-S. Lee, and D.-W. Cho. 2016. Computer-aided multiple-head 3D printing system for printing of heterogeneous organ/tissue constructs. *Scientific Reports* 6 (1):21685. doi: 10.1038/srep21685

Kamalapuram, S. K., H. Handral, and D. Choudhury. 2021. Cultured Meat Prospects for a Billion!. *Foods* 10 (12):2922. doi: 10.3390/foods10122922.

Kang, D.-H., F. Louis, H. Liu, H. Shimoda, Y. Nishiyama, H. Nozawa, M. Kakitani, D. Takagi, D. Kasa, E. Nagamori, et al., 2021. Engineered whole cut meat-like tissue by the assembly of cell fibers using tendon-gel integrated bioprinting. *Nature Communications* 12 (1):5059. doi: 10.1038/s41467-021-25236-9.

Kang, D.-H., F. Louis, H. Liu, H. Shimoda, Y. Nishiyama, H. Nozawa, M. Kakitani, D. Takagi, D. Kasa, E. Nagamori, et al., 2021. Engineered whole cut meat-like tissue by the assembly of cell fibers using tendon-gel integrated bioprinting. *Nature Communications* 12 (1):5059. doi: 10.1038/s41467-021-25236-9

Karnieli, O., O. M. Friedner, J. G. Allickson, N. Zhang, S. Jung, D. Fiorentini, E. Abraham, S. S. Eaker, T. K. Yong, A. Chan, et al., 2017. A consensus introduction to serum replacements and serum-free media for cellular therapies. *Cytotherapy* 19 (2):155–69. doi: 10.1016/j.jcyt.2016.11.011.

Karuppuswamy, P., J. R. Venugopal, B. Navaneethan, A. L. Laiva, S. Sridhar, and S. Ramakrishna. 2014. Functionalized hybrid nanofibers to mimic native ECM for tissue engineering applications. *Applied Surface Science* 322:162–8. doi: 10.1016/j.apsusc. 2014.10.074.

Kattamis, N. T., P. E. Purnick, R. Weiss, and C. B. Arnold. 2007. Thick film laser induced forward transfer for deposition of thermally and mechanically sensitive materials. *Applied Physics Letters* 91 (17):171120. doi: 10.1063/1.2799877.

Keane, T. J., I. T. Swinehart, and S. F. Badylak. 2015. Methods of tissue decellularization used for preparation of biologic scaffolds and *in vivo* relevance. *Methods* (San Diego, Calif.) 84:25–34. doi: 10.1016/j.ymeth.2015.03.005.

Kim, Y., Y. W. Kim, S. B. Lee, K. Kang, S. Yoon, D. Choi, S.-H. Park, and J. Jeong. 2021. Hepatic patch by stacking patient-specific liver progenitor cell sheets formed on multiscale electrospun fibers promotes regenerative therapy for liver injury. *Biomaterials* 274:120899. doi: 10.1016/j.biomaterials.2021.120899.

Knežić, T., L. Janjušević, M. Djisalov, S. Yodmuang, and I. Gadjanski. 2022. Using vertebrate stem and progenitor cells for cellular agriculture-state-of-the-art, challenges, and future perspectives. *Biomolecules* 12 (5):699. doi: 10.3390/biom 12050699.

Ko, H. J., Y. Wen, J. H. Choi, B. R. Park, H. W. Kim, and H. J. Park. Materials Science & Engineering. C, *Materials for Biological Applications* 96:153–65. doi: 10.1016/j. msec.2018.10.086.

Maude, S., E. Ingham, and A. Aggeli. 2013. Biomimetic self-assembling peptides as scaffolds for soft tissue engineering. *Nanomedicine* (London, England) 8 (5):823–47. doi: 10.2217/nnm.13.65

Mendibil, U., R. Ruiz-Hernandez, S. Retegi-Carrion, N. Garcia-Urquia, B. Olalde-Graells, and A. Abarrategi. 2020. Tissue-specific decellularization methods: Rationale and strategies to achieve regenerative compounds. *International Journal of Molecular Sciences* 21 (15):5447. doi: 10.3390/ijms21155447.

Nonoyama, T., Y. W. Lee, K. Ota, K. Fujioka, W. Hong, and J. P. Gong. 2020. Instant thermal switching from soft hydrogel to rigid plastics inspired by thermophile proteins. *Advanced Materials* 32 (4):1905878. doi: 10.1002/adma.201905878.

O'Neill, E. N., Z. A. Cosenza, K. Baar, and D. E. Block. 2021. Considerations for the development of cost-effective cell culture media for cultivated meat production. *Comprehensive Reviews in Food Science and Food Safety* 20 (1):686–709. doi: 10.1111/1541-4337.12678.

Ofek, G., C. M. Revell, J. C. Hu, D. D. Allison, K. J. Grande-Allen, and K. A. Athanasiou. 2008. Matrix development in self-assembly of articular cartilage. *PLoS One* 3 (7):e2795. doi: 10.1371/journal. pone.0002795.

Ogilvie, O., N. Larsen, K. Sutton, L. Domigan, J. Gerrard, N. Demarais, and S. Roberts. 2020. A targeted mass spectrometry method for the accurate label-free quantification of immunogenic gluten peptides produced during simulated digestion of food matrices. *Methods*X 7:101076. doi: 10.1016/j.mex.2020.101076.

Okamura, L. H., P. Cordero, J. Palomino, V. H. Parraguez, C. G. Torres, and O. A. Peralta. 2018. Myogenic differentiation potential of mesenchymal stem cells derived from fetal bovine bone marrow. *Animal Biotechnology* 29 (1):1–11. doi: 10.1080/10495398.2016.1276926.

Ong, S., L. Loo, M. Pang, R. Tan, Y. Teng, X. Lou, S. K. Chin, M. Y. Naik, and H. Yu. 2021. Decompartmentalisation as a simple color manipulation of plant-based marbling meat alternatives. *Biomaterials* 277:121107. doi: 10.1016/j.biomaterials.2021.121107.

Orellana, N., E. Sánchez, D. Benavente, P. Prieto, J. Enrione, and C. A. Acevedo. 2020. A new edible film to produce *in vitro* meat. *Foods* 9 (2):185. doi: 10.3390/foods9020185.

Orellana, N., E. Sánchez, D. Benavente, P. Prieto, J. Enrione, and C. A. Acevedo. 2020. A new edible film to produce *in vitro* meat. *Foods* 9 (2):185. doi: 10.3390/foods9020185.

Oryan, A., A. Kamali, A. Moshiri, H. Baharvand, and H. Daemi. 2018. Chemical crosslinking of biopolymeric scaffolds : Current knowledge and future directions of crosslinked engineered bone scaffolds. *International Journal of Biological Macromolecules* 107 (Pt A):678–88. doi: 10.1016/j.ijbiomac.2017.08.184.

Page, H., P. Flood, and E. G. Reynaud. 2013. Three-dimensional tissue cultures: Current trends and beyond. *Cell and Tissue Research* 352 (1):123–31. doi: 10.1007/s00441-012-1441-5.

Park, S., S. Jung, J. Heo, W.-G. Koh, S. Lee, and J. Hong. 2021. Chitosan/cellulose-based porous nanofilm delivering c-phycocyanin: A novel platform for the production of cost-effective cultured meat. *ACS Applied Materials & Interfaces* 13 (27):32193–204.

Piluso, S., D. Flores Gomez, I. Dokter, L. Moreira Texeira, Y. Li, J. Leijten, R. van Weeren, T. Vermonden, M. Karperien, and J. Malda. 2020. Rapid and cytocompatible cell-laden silk hydrogel formation via riboflavin-mediated crosslinking. *Journal of Materials Chemistry B* 8 (41):9566–75. doi: 10.1039/d0tb01731k.

Ramani, S., D. Ko, B. Kim, C. Cho, W. Kim, C. Jo, C.-K. Lee, J. Kang, S. Hur, and S. Park. 2021. Technical requirements for cultured meat production: A review. *Journal of Animal Science and Technology* 63 (4):681–92. doi: 10.5187/jast.2021.e45.

Ramírez-Espinosa, J. J., L. González-Dávalos, A. Shimada, E. Piña, A. Varela-Echavarria, and O. Mora. 2016. Bovine (bos taurus) bone marrow mesenchymal cell differentiation to adipogenic and myogenic lineages. *Cells Tissues Organs* 201 (1):51–64. doi: 10.1159/000440878.

Reiss, J., S. Robertson, and M. Suzuki. 2021. Cell sources for cultivated meat: Applications and considerations throughout the production workflow. *International Journal of Molecular Sciences* 22 (14):7513. doi: 10.3390/ijms22147513.

Reiss, J., S. Robertson, and M. Suzuki. 2021. Cell sources for cultivated meat: Applications and considerations throughout the production workflow. *International Journal of Molecular Sciences* 22 (14):7513. doi: 10.3390/ijms22147513.

Seah, J. S. H., S. Singh, S. L. P. Tan, and D. Choudhury. 2022. Scaffolds for the manufacture of cultured meat. *Critical Reviews in Biotechnology* 42 (2):311–23. doi: 10.1080/07388551.2021.1931803.

Seah, J. S. H., S. Singh, S. L. P. Tan, and D. Choudhury. 2022. Scaffolds for the manufacture of cultured meat. *Critical Reviews in Biotechnology* 42 (2):311–23. doi: 10.1080/07388551.2021.1931803.

Sealy, M., K. Avegnon, A. Garrett, L. Delbreilh, S. Bapat, and A. Malshe. 2022. Understanding biomanufacturing of soy-based scaffolds for cell-cultured meat by vat polymerization. *CIRP Annals* 71 (1):209–12. doi: 10.1016/j.cirp.2022.04.001.

Seol, Y.-J., T.-Y. Kang, and D.-W. Cho. 2012. Solid freeform fabrication technology applied to tissue engineering with various biomaterials. *Soft Matter* 8 (6):1730–5. doi: 10.1039/C1SM06863F.

Serpooshan, V., M. Mahmoudi, D. A. Hu, J. B. Hu, and S. M. Wu. 2017. Bioengineering cardiac constructs using 3D printing. *Journal of 3D Printing in Medicine* 1 (2):123–39. doi: 10.2217/3dp-2016-0009.

Serrano, D. R., M. C. Terres, and A. Lalats. 2018. Applications of 3D printing in cancer. *Journal of 3D Printing in Medicine* 2 (3):115–27. doi: 10.2217/3dp-2018-0007.

Sha, L., and Y. L. Xiong. 2020. Plant protein-based alternatives of reconstructed meat: Science, technology, and challenges. *Trends in Food Science & Technology* 102:51–61. doi: 10.1016/j.tifs.2020.05.022.

Specht, E. A., D. R. Welch, E. M. R. Clayton, and C. D. Lagally. 2018. Opportunities for applying biomedical production and manufacturing methods to the development of the clean meat industry. *Biochemical Engineering Journal* 132:161–8. doi: 10.1016/j.bej.2018.01.015.

Stout, A. J., A. B. Mirliani, M. L. Rittenberg, M. Shub, E. C. White, J. S. K. Yuen, and D. L. Kaplan. 2022. Simple and effective serum-free medium for sustained expansion of bovine satellite cells for cell cultured meat. *Communications Biology* 5 (1):466. doi: 10.1038/s42003-022-03423-8.

Thyden, R., L. R. Perreault, J. D. Jones, H. Notman, B. M. Varieur, A. A. Patmanidis, T. Dominko, and G. R. Gaudette. 2022. An edible, decellularized plant derived cell carrier for lab grown meat. *Applied Sciences* 12 (10):5155. doi: 10.3390/app12105155.

Tiberius, V., J. Borning, and S. Seeler. 2019. Setting the table for meat consumers: An International Delphi study on *in vitro* meat. *NPJ Science of Food* 3 (1):10–6. doi: 10.1038/s41538-019-0041-0.

Tuomisto, H. L. 2019. The eco-friendly burger: Could cultured meat improve the environmental sustainability of meat products? *EMBO Reports* 20 (1):e47395. doi: 10.15252/embr.201847395.

Vieira, S., A. R. Franco, E. M. Fernandes, S. Amorim, H. Ferreira, R. A. Pires, R. L. Reis, A. Martins, and N. M. Neves. 2018. Fish sarcoplasmic proteins as a high value marine material for wound dressing applications. *Colloids and Surfaces B, Biointerfaces* 167:310– 7. doi: 10.1016/j.colsurfb.2018.04.002.

Wahid, F., T. Khan, Z. Hussain, and H. Ullah. 2018. 30 - Nanocomposite scaffolds for tissue engineering; properties, preparation and applications. In *Applications of nanocomposite materials in drug delivery*, eds. I. A. M. Asiri and A. Mohammad, 701–35. Sawston: Woodhead Publishing.

Xiang, N., Y. Yao, J. S. YuenJr., A. J. Stout, C. Fennelly, R. Sylvia, A. Schnitzler, S. Wong, and D. L. Kaplan. 2022. Edible films for cultivated meat production. *Biomaterials* 287:121659. doi: 10.1016/j.biomaterials.2022.121659.

Xue, X., Y. Hu, S. Wang, X. Chen, Y. Jiang, and J. Su. 2022. Fabrication of physical and chemical crosslinked hydrogels for bone tissue engineering. *Bioactive Materials* 12:327–39. doi: 10.1016/j.bioactmat.2021.10.029.

Chapter 3

The Critical Role of Stem Cells in the Production of Cultured Meat

Abstract

Cellular agriculture is a growing field in agribiotechnology aimed at producing agricultural products using stem cells, without the need for animal sacrifice or crop cultivation. Cultivated meat, a key product of this field, is being researched to address issues related to food insecurity, the environment, and ethics. This chapter explores the use of stem cells for cultivating muscle and adipose tissue to replicate fresh meat. Establishing safety standards and regulations for stem cell culturing is crucial for introducing cultured meat-based biomaterials to the market. Advances in cultivated meat techniques could pave the way for the development of other cellular agriculture industries in the future.

Keywords: stem cells, cultured meat, muscle tissue engineering, cell differentiation

Introduction

Currently, upcoming inevitable challenges are driving traditional agriculture to change. As the world population is estimated to reach approximately 10 billion by 2050, food production from the conventional agriculture industry would not be sufficient to nourish people globally because of the lack of arable land and the crisis of climate change (Eibl et al., 2021). To overcome the problems ahead, cellular agriculture is being considered as one of the potential solutions for food security. Cellular agriculture is an emerging research field of agribiotechnology that aims to produce agricultural products and by-products by culturing microorganisms, plant, and animal cells or tissues (Rischer et al., 2020). The resulting products of cellular agriculture are generally classified into acellular or cellular products. The former includes by-products derived from the cells of microorganisms, plants, and animals, such as proteins, fats, food additives, pigments, flavors, and aroma components, while the latter refers to *in vitro* cultivated cells or tissues themselves, such as meat and leather (Eibl et al., 2021; Rischer et al., 2020). Cellular agriculture is further categorized based on production methods, including tissue engineering or fermentation (Stephens et al., 2018). Although the term "cellular agriculture" was first coined in 2015, its foundational studies began with the discovery of plant cell totipotency in the early 20th century, followed by the development of *in vitro* culturing of animal cells and tissues, fermentation technology, and recombinant bacterial DNA production (Eibl et al., 2021; Haberlandt, 1902). Subsequently, various academic and industrial attempts have been conducted to create artificial agricultural products, such as secondary metabolite production by plant cells and tissues, ginsenosides, red dye shikonin, and vanillin (Eibl et al., 2021).

Cultivated meat, as the representative cellular product of cellular agriculture, is actively researched and its industrial scale is rapidly growing due to worldwide food insecurity, and environmental and ethical issues (Post et al., 2020). The concept of cultivated meat, originally proposed in Winston Churchill's book *50 Years Hence* nearly a century ago, became a reality in 2013 with the production of hamburger patties (Post, 2014; Reiss et al., 2021). These patties were composed of 10,000 individual muscle fibers differentiated from *in vitro* cultured bovine muscle stem cells using tissue engineering technology (Reiss et al., 2021). The manufacturing process of cultivated meat generally involves several steps, including the isolation of muscle stem cells, primary and upscaled culture of these cells, and muscle differentiation and maturation (detailed reviews can be found in previous articles such as Choi et al., 2021; Ong et al., 2021; Post et al., 2020). Numerous efforts have been made to optimize each material and process to produce cost-effective, consumer-satisfying cultivated meat. Along with technological advancements, support from legal and socio-cultural aspects, such as standards and safety regulations, has been concurrently discussed. The fact that meat analogs could be obtained without raising domestic animals has explosively increased multidisciplinary investments and interest in cultivated meat. Eventually, in 2020, the cultivated chicken of Eat Just was approved as a novel food by the Singapore Food Agency (SFA) and has been sold in Singapore (Ng et al., 2021).

Four technologies including cells, culture media, scaffolds, and mass culture system are reportedly fundamental to producing cultivated meat. Stem cells are considered the most important of these, as they serve as the fundamental and structural units of cultivated meat and cellular agriculture, analogous to seeds in conventional agriculture. A deep understanding of stem cell biology is therefore crucial for producing efficient and reliable *in vitro* cultured cell-based food materials. This chapter will comprehensively discuss the basics of stem cell biology and its applications in cultivated meat production. It will then analyze the factors influencing production costs and safety considerations associated with cultivated meat.

Stem Cells for Muscle Tissue

Muscle Development

During embryogenesis, pluripotent inner cell mass (ICM; or epiblasts) sequentially develops toward myogenic progenitor via paraxial mesoderm, thereby generating muscle fibers and muscle tissues by cell-to-cell fusion (Chal and Pourquie, 2017). The myogenic progenitor cells settle beneath the basal lamina of muscle fibers, where they differentiate into quiescent myogenic satellite cells, forming the stem cell reservoir responsible for regenerating muscle tissues throughout life during postnatal development. From the embryo to adulthood, various stem cell types, including ESCs, muscle stem cells (MuSCs), and mesenchymal stem cells (MSCs), contribute to muscle production. Extensive research on muscle tissue regeneration has been conducted in the past to address muscle dysfunction and, more recently, to produce cultivated meat.

Pluripotent Stem Cells

Practically, PSCs represented by ESCs and iPSCs are considered ideal cell sources for producing cultivated meat without sacrificing animals due to their indefinite self-renewal capacity (Post et al., 2020). Authentic ESC lines from domestic animals including pigs and cows have recently been derived, while domestic animal iPSC lines independent of transgene expression have not been reported yet (Bogliotti et al., 2018; Choi et al., 2019). Although cellular reprogramming techniques are meant to be an alternative to create PSCs from somatic cells, genetically modified organism-driven issues related to food safety regulations and consumer acceptance remain unsolved (Sendhil et al., 2022). MuSCs and muscle fibers can be derived from pluripotent stem cells (PSCs) through *in vitro* recapitulation of embryonic myogenesis, as PSCs share molecular biological characteristics with the epiblasts of early embryos

(Chal and Pourquie, 2017). *In vitro* myogenesis, achieved through various combinations of hormones and cytokines, upregulates myogenic determinants such as PAX7, MYOD1, and MYF5 in PSCs, leading to their differentiation into MuSCs. Recent studies report myogenic differentiation rates ranging from 20 to 40% (Chal et al., 2018). A deeper understanding of the *in vivo* environment involved in myogenic precursor specification could improve the efficiency and reliability of the results. However, maintaining stemness and regulating differentiation pathways in highly potent stem cells necessitate advanced cell culture techniques. Therefore, alternatives for producing cultivated meat have been investigated and are analyzed herein.

Muscle Stem Cells

Various types of stem/progenitor cells resident in muscle tissues reportedly participate in skeletal muscle regeneration, such as myogenic satellite cells, MSCs, interstitial cells, fibroadipogenic progenitors, muscle side population cells, and pericytes (Klimczak et al., 2018). Of those, MuSCs including quiescent myogenic satellite cells and proliferating myoblasts are the main contributors to muscle fiber production (Kuang et al., 2007). Therefore, *in vitro* cultured muscle stem cells (MuSCs) exhibit the highest myogenic competency compared to other types of stem cells. MuSCs can be isolated from the muscle tissues of fetuses, adults, and even post-slaughter. The yield and biological characteristics of isolated MuSCs are influenced by factors such as the age, breed, and sex of the donor animals (Choi et al., 2021). Additionally, the number of MuSCs varies depending on the specific location of the muscle tissues.

Male animals have a greater number of MuSCs than female animals, especially proliferating satellite cells (Day et al., 2010). As muscle grows, the total number of stem cells increases while their density decreases, and in turn, satellite cell population declines losing its proliferative potential with aging (Campion et al., 1981). Stem cells from breeds having different genetic backgrounds have distinct myogenic ability features. For example, stem cells from the 'Belgian blue' breed harboring a GDF8 gene mutation exhibit enhanced proliferation capacity compared to those from other cattle breeds (Quinn et al., 1990). In fact, for *in vitro* culture, the stemness of MuSCs is reduced with prolonged culture, even when sourced from fetuses or young animals. This suggests that a constant supply of muscle tissues, such as from biopsies and slaughters, would be necessary for the continuous and sufficient production of cultivated meat. Therefore, alongside research on rejuvenating and extending the culture period of MuSCs, considering various animal conditions is crucial to improving the isolation efficiency of MuSCs while minimizing animal sacrifices. Cellular immortalization may spontaneously occur due to stress-induced disruptions in the gene regulatory network during long-term culture and could be utilized (Soice and Johnston, 2021). However, naturally immortalized cell lines are difficult to obtain due to low yields and unpredictable physiological characteristics resulting from genetic alterations.

Mesenchymal Stem/Stromal Cells

Mesenchymal stem/stromal cells (MSCs) reportedly reside in various tissues such as bone marrow and adipose tissues in a quiescent state, and upon injury, they undergo activation to regenerate various tissues. To date, although their origin remains poorly understood, some research suggests that pluripotent epiblasts differentiate into mesodermal progenitors through lateral-plate mesoderm formation via epithelial-mesenchymal transition (EMT), contributing to body formation (Li et al., 2021). *In vitro*, these progenitors can be isolated from various tissues, including bone marrow, adipose tissue, and the umbilical cord, or obtained through direct differentiation from pluripotent stem cells (PSCs). MSCs as multipotent stem cells could reportedly be differentiated into several mesodermal tissues including adipocytes, osteoblasts, chondrocytes, and myogenic progenitors (Joe et al., 2010). Indeed, the myogenic potential of MSCs has been controversial. *In vitro* cultured MSCs isolated from bone marrow highly expressed myogenic determinant genes by differentiation cues, although their biological functions were poorly characterized along with a low differentiation rate (Okamura et al., 2018). Furthermore, transplant studies using mesenchymal stem cells (MSCs) tagged with reporter genes have demonstrated that MSCs contribute to muscle regeneration by forming muscle fibers (Fukada et al., 2002). However, other studies have indicated that MSCs lack myogenic potential after transplantation, as MSCs and muscle stem cells (MuSCs) originate from different developmental lineages: lateral-plate mesoderm and paraxial mesoderm, respectively (Leinroth et al., 2022; Uezumi et al., 2010). MSCs secrete various cytokines that promote the proliferation and differentiation of MuSCs, as well as facilitate the infiltration of immune cells to induce inflammatory responses, thereby supporting muscle reconstruction in indirect ways (Joe et al., 2010; Leinroth et al., 2022). In this regard, MSCs could be applied to enhance muscle tissue growth or for the production of adipose tissue.

Genetic Modification

Recently, genetically modified animals, including pigs and salmon, were approved as food by the US FDA (Dolgin, 2021; Waltz, 2017). Although both genetically-engineered animal- or cell-derived food materials have a long way to go for approval by the government around the world, the technology would have to be prepared for the potential upcoming food market. It is possible to transdifferentiate non-myogenic cells into MuSCs by ectopic expression of myogenic determinants such as *MYOD1* and *PAX3/7* applying the cellular reprogramming technique used to derive iPSCs (Chal and Pourquie, 2017). The transdifferentiation study using MYOD1 first elucidated that cellular lineage could be altered by regulating a fate determinant (Davis et al., 1987). Fibroblasts, the most abundant cells in the human body, could be utilized for muscle production through the turn-on/off of genes at will. In addition, the upregulation of these genes allows PSCs to directly differentiate into muscle lineage by bypassing the sequential developmental process (Chal and Pourquie, 2017).

As somatic cells could acquire pluripotency by regulating gene expression, genetic manipulation allows for MuSCs to gain new traits, such as enhanced proliferation and differentiation capacities. The immortalization of cells has been widely applied to establish stable cell lines by extending their lifespan, which allows for the maintenance of stemness over an extended period without senescence (Soice and Johnston, 2021). As mentioned earlier, muscle stem cells (MuSCs) gradually lose their proliferation and differentiation abilities during *in vitro* culture. Naturally immortalized cell lines are difficult to obtain and are typically derived by engineering genes involved in cellular senescence and division. In particular, telomere synthesis enzyme, also called telomerase, and cell cycle activators such as *CDK4* and *BMI-1,* have been applied to accomplish the immortalization of MuSCs (Chua et al., 2019; Douillard-Guilloux et al., 2009). It has been proven that the shortening of telomeres, repetitive nucleotide sequences at the ends of chromosomes, is associated with cellular aging (Soice and Johnston, 2021). To extend self-renewal through genetic engineering, careful strategies are required, as constant proliferation hinders myogenic differentiation (Chua et al., 2019). Currently, CRISPR/Cas9, known as gene scissors, is recognized as the next-generation genetic engineering tool. Its use promotes muscle growth by disrupting the myostatin gene without introducing ectopic genes, suggesting that among various genetic engineering methods, gene ablation may be more suitable for regulatory compliance (Wang et al., 2015).

Fat Culture for Cultivated Meat Production

The importance of fat has been emphasized recently, as it has a substantial impact on the juiciness and flavor of cultivated meat (Hausman et al., 2014). Fat could be produced through adipogenesis, in which stem cells differentiate into adipocytes, and lipogenesis, in which triglycerides are accumulated in adipocytes. Adipogenesis begins with the commitment of MSCs or pre-adipocytes to an adipogenic lineage regulated by Zinc Finger Protein 423 (ZFP423), an upstream transcriptional regulator of pre-adipocyte-associated genes (Gupta et al., 2012). For further differentiation, preadipocytes are regulated by the transcription factor CCAAT/enhancer-binding protein (C/EBP) family induced by proliferator-activated receptor gamma (PPARγ) to generate mature adipocytes (Du et al., 2013; Hausman et al., 2014).

Although various cell types are capable of adipogenic differentiation *in vitro*, it is not yet clear which is the optimal source for producing fat for cultivated meat. Characteristics of an ideal cell source include high proliferation and efficient differentiation capacity, low media requirements, homogeneity, stability, and adaptability to mass culture conditions (Fish et al., 2020). With the exception of mass culture adaptability, most of the aforementioned features can potentially be addressed through basic cell characterization. However, due to the transition point between cell regulation and differentiation, efficient cell production is challenging without a comprehensive understanding of these properties. Furthermore, without this understanding, additional negative factors may arise during the cultivated meat production process (Zhang et al., 2020).

Fat Development

Adipogenesis proceeds as MSCs participate in the adipose lineage and differentiate into adipocytes. In this process, delicate regulation of the differentiation process is required. Mesenchymal stem cells (MSCs) are multipotent adult stem cells capable of differentiating into various cell types of mesodermal lineages, including adipocytes and osteoblasts. The interaction between cell cycle regulation and differentiation factors triggers a series of events that ultimately lead to adipocyte production (Avgustinova & Benitah, 2016). Adipogenesis from multipotent stem cells occurs in two stages, each marked by specific gene expression patterns. The first stage, known as the determination process, involves the commitment of stem cells to pre-adipocytes. In the second stage, called terminal differentiation, pre-adipocytes gradually acquire the physiological functions of mature adipocytes, such as lipid transport and synthesis, insulin sensitivity, and the secretion of adipocyte-specific proteins (Zhang et al., 2020). This is a complex and highly regulated process in which gene expression is precisely controlled (Zhang et al., 2020). Generally, adipogenic differentiation is induced by treatment with 1-methyl-3-isobutylxanthine, dexamethasone, insulin, and indomethacin (Fish et al., 2020; Tang et al., 2004). Various chemicals and methods are being studied in different species for a higher differentiation rate and long-term culture of adipocytes. Moreover, the differentiation of MSCs into a mesenchymal lineage is genetically manipulated by promoting specific transcription factors associated with a particular cell lineage (Fish et al., 2020; Zhang et al., 2020). To date, several transcription factors have been identified for the differentiation of MSCs into adipocytes. The adipogenic-specific peroxisome PPARγ is one of the representative transcription factors that regulates the expression of genes responsible for adipogenic differentiation.

Stromal Vascular Fraction (SVF)

The SVF of adipose tissue contains pluripotent cells that can differentiate into adipocytes, chondrocytes, and bone cells (Crossno et al., 2006). However, the use of SVF-origin cells for fat culture is limited due to the short culture period and reduced differentiation potential through subculture, which necessitates continued animal sacrifice. To address these challenges, research into stem cells for stable and sustainable cell supply is essential. Preadipocytes can be obtained from SVF cells using enzymes such as collagenase. SVF contains various cell types, including endothelial cells, pericytes, T cells, and macrophages (Ramakrishnan and Boyd, 2018). Ongoing studies aim to classify preadipocytes to identify cell lines with the potential to differentiate into fat (Yu et al., 2020). The purely isolated preadipocytes can maximize the differentiation yield into adipocytes. Recently, multiple studies have been conducted on adipocyte culture for fat addition to enhance flavor, in addition to muscle culture. Dohmen et al., successfully sorted pre-adipocytes from muscles and applied them into alginate for fat tissue formation, but more research is needed to satisfy the cultivated meat industry (Dohmen et al., 2022).

Dedifferentiated FAT (DFAT)

As previously described, the adipogenic potential of SVF is drastically decreased during *in vitro* culture, which is considered a hurdle to producing cultivated fat. DFAT is supposed to be another candidate cell source for creating cultivated fat to overcome the disadvantages of SVF. Dedifferentiation of fully differentiated adipose can be accomplished through their ceiling culture. During the culture, lipid storage of adipose is reduced and, in turn, reverts to fibroblast-like shaped progenitor cells, so-called DFAT. DFAT reportedly has multiple differentiation capabilities into mesenchymal lineages such as adipocytes and osteoblasts (Matsumoto et al., 2008). Dedifferentiation of FAT has been studied in humans (Kishimoto et al., 2018), rats (Akita et al., 2016), mice (Yagi et al., 2004), and livestocks (Peng et al., 2015; Wei et al., 2013). Various studies have shown that the enhanced adipogenic potential of DFAT is stably maintained with normal chromosomal karyotypes over an extended period (Peng et al., 2015). However, a major obstacle for applying DFAT in the production of cultivated meat is their dependence on high serum concentrations (often 15–20%) to maintain a proliferative state *in vitro*. This is rarely favored for the creation of novel alternative protein foods due to the high cost and animal welfare concerns (Fish et al., 2020). Along with developing serum-free culture conditions for derivation and proliferation, the major features of DFAT remains to be solved for application to create cultivated fat.

Scaffold and Tissue Engineering

For *in vitro* culture, scaffolds as biomimetic materials recapitulate the *in vivo* stem cell niches by providing a 3-dimensional adhesive surface, which supports the *in vitro* organogenesis of stem cells and maturation of tissues, unlike 2-dimensional culture using a tissue culture plate (Ostrovidov et al.,

2014). Hydrogel and porous scaffolds have been generally applied to organize artificial tissues from stem cells *in vitro* for a long time. Hydrogel is a hydrophilic polymer that has been widely used to generate self-organizing artificial tissues from stem or progenitor cells, known as organoids. By mimicking *in vivo* microenvironments, hydrogels enable stem cells to recapitulate their developmental programs, thereby facilitating *in vitro* organogenesis (Hofer and Lutolf, 2021). Stem cells differentiate into organs based on their developmental codes, suggesting that organoid techniques offer a more reliable approach for generating tissues compared to other current differentiation methods. The first brain-like organoids were generated in 2013, and since then, muscle organoids have also been developed using hydrogel scaffolds (Maffioletti et al., 2018). Although organoids have a high similarity to the actual organs in a biological aspect, the texture and shape of cultivated meat are hard to replicate. Porous scaffolds, as biomaterials with sponge-like structures, offer distinct advantages in achieving the texture and shape of cultivated meat by providing a framework for cells to organize in a three-dimensional structure, compared to hydrogel scaffolds (Ben-Arye et al., 2020). However, the tissues produced by porous scaffolds often contain a high proportion of scaffold material and exhibit different histological structures, which distinguishes them from actual meat. This results in a significant gap in terms of nutrition and sensory characteristics when compared to conventional meat.

Likewise, scaffold also plays a crucial role in creating cultivated fat in terms of differentiation efficiency and tissue maturity. Compared to 2D culture systems, 3D fat production improves efficiency and degree of differentiation (Ma et al., 2018). While the 2D protocol generally produces only multi-layered lipid accumulation, the 3D system can produce large unilocular lipid droplets that are characteristic of mature adipocytes. Furthermore, 3D induction can induce a higher differentiation rate as compared to monolayer culture (Dohmen et al., 2022). In addition, scaffolding approaches have been developed to optimize the microenvironment of cultured fat. The physical properties of scaffolds play a crucial role in determining the quality of adipocytes, which is essential for various tissue engineering and regenerative medicine applications. It is clear that the mechanical stiffness of hydrogels significantly influences adipocyte differentiation and function. Soft hydrogels, which mimic the natural elasticity of adipose tissue, promote the formation of mature and functional adipocytes, whereas stiffer hydrogels may impair their development (Ribeiro et al., 2016; Young et al., 2013). Moreover, the porosity and pore size of hydrogels are vital in ensuring optimal diffusion of nutrients, oxygen, and waste products within the gel. An appropriate pore size enables the exchange of essential molecules, thereby supporting healthy adipocyte growth and metabolic activity (Habanjar et al., 2021; Loh and Choong, 2013). Furthermore, the swelling capacity of hydrogels, attributed to their water absorption and retention properties, significantly impacts the hydrated microenvironment required for adipocyte viability and function (Cao et al., 2021). Considering the degradation rate of the hydrogel, it is essential for it to align with the various stages of adipocyte development and maturation, providing sustained support while allowing for extracellular matrix remodeling (Suh and Matthew, 2000). Additionally, the biochemical composition of the hydrogel, often functionalized with bioactive molecules such as growth factors or extracellular matrix components, can regulate adipocyte behavior by promoting adipogenesis, enhancing adipocyte viability, and influencing metabolic activity (Li et al., 2018; Saldin et al., 2017). By thoroughly understanding and optimizing these physical properties, hydrogels can provide an ideal environment to support the growth, maturation, and maintenance of high-quality adipocytes.

In addition, to scaffold types, it is crucial to seek edible scaffold materials to produce cell-based food materials, including cultivated meat. To date, various edible materials, including natural extracellular matrix proteins, decellularized plants, and polysaccharides, have been widely investigated for the creation of artificial muscle tissues (Ben-Arye et al., 2020; Jones et al., 2021; MacQueen et al., 2019). Additionally, to achieve a reliable meat analogue to conventional meats, the maturation of *in vitro*-produced muscles through mechanical stretching and electrical stimulation may be crucial for attaining the desired texture and nutrition (Langelaan et al., 2011; Powell et al., 2002). Therefore, identifying suitable scaffold biomaterials and developing tissue organization techniques are essential for improving the quality of cultivated meat.

Production Cost of Cultivated Meat

In 2013, Dr. Mark J. Post, a professor at Maastricht University in the Netherlands, and his colleagues proved that meat could be served by cultivating animal cells/tissues without slaughtering. They made beef patties using bovine muscle stem cells, which cost approximately 300,000 USD. Subsequently, emergent research teams have released other prototypes of cultivated meat and the cost thereof; however, different standards have been applied to calculate them, which can be confusing to potential consumers and relevant industries. The price of *in vitro*-produced tissues primarily depends on the production costs of cells, culture media, and scaffolds. For cells, this includes the cost of donor animals, cell separation, and the cell content in the final products. For culture media and scaffolds, the production costs and the quantities used during cultivation, along with the culture technique and maintenance costs, must be considered when estimating inputs. However, for an objective evaluation, it is essential to disclose the origin of cells, the ingredients used for culture, and the composition and nutrients of *in vitro* cultured cell-based foods alongside their price. Additionally, providing objective information about cultivated meat, based on conventional meat standards, would enhance consumer awareness and satisfaction. This could potentially lead to the introduction of a cultivated meat grading system, thus expanding the market for this emerging industry.

The high cost of producing cultivated meat is a known factor that hinders the market entry of cultivated meat along with consumer acceptance and government regulations. Examination of potential consumers' willingness to pay (WTP) has shown that consumers have higher WTP for cultivated meat than traditional burgers by approximately two-fold if cultivated meat tastes equal to conventional meat and has environmental benefits (Kantor and Kantor, 2021). As mentioned above, although the production cost of cultivated meat has dropped significantly through extensive research and development, it still remains much higher than that of conventional meat. To produce cost-effective cell culture products, the continuous supply of cells using immortalized cell lines or embryonic stem cells without the need for slaughtering is being considered a key solution to reduce the cost of cell supply (Post et al., 2020). Fetal bovine serum (FBS) and growth factors are reported to be major contributors to the high cost of culture media. Numerous researchers in cell biology have long worked to replace FBS with more affordable alternatives, such as hydrolysates of by-products from animals and plant/microorganism extracts, as well as producing growth factors from plant and insect cell culture systems (Andreassen et al., 2020; Cronin, 2020; Dohmen et al., 2022; Messmer et al., 2022; O'Neill et al., 2021). Furthermore, the price of the culture medium and scaffold can be reduced by replacing the ingredients with lower grade materials or by developing minimal culture conditions through the optimization of their concentration (Lyra-Leite et al., 2021; Ma and Suh, 2019). Automation of the culture process and reduction of culture equipment costs should be implemented alongside the factors mentioned above (Bakmiwewa et al., 2015; Cronin, 2020; Lucendo-Villarin et al., 2020). To produce cost-effective cultivated meat, research on various approaches should be conducted concurrently. Additionally, industrial preparations for scaling up production facilities will be necessary to achieve a substantial cost reduction compared to the current R&D phase.

Safety of Cell-Based Food Materials

The technology development of cellular agriculture, particularly stem cell biotechnology, has led to the commercialization of cell-based food like cultivated meat in the market. However, several issues remain, including nomenclature (labelling), safety management, authenticity, and consumer perception (Post et al., 2020). Particularly, safety management during and after the production of cultivated meat must be ensured, as lab-grown meat technology, if not properly managed, can pose direct risks to consumer health. Potential risks include microbial contamination from inadequate hygiene practices during production, allergenic reactions due to novel ingredients or additives, nutritional imbalances arising from the failure to replicate the optimal nutritional composition of traditional meat, unknown long-term effects that necessitate continuous monitoring and research, and concerns regarding antibiotic use and the development of antibiotic-resistant bacteria, which could pose significant threats to public health. Proper regulation and stringent monitoring are essential to mitigate these concerns.

In the case of food using new materials and technologies, it is necessary to establish safety evaluation processes suitable for the product. Currently, cultivated meat is not a natural sister product of existing meat, and it is difficult to have substantial equivalence with meat because, as stated above, completely different manufacturing processes are used to achieve mass production *in vitro* and to make it similar to real meat (Choi et al., 2021; Lee et al., 2020; Ramani et al., 2021). In this regard, cultivated meat is generally classified as a novel food due to the lack of a historical consumption record. The European Union, Canada, Australia, Singapore, South Korea, and other countries have adopted this principle, while the United States may use the term GRAS (Generally Recognized as Safe) (Sergelidis, 2019). Notably, the regulatory approval for cultivated chicken and shrimp by the Singapore Food Agency in 2020 also adhered to the novel food policy.

In terms of safety considerations, donor animals, origin and stage of the cells, cells before and after storage, manufacturing processes and ingredients, and cultured muscle after harvest should be listed and verified by an effective and reliable management scheme (Lee et al., 2020). Most materials (e.g., chemicals, nutrients, growth factors, scaffolds) for cultivated meat production are not approved yet for food use. In line with safety, the establishment of the cultivated meat standard and manufacturing process will help the industry to further develop other values (e.g., purity and density of muscle stem cells from their origin, cells-scaffolds ratio). In this regard, the establishment and adaptation of harmonized regulations will ensure the safety of cultivated meat after open discussions with various fields, regions, and stakeholders. Recently, organizations within the United Nations, such as the World Health Organization and the Food and Agriculture Organization, have begun addressing the agenda with member countries. The US FDA's recent decision to approve cultivated chicken meat as safe for human consumption may help facilitate these efforts.

Perspectives

We are standing before the door to the cellular agricultural era. Humanity has undergone a paradigm shift across society with remarkable advancements through sequential industrial revolutions. Civilization was built away from hunting and gathering through the 'Agricultural Revolution.' In addition, we have witnessed the establishment of a capitalist economy through the 'Industrial Revolution,' the increase in agricultural production through the 'Green Revolution,' and the advancement of information and communication technology through the recent '4th Industrial Revolution.' Accelerating the trend of research and development of innovative technologies, the world is headed toward significant changes through another technological revolution. Biotechnology is a field that utilizes biological systems and organisms to develop new products or techniques, which started with agriculture, including breeding. Food production has been dramatically increased by the Agricultural Revolution via the utilization of seeds. Likewise, numerous researchers are attempting to shift the paradigm on the production of foods using stem cells, which are known as the seeds of tissues in the body.

Stem cells, as progenitor cells that can differentiate into various cells in the body, have been highlighted as a promising source for curing human diseases and deciphering biological molecular mechanisms for a long time. Recently, advanced stem cell technologies have enabled the organization of artificial tissues *in vitro* using animal muscle stem cells derived from various sources, facilitated by nutrient supplementation. This has made the development of food materials possible. Moreover, these technologies have been extended to create new cellular agriculture products, including leather, fur, and milk, without the need to raise domestic animals. In this context, the shift toward the production of sustainable agricultural products using stem cells can be considered the 'Cell Revolution.' Much like how human life was transformed by the Agricultural and Green Revolutions, cultivated meat, as the starting point of the 'Cell Revolution,' has the potential to diversify our lives through expansion and integration across various industries. Accordingly, in order to pave the way for revitalizing cellular agriculture, it is crucial to further advance the basic technology derived from cultivated meat research and establish an industrial ecosystem and infrastructure. The next wave of stem cell research is coming along with the development of the cellular agricultural industry., which will open the ('Brave') new world.

Conclusion

This chapter has highlighted the critical role of stem cells in the production of cultured meat. Stem cells, particularly myosatellite and pluripotent stem cells, provide the foundation for developing muscle and fat tissues, which are essential for creating meat that closely mimics the texture, flavor, and nutritional profile of conventional meat. The successful application of stem cells in cultured meat production offers significant potential for improving food security, sustainability, and animal welfare. However, challenges remain in optimizing stem cell sources, scaling production, and ensuring the efficient differentiation of stem cells into functional muscle and fat tissues. The development of suitable growth media, scaffolding techniques, and bioreactor systems remains central to overcoming these hurdles. Furthermore, ensuring the safety and cost-effectiveness of cultured meat products will require continued research and innovation. In conclusion, while the use of stem cells in cultured meat production is a promising area of biotechnology, it is still in the early stages of development. The next steps involve refining stem cell techniques, improving scalability, and addressing regulatory and safety concerns. With continued advancements, stem cell-based cultured meat could become a viable and sustainable solution to meet global protein demands, offering a more ethical and environmentally friendly alternative to traditional meat production.

References

Akita, D., Kano, K., Saito-Tamura, Y., Mashimo, T., Sato-Shionome, M., Tsurumachi, N., Yamanaka, K., Kaneko, T., Toriumi, T., Arai, Y., Tsukimura, N., Matsumoto, T., Ishigami, T., Isokawa, K., Honda, M., 2016. Use of rat mature adipocyte-derived dedifferentiated fat cells as a cell source for periodontal tissue regeneration. *Front. Physiol.* 7, 50. https://doi.org/10.3389/fphys.2016.00050.

Andreassen, R. C., Pedersen, M. E., Kristoffersen, K. A., Rønning, S. B., 2020. Screening of by-products from the food industry as growth promoting agents in serum-free media for skeletal muscle cell culture. *Food Funct.* 11 (3), 2477–2488.

Avgustinova, A., Benitah, S. A., 2016. Epigenetic control of adult stem cell function. *Oct Nat. Rev. Mol. Cell Biol.* 17 (10), 643–658. https://doi.org/10.1038/nrm.2016.76.

Bakmiwewa, S. M., Heng, B., Guillemin, G. J., Ball, H. J., Hunt, N. H., 2015. An effective, low-cost method for achieving and maintaining hypoxia during cell culture studies. *Biotechniques* 59 (4), 223–229.

Ben-Arye, T., Shandalov, Y., Ben-Shaul, S., Landau, S., Zagury, Y., Ianovici, I., Lavon, N., Levenberg, S., 2020. Textured soy protein scaffolds enable the generation of three-dimensional bovine skeletal muscle tissue for cell-based meat. *Nature Food* 1 (4), 210–220.

Bogliotti, Y. S., Wu, J., Vilarino, M., Okamura, D., Soto, D. A., Zhong, C., Sakurai, M., Sampaio, R. V., Suzuki, K., Izpisua Belmonte, J. C., Ross, P. J., 2018. Feb 27). Efficient derivation of stable primed pluripotent embryonic stem cells from bovine blastocysts. *Proc. Natl. Acad. Sci.* U. S. A. 115 (9), 2090–2095. https://doi.org/10.1073/pnas.1716161115.

Campion, D. R., Richardson, R. L., Reagan, J. O., Kraeling, R. R., 1981. Changes in the satellite cell population during postnatal growth of pig skeletal muscle. *May J. Anim. Sci.* 52 (5), 1014–1018. https://doi.org/10.2527/jas1981.5251014x.

Cao, H., Duan, L., Zhang, Y., Cao, J., Zhang, K., 2021. Current hydrogel advances in physicochemical and biological response-driven biomedical application diversity. Signal Transduct. *Targeted Ther.* 6 (1), 426.

Chal, J., Pourquie, O., 2017. Making muscle: skeletal myogenesis *in vivo* and *in vitro.* Jun 15 *Development* 144 (12), 2104–2122. https://doi.org/10.1242/dev.151035.

Choi, K. H., Lee, C. K., 2019. Pig pluripotent stem cells as a candidate for biomedical application. *Journal of Animal Reproduction and Biotechnology* 34 (3), 139–147.

Choi, K. H., Lee, D. K., Kim, S. W., Woo, S. H., Kim, D. Y., Lee, C. K., 2019. Chemically defined media can maintain pig pluripotency network *in vitro. Jul 9 Stem Cell Rep.* 13 (1), 221–234. https://doi.org/10.1016/j.stemcr.2019.05.028.

Choi, K. H., Yoon, J. W., Kim, M., Lee, H. J., Jeong, J., Ryu, M., Jo, C., Lee, C. K., 2021. Muscle stem cell isolation and *in vitro* culture for meat production: a methodological review. *Jan Compr. Rev. Food Sci. Food Saf.* 20 (1), 429–457. https://doi.org/10.1111/1541-4337.12661.

Chua, M. W. J., Yildirim, E. D., Tan, J. H. E., Chua, Y. J. B., Low, S. M. C., Ding, S. L. S., Li, C.w., Jiang, Z., Teh, B. T., Yu, K., 2019. Assessment of different strategies for scalable production and proliferation of human myoblasts. *Cell Prolif.* 52 (3), e12602.

Du, M., Huang, Y., Das, A. K., Yang, Q., Duarte, M. S., Dodson, M. V., Zhu, M. J., 2013. Meat Science and Muscle Biology Symposium: manipulating mesenchymal progenitor cell differentiation to optimize performance and carcass value of beef cattle. *Mar J. Anim. Sci.* 91 (3), 1419–1427. https://doi.org/10.2527/jas.2012-5670.

Eibl, R., Senn, Y., Gubser, G., Jossen, V., van den Bos, C., Eibl, D., 2021. Mar 25). Cellular agriculture: opportunities and challenges. *Annu. Rev. Food Sci. Technol.* 12, 51–73. https://doi.org/10.1146/annurev-food-063020-123940.

Engineered whole cut meat-like tissue by the assembly of cell fibers using tendon-gel integrated bioprinting. *Aug 24 Nat. Commun.* 12 (1). https://doi.org/10.1038/s41467-021-25236-9.

Evans, M. J., Kaufman, M. H., 1981. Establishment in culture of pluripotential cells from mouse embryos. *Nature* 292 (5819), 154–156.

Fish, K. D., Rubio, N. R., Stout, A. J., Yuen, J. S. K., Kaplan, D. L., 2020. Prospects and challenges for cell-cultured fat as a novel food ingredient. *Apr Trends Food Sci. Technol.* 98, 53–67. https://doi.org/10.1016/j.tifs.2020.02.005.

Gupta, R. K., Mepani, R. J., Kleiner, S., Lo, J. C., Khandekar, M. J., Cohen, P., Frontini, A., Bhowmick, D. C., Ye, L., Cinti, S., Spiegelman, B. M., 2012. Feb 8). Zfp423 expression identifies committed preadipocytes and localizes to adipose endothelial and perivascular cells. *Cell Metabol.* 15 (2), 230–239. https://doi.org/10.1016/j.cmet.2012.01.010.

Habanjar, O., Diab-Assaf, M., Caldefie-Chezet, F., Delort, L., 2021. 3D cell culture systems: tumor application, advantages, and disadvantages. *Int. J. Mol. Sci.* 22 (22), 12200.

Haberlandt, G., 1902. Kulturversuche mit Isolierten Pflanzenzellen, Sitzungsberg. Kais. Akad Wiss Wien Mat-Naturwiss. *Kl Abt* 111, 69–92.

Joe, A. W., Yi, L., Natarajan, A., Le Grand, F., So, L., Wang, J., Rudnicki, M. A., Rossi, F., 2010. Muscle injury activates resident fibro/adipogenic progenitors that facilitate myogenesis. *Nat. Cell Biol.* 12 (2), 153–163.

Jones, J. D., Rebello, A. S., Gaudette, G. R., 2021. Decellularized spinach: an edible scaffold for laboratory-grown meat. *Food Biosci.* 41, 100986.

Joo, S., Kim, G., Hwang, Y., Ryu, Y., 2013. Control of fresh meat quality through manipulation of muscle fiber characteristics. *Meat Sci.* 95 (4), 828–836.

Joshi, V., Kumar, S., 2015. Meat Analogues: plant based alternatives to meat products-A review. *Int. J. Food Ferment. Technol.* 5 (2), 107–119.

Kang, D. H., Louis, F., Liu, H., Shimoda, H., Nishiyama, Y., Nozawa, H., Kakitani, M., Takagi, D., Kasa, D., Nagamori, E., Irie, S., Kitano, S., Matsusaki, M., 2021. Recapitulating early development of mouse musculoskeletal precursors of the paraxial mesoderm *in vitro*. *Development* 145 (6). https://doi.org/10.1242/dev.157339.

Kantor, B. N., Kantor, J., 2021. Public attitudes and willingness to pay for cultured meat: a cross-sectional experimental study. Front. Sustain. *Food Syst.* 26.

Langelaan, M. L., Boonen, K. J., Rosaria-Chak, K. Y., van der Schaft, D. W., Post, M. J., Baaijens, F.P., 2011. Advanced maturation by electrical stimulation: differences in response between C2C12 and primary muscle progenitor cells. *Journal of tissue engineering and regenerative medicine* 5 (7), 529–539.

Lee, H. J., Yong, H. I., Kim, M., Choi, Y. S., Jo, C., 2020. Status of meat alternatives and their potential role in the future meat market - a review. Oct Asian-Australas. *J. Anim. Sci.* 33 (10), 1533–1543. https://doi.org/10.5713/ajas.20.0419.

Lee, Y. E., Lee, H. J., Kim, M., Yoon, J. W., Ryu, M., Jo, C., 2021. Analysis on difference of consumer's evaluation on visual features of pork cuts. *J. Anim. Sci. Technol.* 63 (3), 614.

Leinroth, A. P., Mirando, A. J., Rouse, D., Kobayahsi, Y., Tata, P. R., Rueckert, H. E., Liao, Y., Long, J. T., Chakkalakal, J. V., Hilton, M.J., 2022. Identification of distinct non-myogenic skeletal-muscle-resident mesenchymal cell populations. *Cell Rep.* 39 (6), 110785.

Li, C., Zhao, H., Wang, B., 2021. Mesenchymal stem/stromal cells : developmental origin, tumorigenesis and translational cancer therapeutics. *Jan Transl Oncol* 14 (1), 100948. https://doi.org/10.1016/j.tranon.2020.100948.

Li, S., Poche, J. N., Liu, Y., Scherr, T., McCann, J., Forghani, A., Smoak, M., Muir, M., Berntsen, L., Chen, C., 2018. Hybrid Synthetic-Biological hydrogel system for adipose tissue regeneration. *Macromol. Biosci.* 18 (11), 1800122.

Ma, T., Xie, M., Laurent, T., Ding, S., 2013. Progress in the reprogramming of somatic cells. *Feb 1 Circ. Res.* 112 (3), 562–574. https://doi.org/10.1161/CIRCRESAHA.111.249235.

Ma, W., Suh, W. H., 2019. Cost-effective cosmetic-grade hyaluronan hydrogels for ReNcell VM human neural stem cell culture. *Biomolecules* 9 (10), 515.

Ma, Y. N., Wang, B., Wang, Z. X., Gomez, N. A., Zhu, M. J., Du, M., 2018. Three-dimensional spheroid culture of adipose stromal vascular cells for studying adipogenesis in beef cattle. *Feb 22 Animal* 1–7. https://doi.org/10.1017/S1751731118000150.

MacQueen, L. A., Alver, C. G., Chantre, C. O., Ahn, S., Cera, L., Gonzalez, G. M., O'Connor, B. B., Drennan, D. J., Peters, M. M., Motta, S. E., Zimmerman, J. F., Parker, K. K., 2019. Muscle tissue engineering in fibrous gelatin: implications for meat analogs. *NPJ Sci Food* 3, 20. https://doi.org/10.1038/s41538-019-0054-8.

Maffioletti, S. M., Sarcar, S., Henderson, A. B., Mannhardt, I., Pinton, L., Moyle, L. A., Steele-Stallard, H., Cappellari, O., Wells, K. E., Ferrari, G., 2018. Three-dimensional human iPSC-derived artificial skeletal muscles model muscular dystrophies and enable multilineage tissue engineering. *Cell Rep.* 23 (3), 899–908.

Matsumoto, T., Kano, K., Kondo, D., Fukuda, N., Iribe, Y., Tanaka, N., Matsubara, Y., Sakuma, T., Satomi, A., Otaki, M., Ryu, J., Mugishima, H., 2008. Mature adipocyte-derived dedifferentiated fat cells exhibit multilineage potential. *Apr J. Cell. Physiol.* 215 (1), 210–222. https://doi.org/10.1002/jcp.21304.

Messmer, T., Klevernic, I., Furquim, C., Ovchinnikova, E., Dogan, A., Cruz, H., Post, M. J., Flack, J. E., 2022. A serum-free media formulation for cultured meat production supports bovine satellite cell differentiation in the absence of serum starvation. *Nature Food* 3 (1), 74–85.

Peng, X., Song, T., Hu, X., Zhou, Y., Wei, H., Peng, J., Jiang, S., 2015. Phenotypic and functional properties of porcine dedifferentiated fat cells during the long-term culture *in vitro*. *BioMed Res. Int.*, 673651 https://doi.org/10.1155/2015/673651, 2015.

Post, M. J., 2014. Cultured beef: medical technology to produce food. *Apr J. Sci. Food Agric.* 94 (6), 1039–1041. https://doi.org/10.1002/jsfa.6474.

Post, M. J., Levenberg, S., Kaplan, D. L., Genovese, N., Fu, J. A., Bryant, C. J., Negowetti, N., Verzijden, K., Moutsatsou, P., 2020. Scientific, sustainability and regulatory challenges of cultured meat. *Jul Nature Food* 1 (7), 403–415. https://doi.org/10.1038/s43016-020-0112-z.

Powell, C. A., Smiley, B. L., Mills, J., Vandenburgh, H. H., 2002. Mechanical stimulation improves tissue-engineered human skeletal muscle. *Am. J. Physiol. Cell Physiol.* 283 (5), C1557–C1565.

Quinn, L. S., Ong, L. D., Roeder, R. A., 1990. Paracrine control of myoblast proliferation and differentiation by fibroblasts. *Jul Dev. Biol.* 140 (1), 8–19. https://doi.org/10.1016/0012-1606(90)90048-n.

Ramakrishnan, V. M., Boyd, N. L., 2018. The adipose stromal vascular fraction as a complex cellular source for tissue engineering applications. *Aug Tissue Eng. B Rev.* 24 (4), 289–299. https://doi.org/10.1089/ten.TEB.2017.0061.

Renerre, M., 1990. Factors involved in the discoloration of beef meat. *Int. J. Food Sci. Technol.* 25 (6), 613–630.

Ribeiro, A. J., Denisin, A. K., Wilson, R. E., Pruitt, B. L., 2016. For whom the cells pull: hydrogel and micropost devices for measuring traction forces. *Methods* 94, 51–64.

Rischer, H., Szilvay, G. R., Oksman-Caldentey, K. M., 2020. Cellular agriculture - industrial biotechnology for food and materials. *Feb Curr. Opin. Biotechnol.* 61, 128–134. https://doi.org/10.1016/j.copbio.2019.12.003.

Rossant, J., Tam, P. P. L., 2017. Jan 5). New insights into early human development: lessons for stem cell derivation and differentiation. *Cell Stem Cell* 20 (1), 18–28. https://doi.org/10.1016/j.stem.2016.12.004.

Saldin, L. T., Cramer, M. C., Velankar, S. S., White, L. J., Badylak, S. F., 2017. Extracellular matrix hydrogels from decellularized tissues: structure and function. *Acta Biomater.* 49, 1–15.

Sergelidis, D., 2019. Lab grown meat: the future sustainable alternative to meat or a novel functional food? *Biomedical Journal of Scientific & Technical Research* 17 (1), 12440–12444.

Sendhil, R., Nyika, J., Yadav, S., Mackolil, J., G, R. P., Workie, E., Ragupathy, R., Ramasundaram, P., 2022. Genetically modified foods: bibliometric analysis on consumer perception and preference. *Dec 31 GM Food* 13 (1), 65–85. https://doi.org/10.1080/21645698.2022.2038525.

Sieweke, M. H., 2015. Waddington's valleys and Captain Cook's islands. *Jan 8 Cell Stem Cell* 16 (1), 7–8. https://doi.org/10.1016/j.stem.2014.12.009.

Suh, J. K. F., Matthew, H. W., 2000. Application of chitosan-based polysaccharide biomaterials in cartilage tissue engineering: a review. *Biomaterials* 21 (24), 2589–2598.

Takahashi, K., Yamanaka, S., 2006. Induction of pluripotent stem cells from mouse embryonic and adult fibroblast cultures by defined factors. *Aug 25 Cell* 126 (4), 663–676. https://doi.org/10.1016/j.cell.2006.07.024.

Tang, Q. Q., Otto, T. C., Lane, M. D., 2004. Commitment of C3H10T1/2 pluripotent stem cells to the adipocyte lineage. *Jun 29 Proc. Natl. Acad. Sci. U. S. A.* 101 (26), 9607–9611. https://doi.org/10.1073/pnas.0403100101.

Chapter 4

Packaging Requirements and Labeling Design of Cultured Meat

Abstract

The growing consumer concern over sustainability has increased the demand for alternative protein products to replace animal-based meat. Cultured meat, as a new market product, faces consumer skepticism, making improvements in its functionality and sensory quality crucial for better acceptance. Key consumer considerations include unnaturalness, healthiness, texture, price, and safety. Tissue-engineering technology enables cultured meat to meet consumer demands for acceptable quality. Packaging plays a crucial role in preserving food quality and extending shelf life, thereby influencing consumer acceptance in various countries. This chapter discusses processing technologies, packaging, shelf life, and the future development of cultured meat.

Keywords: cultured meat, packaging technology, labeling design, consumer acceptance

Introduction

Consumer demand for sustainable food has forced food companies to find new alternative protein-based products to alter muscle meat in the market (Niva & Vainio, 2021). The muscle meat from livestock animals contribute to global environmental degradation. These livestock animals utilize 30% of global land area and 8% of water and contribute to 14.5% of gas emissions (over global transportation emission) (Tuomisto & Joost Teixeira de Mattos, 2011). Other rising issues due to livestock production are eutrophication and deforestation, which contribute to 34% of greenhouse gas (GHG) emissions (Lamb et al., 2021). From 1961 to 2013, production has doubled. According to eating trends in different nations, 40–45% of total meat production was consumed by Asians, while 19% and 15% were consumed by Europeans and North

Americans, respectively, in 2018 (Tuomisto & Joost Teixeira de Mattos, 2011). Global meat consumption is expected to increase in the coming decades, with the global population projected to rise by 70% by 2050 (Gerber, Steinfeld, Henderson, Mottet, et al., 2013). The increase in meat consumption and production will impact the global environment in the future, highlighting the need for more efficient meat alternatives.

Plant-based products offer a promising alternative, but they currently lack a complete resemblance to meat, which many consumers consider a barrier (Escribano et al., 2021; Michel, Hartmann, & Siegrist, 2021; Onwezen, Bouwman, Reinders, & Dagevos, 2021; Van der Weele C. et al., 2019). Cultured meat is another option, as lab-grown meat has recently gained attention as a meat alternative to challenge traditional livestock in future global markets, with its supply chain shown in Figure 1 (Santo et al., 2020). These factors are likely to significantly affect global cultured meat consumption, addressing a long-standing issue.

Cultured meat called laboratory-growing meat is meat produced in a controlled environment with tissue engineering technology (Arshad et al., 2017; Ben-Arye & Levenberg, 2019). Muscle stem cells are cultured in a suitable medium, e.g., crude cell extracts. The muscle tissue is produced by tissue engineers for the selection and placement of adult cells on a scaffold and is then grown in a bioreactor (Arshad et al., 2017).

Cultured meat as well as traditional meat is regarded as a nutritious product, containing amino acid compositions, minerals, and vitamins (Fraeye, Kratka, Vandenburgh, & Thorrez, 2020). However, it is not clear when the type of cultured meat is significantly different from meat of animal origin. The scaffold material also influences the macronutrient composition of the final product. In muscle tissue engineering, proteins such as collagen and fibrin are commonly used. Collagen is primarily composed of non-essential amino acids, though it also contains a modest amount of the essential amino acid lysine (Listrat et al., 2016). Fat in meat can be classified nutritionally based on its percentage concentration and fatty acid composition. In muscle tissue, myoglobin contains iron, either as part of a heme group or in a non-heme complex with ferritin. Meat is rich in B-group vitamins, particularly vitamin B12 (Kausar, Hanan, Ayob, Praween, & Azad, 2019). In addition to essential nutrients like vitamins, minerals, and amino and fatty acids, meat also contains several bioactive compounds that are beneficial to human health.

Compared to traditional meat production, cultured meat is regarded as healthier, disease-free, environmental-friendly, and safe for consumers (Arshad et al., 2017a). It is important to note that the production of cultured meat eliminates the need to breed, maintain, and slaughter large numbers of animals, significantly reducing material production costs, land use, and the consumption of agricultural crops typically used as animal feed. Additionally, the reduction in the number of farm animals helps address several environmental issues, particularly the release of substantial amounts of methane, ammonia, and carbon dioxide. There are no religious restrictions on the consumption of cultured meat, which exist for different types of meat of animal origin in several countries. The advantage of cultured meat is the possibility of increasing its biological value by adding various vitamins, trace elements, amino acids, unsaturated fatty acids, etc. Cultured meat also cannot be a source of helminths (Bhat, Morton, Bekhit, Kumar, & Bhat, 2022; Handral et al., 2022; Nobre, 2022).

The quality of cultured meat needs to be designed like that of conventional meat. Factors such as aroma, texture, taste, and appearance play a crucial role in their acceptance (Fraeye et al., 2020). One of the problems of cultured meat is the absence of myoglobin protein in its composition, which forms the red color of meat of animal origin. To solve this problem, natural dyes (sugar beet or saffron) or hemoglobin isolated from animal blood, or its derivatives can be used in cultured meat production process (Mateti, Laha, & Shenoy, 2022). The addition of tranquil muscle cells has recently been shown to enhance the functionality of muscle stem cells in scaffolds and improve the sensory quality and toughness of cultured meat (Ben-Arye & Levenberg, 2019; Zheng et al., 2021). However, cultured meat is often viewed as unnatural and unappealing by consumers worldwide (Bryant & Sanctorum, 2021a; Siegrist & Hartmann, 2020). Therefore, improving functionality and sensory quality is crucial to increasing consumer acceptance by emphasizing the positive aspects of the technology.

The consumer acceptance of cultured meat is also influenced by marketing strategy with the fact that cultured meat is not muscle meat (Treich, 2021). Unnaturalness, healthiness, texture, price, and safety are the considered attributes for the consumer to purchase cultured meat (Siddiqui et al., 2022a, 2022b). These attributes need to be well-promoted to increase the acceptability of cultured meat as modified products, since cultured meat is reported to have more benefits, in terms of sustainability and safety (Jairath, Mal, Gopinath, & Singh, 2021). The reported benefits of cultured meat need to be effectively addressed through labeling and well-presented marketing strategies. This chapter explores processing, packaging, shelf life, labeling design, and presentation methods to enhance consumer acceptance of cultured meat. Labeling plays a crucial role in helping consumers understand the information about packaged products and ensures compliance with regulatory standards. While no reports on potential packaging have been published yet, it is important to investigate its potential application in cultured meat to protect the product during storage and distribution, as cultured meat has a shorter shelf life and is at risk of contamination after processing and packaging. This chapter also aims to provide food companies with insights into techniques for achieving consumer acceptance through appropriate packaging, label design, and product presentation.

Processing and Packaging Requirements of Cultured Meat

Cultured meat is generated with the technology of tissue engineering, where stem cell lines can multiply and distinguished into meat elements used to produce cultured meat (Arshad et al., 2017). These cells, extracted either from embryos or through cattle biopsies, are initially cultivated in small cell cultures before being expanded in a seed simulation system that provides the necessary medium and optimal conditions for cell proliferation (Ben-Arye & Levenberg, 2019). Subsequently, the cells are transferred to bioreactors, where they develop into muscle tissue by growing on scaffolds. The resulting tissue is typically collected in thin layers, which must be stacked to replicate meat products of a desired thickness. Additionally, multiple cell types can be co-cultured to create three-dimensional cultured meat (Young & Skrivergaard, 2020).

Although biotherapy production and clinical finding provide significant equipment and methods for producing cultured meat, any feature of cultured meat production is uncommon, necessitating additional development (Ben-Arye & Levenberg, 2019). Optimization and innovation are essential throughout the bioprocessing stages of cultured meat production, from initial cell line development to end-product manufacturing and harvesting. Significant investments and collaborations are particularly important for advancing technologies related to cell line creation, cell culture media, scaffolding, and bioreactor systems (Djisalov et al., 2021). The development of packaging for cultured meat should be tailored to its unique origin and compositional properties. Specifically, packaging strategies should take into account factors such as muscle and fat content, which influence key quality attributes like oxidation and proteolysis. Unlike conventional meat, cultured meat does not undergo the typical post-slaughter biochemical transformations. Therefore, specific quality characteristics absent in conventional meat must be considered when designing appropriate packaging solutions for cultured meat products, as discussed below. The red color of conventional muscle meat is related to the heme proteins, called myoglobin. The absence of myoglobin generates a pale color due to its expression suppressed at ambient oxygen conditions (Gholobova et al., 2018; Post & Hocquette, 2017). For the cultured meat, extracellular heme proteins, such as myoglobin, are supplemented to improve the bloody flavor and the red color of the cultured meat. Myoglobin expression can be enhanced by adapting culturing conditions, such as low oxygen environments, stimulating media additives (e.g., lipids and acetic acid), promoting myoglobin protein synthesis, and ensuring a sufficient quantity of iron in the cells (Fraeye et al., 2020; Schlater, de Miranda, Frye, Trumble, & Kanatous, 2014; Simsa et al., 2019). The color change in cultured meat, attributed to the presence of extracellular heme proteins like myoglobin, may occur as a result of the oxidation process. Myoglobin has physiological functions to bind and store oxygen, therefore the reactions with oxygen and other ligands are same, which cause the color changes. The change of the color could be followed by the change of other meat product properties, e.g., texture and pH. Therefore, it is essential to maintain the presence of cultured meat, since the acceptability by the consumers and the consumer decisions to purchase the cultured products are influenced by the appearance.

The cultured techniques are used to produce cell layers of cultured meat, but only few layers are produced due to the absence of the blood, hence causing the lack of nutrients and oxygen distribution throughout the tissue (Gholobova et al., 2018). Cultured meat products have more feasibility to be produced as minced beef, e.g., hamburgers, because product texture is expected to resemble processed burgers made from the conventional meat. The texture of the culture meat can be affected both by supportive scaffolding materials and cultured cells (Rubio, Xiang, & Kaplan, 2020). Structure formation depends on the characteristics of technological function in the dissolved proteins, actin and myosin, during pasteurization (Fraeye et al., 2020). In terms of biochemical composition, cultured meat contains muscle fibers similar to conventional meat (Bhat, Morton, Mason, Bekhit, & Bhat, 2019a). The production of hamburgers involving a heating process induces thermal reactions, leading to the formation of volatile compounds that contribute to the characteristic flavor. This process triggers the Maillard reaction, which involves a reaction between an amino compound and a reducing sugar, as well as lipid degradation due to the susceptibility of polyunsaturated fatty acids in phospholipids and intracellular lipids to oxidation. The oxidation products influence the aroma of the meat, causing off-flavors and spoilage. A sensory panel described the typical bite and texture of cultured meat as having a pleasant, meaty, and juicy flavor (Rubio et al., 2020). These properties of cultured meat products are crucial for considering its shelf life during storage and distribution.

For cultured meat, there is no risk of infection with bacteria at the stages of cell proliferation, post-processing and packaging, which is due to the fact that these technological operations are carried out in sterile conditions (Ketelings, Kremers, & de Boer, 2021). However, it is important to note that bacterial infection can occur during transportation and distribution of cultured meat or products due to poor-quality packaging material. In this regard, the quality of the packaging material and the packaging process itself plays a significant role in the consumption of cultured meat.

Maintaining Quality and Shelf Life of Cultured Meat Products

In terms of shelf life, cultured meat has an advantage over conventional meat, as it is produced in sterile conditions. The sterility of cultured meat enhances its use as a safe meat alternative with a long shelf life, thereby reducing food loss (Furuhashi et al., 2021). However, the manufacturing process of cultured meat involves not only the production of cells and tissues but also the collection and purification of cells post-production, along with storage, transportation, standardization, quality control, and food processing technologies for the creation of meat products based on it (Post et al., 2020). To ensure the quality and safety of cultured meat throughout its entire product life cycle, it is essential to adhere to the quality management system requirements set by the manufacturer.

Food packaging is essential for a new product to reach potential and the right market by maintaining quality and shelf life of the cultured meat products. Cultured meat products require a packaging to protect the product from any physical, chemical, and biological contamination, and to interact with the consumers by informing the condition of the packaged foods.

Despite the fact that the market of cultured meat products is just being formed, and the acceptability of these products by consumers is still being studied, research and development of packaging methods and materials for cultured meat is an important area, which will provide safety and quality of cultured meat products and extend their shelf life. It is important to note that relevant scientific information regarding specific methods of cultured meat packaging is limited in literature. Thus, this review considered packaging for conventional meat and meat products that may be applicable for cultured meat.

Although the cultured meat products are limited in the market and the acceptance of the products by the consumers are being still explored, this chapter explored the appropriate packaging that can be implemented to the cultured meat to keep the quality of the products and extend the shelf life during storage and distribution. The packaging ensures high quality and protection of the products for consumers. The proposed packaging design, as shown in Figure 1, follows the conventional meat packaging approach.

Proposed Packaging of the Cultured Meat

Modified Atmosphere Packaging

Modified atmosphere packaging (MAP) can be applied for the cultured meat product to prevent the oxidation of heme proteins in the cultured meat and color changes during the storage. MAP successfully prolongs the shelf life of the conventional muscle meat. MAP is developed by modifying the atmosphere, e.g., gas content, inside the packaging system as shown in Figure 2a. Oxygen gas inducing the oxidation process inside the packaging is reduced and/or removed from the packaging headspace by changing the atmosphere gas with nitrogen gas and carbon dioxide at certain level of concentrations (Esmer et al., 2011). The gas composition for minced beef was investigated, and it was found that a combination of 50% O2, 30% CO2, and 20% N2 effectively maintained an acceptable color and oxidation stability. Additionally, the use of atomic nitrogen for storing cultured meat is discussed in the work (Hubalek, Post, & Moutsatsou, 2022). There has not been a study reporting the optimum concentration level of both nitrogen and carbon dioxide to extend the shelf life of the muscle meat products. The presence of carbon dioxide and nitrogen can prevent oxidation during storage. The plastic material used to package cultured meat should also have high permeability to prevent the sorption of gases. Low permeability in the packaging material leads to high sorption rates, causing the penetration of nitrogen or carbon dioxide out of the packaging system, which ultimately results in a reduction of the gas content (Hutchings, Smyth, Cunningham, & Mangwandi, 2021). Furthermore, oxygen concentration should also be taken into account, since complete removal of oxygen can cause the development of anaerobic bacteria. This anaerobic reaction could produce off flavor to the beef due microbial growth. In addition, MAP can reduce microbial growth of contaminant

Salmonella spp. in minced beef when packaged in a mixture of 50% CO2, 20% O2, and 30% N2 (Djordjević et al., 2018). Karabagias, Badeka, and Kontominas (2011) found that the use of MAP along with 0.1 wt% thyme oil extends the shelf life of lamb meat by 14–15 days compared to the control sample and by 11–13 days compared to the sample treated with 0.1 wt% thyme oil alone. The study by Meredith et al., (2014) demonstrated that a 40:30:30 CO2:O2:N2 gas mixture can extend the shelf life of chicken fillets by 9 days. Furthermore, Höll, Behr, and Vogel (2016) discovered that increasing the oxygen content from 20% to 35% leads to significant changes in the development of pathogenic microorganisms. The authors have shown that the use of 35% oxygen in the medium makes it possible to suppress the growth of Yersinia culture bacteria. Nauman et al., (2022) reported that MAP extends the shelf life of poultry meat due to oxygen-permeable packaging for at least 5 days in comparison with aerobic packaging. Meat in MAP (Modified Atmosphere Packaging) exhibited a lower total number of viable microorganisms, as well as a reduced count of *Pseudomonas aeruginosa* and coliforms, compared to the control samples.

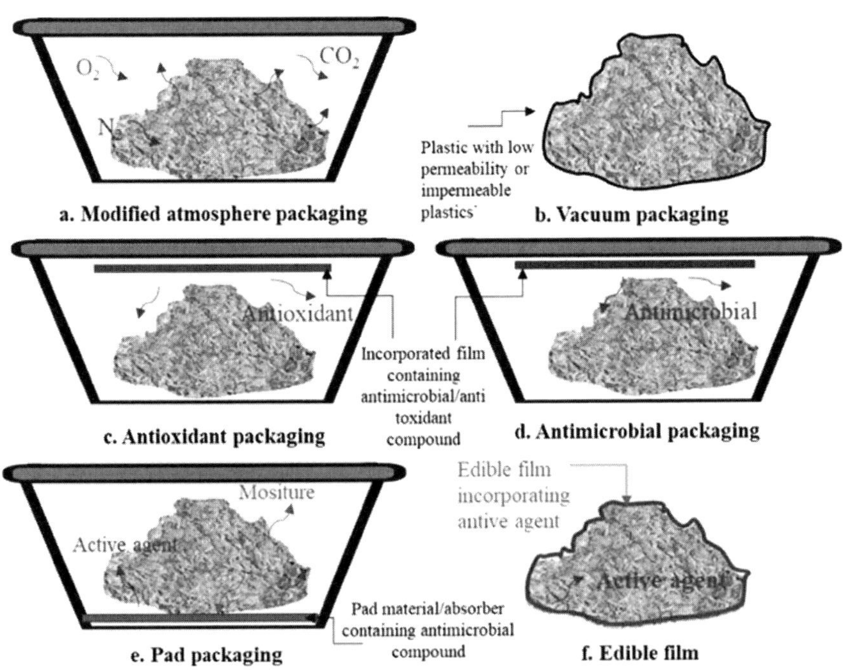

Figure 1. Suggested packaging design for cultured meat.

Thus, modified atmosphere packaging (MAP) can significantly affect lipid oxidation, color changes, and microbial growth, offering potential applications in cultured meat. By preventing lipid oxidation, color changes, and microbial growth, MAP helps maintain the appealing appearance of cultured meat for an extended period, potentially increasing consumer interest and encouraging product purchases.

Vacuum Packaging

Another potential packaging system is vacuum packaging as shown in Figure 1b. Several studies have focused on investigating the effect of vacuum packaging on the shelf life of meat and meat products. Lorenzo and Gómez (2012) found that vacuum packaging extended the shelf life of fresh foal meat by 7 days compared to samples packaged in modified atmosphere packaging and wrapping. Similarly, Devatkal, Thorat, and Manjunatha (2014) demonstrated that vacuum packaging prolonged the shelf life of goat meat and nuggets when stored in a refrigerator. In a similar study, it was found that ostrich meat under vacuum packaging can be stored for 21 days (Brenesselová et al., 2015). In this regard, it is logical to assume that vacuum packaging can also be used for cultured meat.

The packaging systems would be effective to prevent the color change and oxidation process in culture meat as revised atmosphere packaging, although the MAP is more effective in dropping the pathogenic and spoilage bacteria in packaged minced beef (Djordjević et al., 2018). The removal of oxygen helps prevent fat oxidation, discoloration, and the formation of oxidation products that cause off-flavors and spoilage. The plastic material used for packaging cultured meat should be impermeable or have low permeability to prevent the absorption of oxygen from both outside and inside the packaging system. Polypropylene (BOPP) coated with acrylic/polyvinylidene chloride (BOPPAcPVDC) has good properties in protection of fresh pork from physico-chemical, sensory and microbiological factors and improved the food shelf life by double – which was up to 14 days at 4°C. However, the absence of oxygen causes the formation of deoxymyoglobin (purple colors) converted from metmyoglobin in the oxygen-free storage (vacuum package) (Avilés, Juárez, Larsen, Rodas-González, & Aalhus, 2013).

Active Packaging

Active packaging is a recently established packaging proving to be an active agent that can interact with packaged food and headspace gas inside a packaging system. The biochemical reaction and microbial growth in the cultured meat could be interacted with the active agent inside the packaging system to inhibit the reaction and microbial growth, consequently extending the shelf life of cultured meat. The potential of active packaging, e.g., oxygen scavenger, antioxidant packaging, antimicrobial packaging, moisture absorber, etc, are discussed below.

Fat oxidation and discoloration due to the appearance of oxygen in the packaging system can be handled by presenting an antioxidant agent or oxygen scavenger, as shown in Figure 1c. Since antioxidants are highly sensitive to ambient conditions, they can be encapsulated in packaging materials to control their release into the packaging system (Gvozdenko et al., 2022). Wrona et al., (2017b) reported the encapsulation of green tea extract in polyethylene to control the release of catechins and caffeine into packaged minced pork meat. The results demonstrated an extension of shelf life, with significant differences in sensory evaluation, metmyoglobin levels, and a* (redness) after 9 days of storage. Oxygen scavenging packaging, based on iron incorporated into the food packaging, can reduce oxygen levels to inhibit lipid oxidation and maintain oxidative stability (Johnson et al., 2018). By applying oxygen scavenging packaging, the nutritional quality and appearance of cultured meat can be preserved and enhanced by preventing fat oxidation and discoloration.

The risk of microbial contamination and spoilage on cultured meat is the main issue since cultured meat has a short shelf life in its' native form (Hadi & Brightwell, 2021). No research related to the inhibition of microbial growth in cultured meat was found in references. The cultured meat might be prone to contamination from pathogenic bacteria, e.g., *Salmonella, Listeria*, and *Escherichia coli* (Ong et al., 2021), and spoilage bacteria, e.g., *Pseudomonas* spp. Antimicrobial packaging for cultured meat can be achieved by either adding an antimicrobial agent inside the packaging system or incorporating the agent into the packaging polymer. A recent advancement in antimicrobial packaging involves systems that trigger and maintain the release of antimicrobial compounds, as illustrated in Figure 1d (Bahmid et al., 2020; Zaitoon, Luo, & Lim, 2021). Bahmid, Dekker, Fogliano, and Heising (2021) improved a new packaging system by incorporating natural seeds, e.g., ground mustard seeds, into a packaging polymer that is attached to the packaging system. This packaging absorbs the moisture desorbed from the beef and then sustains releasing the antimicrobial compounds to the ground beef, consequently the antimicrobial growth of the bacteria in the beef can be inhibited to extend the shelf life of the ground beef. The encapsulation of gases, such as carbon dioxide, ethylene, and 1-methylcyclopropene, along with the incorporation of volatiles like ethanol, essential oils, and allyl isothiocyanates into the packaging material (Zaitoon et al., 2021), holds potential for application in cultured meat to inhibit bacterial growth. The packaging material can also be incorporated into an edible coating, as shown in Figure 1f, allowing consumers to easily consume the products while extending their shelf life (Umaraw et al., 2020).

The microbial growth of the packaged cultured meat is potentially inhibited by moisture absorbers, such as the pad shown in Figure 1e. The high humidity inside the packaging system induces bacterial growth due to high water activity in the beef products (Robertson, 2009). Cellulosic pads containing emulsions of essential oils were able to inhibit bacterial species specifically and several common foodborne pathogens to prolong the storage time of hamburger and minced meat for 15 and 12 days at refrigerated temperatures. The cellulosic pads are typically placed at the bottom of the packaging trays (Agrimonti et al., 2019). Similar results have been observed with absorbent pads used in conjunction with oregano essential oil, which extends the shelf life of over-stored chicken drumsticks at refrigerated temperatures by releasing antimicrobial compounds and absorbing moisture to inhibit bacterial growth (Oral et al., 2009; Sisilia Yolanda, Dirpan, Nur Faidah Rahman, Djalal, & Hatul Hidayat, 2020). The role of active packaging in preventing color changes and microbial growth, along with the moisture absorption benefits, is crucial in maintaining the quality and consumer acceptability of meat products, particularly cultured meat.

Labeling Design to Increase Intention to Purchase Cultured Meat Products

Labeling is one of the most important factors to attract consumers to consume a product. Labeling should, in theory, present consumers with the type of information that might influence their purchasing decisions (Muller, Lacroix, & Ruffieux, 2019; Zander et al., 2022). Labels are a method of informing consumers and influencing their purchase decisions. Labeling presents a unique problem wherein consumers must be aware of the concerns being discussed and that these issues are important to them. Only then labeling will have an influence on consumers' product impressions and judgments, as well as their purchase decisions (Solomon, 2010). These impressions can be captured by colors and symbols (Muller et al., 2019). However, how cultured meat is labeled is debatable, concerning that the consumers might have been misled to purchase conventional meat products (Rubio et al., 2020; Siddiqui, Khan, Farooqi, et al., 2022). In October 2019, Federal legislation was established to limit the widespread usage in a packaging label for the term "meat" for cell-based meat and plant-based products. Each product without conventional "real" meat content are required to be labeled as an "imitation"(Janet Riley & Eric Mittenthal, 2019; Santo et al., 2020). At least 25 countries approve this restriction for the usage of the word "meat' for cell-based meat or plant-based products (Santo et al., 2020). Despite the negative connotations associated with the term "imitation" in relation to both flavor and quality, these regulations are likely intended more to restrict marketing than to provide consumers with informative insights, as consumers are generally aware of the nature of meat substitutes (Rubio et al., 2020; Siddiqui, Khan, Murid, et al., 2022). It could be argued that some cell-based meats contain comparable levels of all essential nutrients to their conventional counterparts and, therefore, should not be considered "imitations." This discussion emphasizes the importance of labeling design in terms of policy, nutrition, sensory attributes, health claims, and environmental factors, as these elements significantly influence consumer purchase intentions and willingness to pay (WTP) for cultured meat products (Silva & Semprebon, 2021).

Policy and Regulations-Based Labeling Design

Design labeling textured meat should fulfill the policy and regulations. To enhance individual and public health, regulation is necessary to assure the safety and efficacy the cultured meat. There is presently no structure to adequately regulate cultured meat products. However, it is necessary to strike a balance between safeguarding consumers and ensuring that rules do not erect large obstacles to the market entrance for producers. Some argue that cell-based meat falls outside the range of significance comparable to genuine meat due to the technical methods and additives required to achieve the desired flavor, texture, and color (Wendin & Undeland, 2020). The range of chemical, physical, microbiological, and biological qualities required for quality assessment might be further defined by establishing control quality attributes (CQAs) (Chen et al., 2022). The nutritional characteristics of cultured meat, including levels of vitamin B12, selenium, iron, zinc, and amino acids, should either match or improve the bioavailability of conventional meat (Verbeke, Sans, & van Loo, 2015). Identifying contaminants related to the production process, monitoring bioreactor conditions (e.g., pressure, viscosity, flow rate, temperature), and assessing physical product characteristics (e.g., texture, color, density) are all crucial CQAs (Wendt et al., 2009). This information should be stated clearly on the packaging label of the cultured meat products in order to not mislead the consumers.

The labeling of cultured meat should have accreditation to guarantee the safety of the products. The use of proper controls, e.g., assays to evaluate process conditions, should guarantee that the finished products meet the desired quality specifications. They can also assist to reduce the danger of contamination by unintended microorganisms (e.g., bacteria, fungus, and viruses) throughout the operations of the life cycle of the product including packaging, transportation and storage. Improvement and control of culture process conditions rely heavily on sensor systems. Preventive methods utilized in adjacent sectors, such as biomanufacturing, can also be applied in the manufacture of cultured meat. Cross-contamination and exposure to hazardous materials can be prevented by adopting excellent production methods (e.g., cleanrooms) (Chen et al., 2022).

The Food and Drug Administration and the United States Department of Agriculture (USDA) in 2019 accepted to create a collaborative framework of regulation for cultured meat in the United States (Food and Drug Administration, 2020). The FDA will oversee the upstream elements of the value chain, including cell banks, differentiation, cell collection, and cell growth, while the USDA will regulate the downstream processes, such as product finishing, transportation, and labeling. Additional regulations are necessary to effectively control products imported from overseas producers, especially since these producers may include unauthorized additives or other contaminants (Chen et al., 2022). In the European Union, cultured meat could fall under the Novel Food Regulation (EU Regulation No. 2015/2283) and is expected to be as nutritious as conventional meat (European Parlement, 2018).

Nutrition-Based Labeling

Just as shown on a normal packaging label, the nutrition of the cultured meat should be stated on a packaging label. Laboratory tests can be used to determine the nutritional content of cell cultures using minimal sample volumes. Differentiated muscle meat cells will possibly be the predominant protein source, whereas mature adipocytes can influence the composition of fatty acids (Rubio et al., 2020). Another consideration is a genetic alteration for nutritional benefit, yet genetic approaches may cause issues with regulatory strategy and customer acceptance (Bryant & Barnett, 2020). The nutritional characteristics and cultured meat safety cause an increased 'willingness to pay (WTP)' albeit not a 'willingness to try' because of the dependence on subtle elements, e.g., such as texture and flavor (Chen et al., 2022; Palmieri, Perito, & Lupi, 2021a).

Sensory Quality-Based Labeling

For improved customer acceptance, cultured meat should ideally have flavor and taste like conventional meat (Arshad et al., 2017; Verbeke, Sans, & van Loo, 2015). However, the labeling should accurately reflect the sensory characteristics of the products to ensure consumers are informed about the true taste and flavor of cultured meat. Technically, through the customized engineering of cell culture, researchers are working to enhance the structural and sensory characteristics of cultured meat (Arshad et al., 2017). One approach involves co-culturing a mix of cell types, including myoblasts, adipocytes, and fibroblasts, to generate a 3D structure of meat with connective tissues (Ben-Arye & Levenberg, 2019). Another strategy involves the selection and application of external stimuli (e.g., electrical stimulation) and scaffolds to enable muscle cells to mature sufficiently *in vitro* (Chen et al., 2022). A recent study showed that such a technique may be used to generate cow and rabbit muscle cells that have the taste and flavor of meat (Chen et al., 2022). Extracellular heme proteins have also been found to increase a meat-like appearance and cell proliferation. Upcoming solutions are expected to involve a mix of techniques, following the ultimate meat characteristics. With the flavor and taste close to conventional meat, this could improve the promotion of the cured meat to attract more consumers to purchase the products.

Health Claim-Based Labeling

Health claims in food labeling for cultured meat products should follow the regulations, in which the claims are reviewed by FDA. The health claims should show a risk reduction of a disease or other health condition (Ketelings et al., 2021). Cultured meat products are expected to be safer than conventional meat due to their production in a controlled environment, free from the presence of other organisms, whereas natural meat is obtained from animals exposed to the environment, despite each tissue being protected by mucosa (Chriki & Hocquette, 2020; Santo et al., 2020). While the production process in cultured cells may enhance the positive impacts of micronutrients for human health, this remains unclear in the case of *in vitro* meat. The uptake of micronutrients, such as iron, by cultured cells must therefore be thoroughly understood. Additionally, depending on the composition of the culture medium, there is a potential for a reduction in the health benefits of micronutrients. Furthermore, the inclusion of chemicals in the medium may transform cultured beef into a more "chemical" product with a less desirable label.

Environmental-Based Labeling

Cultured meat products are well known as the current environmentally friendly meat products (Chriki & Hocquette, 2020). Meat production consumes a large number of resources, environmental implications both upstream, such as emissions from cattle systems, and downstream, such as waste disposal and transportation, must be considered (Chriki & Hocquette, 2020). Emissions derived from conventional meat production and cultured meat production are different in concentration, global warming potential, type (e.g., N_2O, CH_4, and CO_2), half-life, and among other things, it's crucial to understand how various gases behave (Chen et al., 2022). Hadi and Brightwell (2021) note that the environmental impact of cultured meat may vary depending on the nutrient medium used. Smetana, Mathys, Knoch, and Heinz (2015; 2018) compared the environmental impact of alternative protein sources, including cultured meat. The authors found that the production of cultured meat has higher carbon emissions and water consumption compared to the production of mycoprotein and insect-based proteins. However, the production of protein from insects requires the largest amount of agricultural land compared to mycoprotein and cultured meat.

Research on Belgian consumers revealed that when given more information related to potential benefits of the environmental aspects of cultured meat, they were more likely to try it (Verbeke, Sans, & van Loo, 2015). However, there were still reservations about the products' capacity to replicate the sensory excellence and low cost of conventional beef. Related to consumer acceptance, since the cultured meat are developed for environmental issues, it is important to address the environmental issue in the labeling by using the green color, for example, known as green and sustainability (Muller et al., 2019). The moderating role of packaging color induces the marketing cues influencing consumers' perception of cultured meat products (Sucapane, Roux, & Sobol, 2021).

Conclusion

This chapter has highlighted the crucial aspects of packaging and labeling design for cultured meat, focusing on the importance of maintaining product quality, ensuring safety, and meeting consumer expectations. As the cultured meat industry moves towards commercialization, effective packaging solutions will be key in preserving the sensory and nutritional qualities of the product while ensuring shelf stability. Additionally, innovative labeling practices, which provide clear and transparent information, are essential to gaining consumer trust and acceptance. The chapter also discussed the regulatory landscape and the need for standardized labeling to ensure that cultured meat products are accurately represented and safe for consumption. With the development of clear policies and regulations, the labeling of cultured meat can effectively address consumer concerns about nutrition, environmental impact, and ethical considerations. In conclusion, as the market for cultured meat continues to expand, addressing packaging and labeling challenges will be essential for its success. By providing consumers with well-designed, informative, and environmentally friendly packaging, and through the establishment of clear regulatory frameworks, the industry can move towards achieving widespread acceptance and market growth.

References

A review of trends and drivers of greenhouse gas emissions by sector from 1990 to 2018. *Environmental Research Letters,* 16(7), Article 073005. https://doi.org/10.1088/1748-9326/abee4e.

Agrimonti, C., White, J. C., Tonetti, S., & Marmiroli, N. (2019). Antimicrobial activity of cellulosic pads amended with emulsions of essential oils of oregano, thyme and cinnamon against microorganisms in minced beef meat. *International Journal of Food Microbiology,* 305, Article 108246. https://doi.org/10.1016/J.IJFOODMICRO.2019.108246.

Arshad, M. S., Javed, M., Sohaib, M., Saeed, F., Imran, A., & Amjad, Z. (2017). Tissue engineering approaches to develop cultured meat from cells: A mini review. Cogent

Food & Agriculture, 3(1), Article 1320814. https://doi.org/10.1080/23311932.2017.1320814.

Aviles, C., Ju´arez, M., Larsen, I. L., Rodas-Gonz´alez, A., & Aalhus, J. L. (2013). Effect of multiple vacuum packs on colour development and stability in beef steaks. *Canadian Journal of Animal Science,* 94(1), 63–69. https://doi.org/10.4141/cjas2013-037.

Bahmid, N. A., Dekker, M., Fogliano, V., & Heising, J. (2021). Development of a moisture-activated antimicrobial film containing ground mustard seeds and its application on meat in active packaging system. *Food Packaging and Shelf Life,* 30, Article 100753. https://doi.org/10.1016/J.FPSL.2021.100753.

Ben-Arye, T., & Levenberg, S. (2019). Tissue engineering for clean meat production. *Frontiers in Sustainable Food Systems*, 3, 46. https://www.frontiersin.org/article/10.3389/fsufs.2019.00046.

Bhat, Z. F., Morton, J. D., Bekhit, A. E. D. A., Kumar, S., & Bhat, H. F. (2022). Cultured meat : Challenges in the path of production and 3D food printing as an option to develop cultured meat-based products. In *Alternative proteins* (pp. 271–295). CRC Press.

Bhat, Z., & Bhat, H. (2011). Animal-free meat biofabrication. *American Journal of Food Technology,* 6. https://doi.org/10.3923/ajft.2011.441.459.

Blanchette, A. (2020). Porkopolis: American animality, standardized life, and the factory farm. Duke University Press. https://doi.org/10.1515/9781478012047.

Brenesselova, M., Kore´nekova, B., Macanga, J., Marcincak, S., Jevinova, P., Pipova, M., et al., (2015). Effects of vacuum packaging conditions on the quality, biochemical changes and the durability of ostrich meat. *Meat Science,* 101, 42–47. https://doi.org/10.1016/j.meatsci.2014.11.003.

Bringing cultured meat to market: Technical, socio-political, and regulatory challenges in cellular agriculture. Trends In *Food Science & Technology,* 78, 155–166. https://doi.org/10.1016/J.TIFS.2018.04.010.

Bryant, C., & Barnett, J. (2020). Consumer acceptance of cultured meat: An updated review (2018–2020). *Applied Sciences,* 10(15). https://doi.org/10.3390/app10155201.

Bryant, C., & Sanctorum, H. (2021). Alternative proteins, evolving attitudes: Comparing consumer attitudes to plant-based and cultured meat in Belgium in two consecutive years. *Appetite,* 161, Article 105161. https://doi.org/10.1016/J. APPET.2021.105161.

Bryant, C., & Sanctorum, H. (2021). Alternative proteins, evolving attitudes: Comparing consumer attitudes to plant-based and cultured meat in Belgium in two consecutive years. *Appetite*, 161, Article 105161. https://doi.org/10.1016/J. APPET.2021.105161.

Chen, L., Guttieres, D., Koenigsberg, A., Barone, P. W., Sinskey, A. J., & Springs, S. L. (2022). Large-scale cultured meat production: Trends, challenges and promising biomanufacturing technologies. *Biomaterials,* 280, Article 121274. https://doi.org/10.1016/J.BIOMATERIALS.2021.121274.

de Oliveira, G. A., Domingues, C. H. D. F., & Borges, J. A. R. (2021). Analyzing the importance of attributes for Brazilian consumers to replace conventional beef with cultured meat. *PLoS One,* 16(5), Article e0251432. https://doi.org/10.1371/JOURNAL.PONE.0251432.

Djisalov, M., Kne˘zi´c, T., Podunavac, I., Z˘ivojevi´c, K., Radonic, V., Kne˘zevi´c, N., et al., (2021). Cultivating multidisciplinarity: Manufacturing and sensing challenges in cultured meat production, 2021, Vol. 10, Page 204 *Biology,* 10(3), 204. https://doi.org/10.3390/BIOLOGY10030204.

Djordjevic, J., Boskovic, M., Starcevic, M., Ivanovic, J., Karabasil, N., Dimitrijevi´c, M., et al., (2018). Survival of Salmonella spp. in minced meat packaged under vacuum and modified atmosphere. *Brazilian Journal of Microbiology,* 49(3), 607–613. https://doi.org/10.1016/J.BJM.2017.09.009.

Escribano, A. J., Pen˜a, M. B., Díaz-Caro, C., Elghannam, A., Crespo-Cebada, E., & Mesías, F. J. (2021). Stated preferences for plant-based and cultured meat: A choice experiment study of Spanish consumers. *Sustainability,* 13(15). https://doi.org/10.3390/su13158235.

Esmer, O. K., Irkin, R., Degirmencioglu, N., & Degirmencioglu, A. (2011). The effects of modified atmosphere gas composition on microbiological criteria, color and oxidation values of minced beef meat. *Meat Science,* 88(2), 221–226. https://doi.org/10.1016/J.MEATSCI.2010.12.021.

Food and Drug Administration (FDA). *Food made with cultured animal cells.* Retrieved from https://www.Fda.Gov/Food/Food-Ingredients-Packaging/Food-Made-Cultur ed-Animal-Cells. (Accessed 20 December 2021) https://www.fda.gov/food/food-ingredients-packaging/food-made-cultured-animal-cells.

Fraeye, I., Kratka, M., Vandenburgh, H., & Thorrez, L. (2020). Sensorial and nutritional aspects of cultured meat in comparison to traditional meat: Much to Be inferred. *Frontiers in Nutrition,* 7, 35. https://www.frontiersin.org/article/10.3389/fnut.2020.00035.

Gesell, C., Glas, A. H., & Essig, M. (2021). The impact of buyer–supplier communication on production ramp-up performance. *Journal of Business & Industrial Marketing.* https://doi.org/10.1108/JBIM-06-2020-0284.

Gholobova, D., Gerard, M., Decroix, L., Desender, L., Callewaert, N., Annaert, P., et al., (2018). Human tissue-engineered skeletal muscle: A novel 3D *in vitro* model for drug disposition and toxicity after intramuscular injection. *Scientific Reports,* 8(1), Article 12206. https://doi.org/10.1038/s41598-018-30123-3.

Go´mez-Luciano, C. A., de Aguiar, L. K., Vriesekoop, F., & Urbano, B. (2019). Consumers' willingness to purchase three alternatives to meat proteins in the United Kingdom, Spain, Brazil, and the Dominican Republic. *Food Quality and Preference,* 78, Article 103732. https://doi.org/10.1016/J.FOODQUAL.2019.103732.

Hadi, J., & Brightwell, G. (2021). Safety of alternative proteins: Technological, environmental and regulatory aspects of cultured meat, plant-based meat, insect protein, and single-cell protein. *Foods,* 10(6). https://doi.org/10.3390/foods10061226.

Hamdan, M. N., Post, M., Ramli, M. A., Kamarudin, M. K., Ariffin, M. F. M., & Huri, N. M. F. Z. (2021). Cultured meat : Islamic and other religious perspectives. *UMRAN-International Journal of Islamic and Civilizational Studies,* 8(2), 11–19. https://doi.org/10.11113/umran2021.8n2.475.

Han, J., Liu, Y., Zhu, L., Liang, R., Dong, P., Niu, L., et al., (2021). Effects of spraying lactic acid and peroxyacetic acid on the quality and microbial community dynamics of vacuum skin-packaged chilled beef during storage. *Food Research International*, 142, Article 110205. https://doi.org/10.1016/J.FOODRES.2021.110205.

Harland, P. E., Uddin, Z., & Laudien, S. (2020). Product platforms as a lever of competitive advantage on a company-wide level: A resource management perspective. *Review of managerial science*, 14(1), 137–158. https://doi.org/10.1007/s11846-018-0289-9.

Ho¨ll, L., Behr, J., & Vogel, R. F. (2016). Identification and growth dynamics of meat spoilage microorganisms in modified atmosphere packaged poultry meat by MALDI-TOF MS. *Food Microbiology*, 60, 84–91. https://doi.org/10.1016/j.fm.2016.07.003.

Hocquette, E., Liu, J., Ellies-Oury, M. P., Chriki, S., & Hocquette, J. F. (2022). Does the future of meat in France depend on cultured muscle cells? Answers from different consumer segments. *Meat Science*, 188, Article 108776. https://doi.org/10.1016/J.MEATSCI.2022.108776.

Hwang, J., You, J., Moon, J., & Jeong, J. (2020). Factors affecting consumers' alternative meats buying intentions: Plant-based meat alternative and cultured meat. *Sustainability*, 12(14). https://doi.org/10.3390/su12145662.

Investigation of the physicochemical, bacteriological, and sensory quality of beef steaks held under modified atmosphere packaging and representative of different ultimate pH values. *Meat Science*, 174, Article 108416. https://doi.org/10.1016/J.MEATSCI.2020.108416.

Jairath, G., Mal, G., Gopinath, D., & Singh, B. (2021). A holistic approach to access the viability of cultured meat: A review. *Trends in Food Science & Technology*, 110, 700–710. https://doi.org/10.1016/J.TIFS.2021.02.024.

Kadim, I. T., Mahgoub, O., Baqir, S., Faye, B., & Purchas, R. (2015). Cultured meat from muscle stem cells: A review of challenges and prospects. *Journal of Integrative Agriculture*, 14(2), 222–233. https://doi.org/10.1016/S2095-3119(14)60881-9.

Karabagias, I., Badeka, A., & Kontominas, M. G. (2011). Shelf life extension of lamb meat using thyme or oregano essential oils and modified atmosphere packaging. *Meat Science*, 88(1), 109–116. https://doi.org/10.1016/j.meatsci.2010.12.010.

Karam, L., Chehab, R., Osaili, T. M., & Savvaidis, I. N. (2020). Antimicrobial effect of thymol and carvacrol added to a vinegar-based marinade for controlling spoilage of marinated beef (Shawarma) stored in air or vacuum packaging. *International Journal of Food Microbiology*, 332, Article 108769. https://doi.org/10.1016/J.IJFOODMICRO.2020.108769.

Kausar, T., Hanan, E., Ayob, O., Praween, B., & Azad, Z. R. A. A. (2019). A review on functional ingredients in red meat products. *Bioinformation*, 15(5), 358. https://doi.org/10.6026/97320630015358.

Kavuncuoglu, H., Yalcin, H., & Dogan, M. (2019). *Production of polyhedral oligomeric silsesquioxane (POSS) containing low-density polyethylene (LDPE) based nanocomposite films for minced beef packaging for extension of shelf life.*

Ketelings, L., Kremers, S., & de Boer, A. (2021). The barriers and drivers of a safe market introduction of cultured meat: A qualitative study. *Food Control*, 130, Article 108299. https://doi.org/10.1016/J.FOODCONT.2021.108299.

Liang, Z., Veronica, V., Huang, J., Zhang, P., & Fang, Z. (2022). Combined effects of plant food processing by-products and high oxygen modified atmosphere packaging on the storage stability of beef patties. *Food Control,* 133, Article 108586. https://doi.org/10.1016/J.FOODCONT.2021.108586.

Listrat, A., Lebret, B., Louveau, I., Astruc, T., Bonnet, M., Lefaucheur, L., et al., (2016). How muscle structure and composition influence meat and flesh quality. *The Scientific World Journal.* https://doi.org/10.1155/2016/3182746, 2016.

Mancini, M. C., & Antonioli, F. (2019). Exploring consumers' attitude towards cultured meat in Italy. *Meat Science*, 150, 101–110. https://doi.org/10.1016/J.MEATSCI.2018.12.014.

Michel, F., Hartmann, C., & Siegrist, M. (2021). Consumers' associations, perceptions and acceptance of meat and plant-based meat alternatives. *Food Quality and Preference*, 87, Article 104063. https://doi.org/10.1016/J.FOODQUAL.2020.104063.

Moritz, J., Tuomisto, H. L., & Ryyna¨nen, T. (2022). The transformative innovation potential of cellular agriculture: Political and policy stakeholders' perceptions of cultured meat in Germany. *Journal of Rural Studies,* 89, 54–65. https://doi.org/10.1016/j.jrurstud.2021.11.018.

Muller, L., Lacroix, A., & Ruffieux, B. (2019). Environmental labelling and consumption changes: A food choice experiment. *Environmental and Resource Economics*, 73(3), 871–897. https://doi.org/10.1007/s10640-019-00328-9.

Onwezen, M. C., Bouwman, E. P., Reinders, M. J., & Dagevos, H. (2021). A systematic review on consumer acceptance of alternative proteins: Pulses, algae, insects, plant-based meat alternatives, and cultured meat. *Appetite,* 159, Article 105058. https://doi.org/10.1016/j.appet.2020.105058. Academic Press.

Palmieri, N., Perito, M. A., & Lupi, C. (2021). Consumer acceptance of cultured meat: Some hints from Italy. *British Food Journal,* 123(1), 109–123. https://doi.org/10.1108/BFJ-02-2020-0092.

Palmieri, N., Perito, M. A., & Lupi, C. (2021). Consumer acceptance of cultured meat: Some hints from Italy. *British Food Journal,* 123(1), 109–123. https://doi.org/10.1108/BFJ-02-2020-0092/FULL/PDF.

Post, M. J., Levenberg, S., Kaplan, D. L., et al., (2020). Scientific, sustainability and regulatory challenges of cultured meat. *Nat Food,* 1, 403–415. https://doi.org/10.1038/s43016-020-0112-z.

Rolland, N. C., Markus, C. R., & Post, M. J. (2020). The effect of information content on acceptance of cultured meat in a tasting context. *PLoS One,* 15(4), Article e0231176. https://doi.org/10.1371/journal.pone.0231176.

Rollin, F., Kennedy, J., & Wills, J. (2011). Consumers and new food technologies. *Trends in Food Science & Technology - Trends Food Sci Technol,* 22, 99–111. https://doi.org/10.1016/j.tifs.2010.09.001.

Rosenfeld, D. L., & Tomiyama, A. J. (2022). Would you eat a burger made in a petri dish? Why people feel disgusted by cultured meat. *Journal of Environmental Psychology,* 80, Article 101758. https://doi.org/10.1016/J.JENVP.2022.101758.

Rubio, N. R., Xiang, N., & Kaplan, D. L. (2020). Plant-based and cell-based approaches to meat production. *Nature Communications,* 11(1), 6276. https://doi.org/10.1038/s41467-020-20061-y.

Sergelidis, D. (2019). Lab grown meat: The future sustainable alternative to meat or a novel functional food? *Biomedical Journal of Scientific & Technical Research,* 17. https://doi.org/10.26717/BJSTR.2019.17.002930.

Smetana, S., Aganovic, K., Irmscher, S., & Heinz, V. (2018). Agri-food waste streams utilization for development of more sustainable food substitutes. *Designing Sustainable Technologies, Products and Policies*, 20, 145–155.

Smetana, S., Mathys, A., Knoch, A., & Heinz, V. (2015). Meat alternatives: Life cycle assessment of most known meat substitutes. *International Journal of Life Cycle Assessment,* 20(9), 1254–1267. https://doi.org/10.1007/s11367-015-0931-6.

Stephens, N., & Ruivenkamp, M. (2016). Promise and ontological ambiguity in the *in vitro* meat imagescape: From laboratory myotubes to the cultured burger. *Science As Culture*, 25(3), 327–355. https://doi.org/10.1080/09505431.2016.1171836.

Sucapane, D., Roux, C., & Sobol, K. (2021). Exploring how product descriptors and packaging colors impact consumers' perceptions of plant-based meat alternative products. *Appetite,* 167, Article 105590. https://doi.org/10.1016/J.APPET.2021.105590.

Technological, regulatory, and ethical aspects of *in vitro* meat: A future slaughter- free harvest. *Comprehensive Reviews in Food Science and Food Safety*, 18(4), 1192–1208. https://doi.org/10.1111/1541-4337.12473.

Technological, regulatory, and ethical aspects of *in vitro* meat: A future slaughter- free harvest. *Comprehensive Reviews in Food Science and Food Safety*, 18(4), 1192–1208. https://doi.org/10.1111/1541-4337.12473.

The effect of barrier properties of polymeric films on the shelf-life of vacuum packaged fresh pork meat. *Meat Science,* 158, Article 107880. https://doi.org/10.1016/J.MEATSCI.2019.107880.

Umaraw, P., Munekata, P. E. S., Verma, A. K., Barba, F. J., Singh, V. P., Kumar, P., et al., (2020). Edible films/coating with tailored properties for active packaging of meat, fish, and derived products. *Trends in Food Science & Technology*, 98, 10–24. https://doi.org/10.1016/J.TIFS.2020.01.032.

Van der Weele, C., Feindt, P., van der Goot, A. J., van Mierlo, B., & van Boekel, M. (2019). Meat alternatives: An integrative comparison. *Trends in Food Science & Technology,* 88, 505–512. https://doi.org/10.1016/j.tifs.2019.04.018.

Verbeke, W., Marcu, A., Rutsaert, P., Gaspar, R., Seibt, B., Fletcher, D., et al., (2015). 'Would you eat cultured meat?': Consumers' reactions and attitude formation in Belgium, Portugal, and the United Kingdom. *Meat Science,* 102, 49–58. https://doi.org/10.1016/J.MEATSCI.2014.11.013.

Verbeke, W., Sans, P., & van Loo, E. J. (2015). Challenges and prospects for consumer acceptance of cultured meat. *Journal of Integrative Agriculture*, 14(2), 285–294. https://doi.org/10.1016/S2095-3119(14)60884-4.

Wang, H., Luo, Y., & Ertbjerg, P. (2017). Myofibrillar protein gel properties are influenced by oxygen concentration in modified atmosphere packaged minced beef. *Food Chemistry,* 230, 475–481. https://doi.org/10.1016/J.FOODCHEM.2017.03.073.

Weinrich, R., Strack, M., & Neugebauer, F. (2020). Consumer acceptance of cultured meat in Germany. *Meat Science,* 162, Article 107924. https://doi.org/10.1016/J.MEATSCI.2019.107924.

Wendin, K., & Undeland, I. (2020). Seaweed as food – attitudes and preferences among Swedish consumers. A pilot study. *International Journal of Gastronomy and Food Science,* 22, Article 100265. https://doi.org/10.1016/J.IJGFS.2020.100265.

Young, J. F., & Skrivergaard, S. (2020). *Cultured meat on a plant-based frame*, 2020 1:4.

Chapter 5

Risk Assessment of Cultured Meat

Abstract

Despite some controversies, the development of cell culture and tissue engineering, coupled with the increasing demand for sustainable animal-based protein, positions cultured meat as a potential future source of nutritional protein. As a novel food, the production of cultured meat involves advanced technologies and new ingredients, which may introduce hazards and risks at various stages. These must be reviewed and approved by regulatory authorities to ensure food safety and promote health. While research on the safety and risk assessment of cultured meat is increasing, a lack of systematic, in-depth risk evaluation remains a challenge for its global commercialization. This chapter summarizes potential hazards, proposes countermeasures, and highlights the need for context-specific risk assessments and regulatory harmonization.

Keywords: risk assessment, cultured meat safety, cellular agriculture, food safety standards

Introduction

Livestock farming is an important industry that contributes to national economy and people's livelihood. The positive impacts and benefits of livestock farming include the provision of protein-rich food, the creation of employment opportunities, and the preservation of biodiversity and cultural heritage (Dumont et al., 2019; Dumont, Groot, & Tichit, 2018). However, global livestock farming has recently been constrained by its negative effects on the environment (e.g., greenhouse gas emissions, water consumption and pollution, species extinction), ethics (e.g., animal welfare), and human health (e.g., antibiotic overuse, foodborne illnesses, zoonotic diseases, occupational diseases) (Rubio, Xiang, & Kaplan, 2020). The demand for animal-based proteins is predicted to double with the world population forecast to increase to 9.7 billion by 2050 (Hadidi et al., 2022). With the development of cell culture and tissue engineering techniques, some researchers suggest that it is becoming increasingly feasible to obtain sustainable animal proteins using fewer animal-derived materials than traditional animal husbandry, while avoiding certain drawbacks associated with conventional animal protein consumption (Kumar, Sood, & Han, 2023). The first cultured beef burger in the world was created in the laboratory of Professor Mark Post at Maastricht University on August 5, 2013, sparking a trend in cultured meat research (Mattick, Landis, & Allenby, 2015). According to the 2022 State of the Industry Reports from the Good Food Institute, 156 cultured meat companies across 29 countries have been established, with 19 of them launched in 2022 (Claire et al., 2022). Notably, Singapore is the only country currently permitting the regulated sale of cultured meat products.

Additionally, various terms such as cultured meat, clean meat, *in vitro* meat, lab-grown meat, cell-based meat, cultivated meat and animal-free meat are used to describe this product (Hallman, Hallman, & Hallman, 2023). Its name is actually a matter of debate and controversy and may have great significance in the European or Arabic countries due to legislation and religious beliefs (Chriki, Ellies-Oury, & Hocquette, 2022; FAO, 2022). The priorities of these terms are not discussed in this review and for consistency, the term "cultured meat" is used. Cultured meat refers to the edible muscle tissue that is produced by cultivating cells but is not directly obtained from farming animals (Kumar et al., 2023). As one of the most important fields of cellular agriculture, the production of cultured meat relies on cell biology and tissue engineering technologies. These processes primarily include the collection of targeted cells or tissues from living or freshly slaughtered animals, cell isolation and culture, biomass production (including cell proliferation, expansion, differentiation, and maturation), and product harvesting (Chen et al., 2022). Although life cycle analysis indicates that the energy consumption of cultured meat may be higher than that of poultry meat, the production of cultured meat requires less water, land, and energy, and emits fewer greenhouse gases compared to the production of agricultural beef, sheep, and pork meat. Despite high un- certainty, the authors concluded that the overall environmental impacts of cultured meat production are substantially lower than those of conventionally produced meat (Tuomisto & Teixeira, 2011). Nevertheless, with greater clarity in cultured meat technology in recent years, there may be some inaccuracies in the above conclusion (Mattick et al., 2015). More recent research presents an opposing viewpoint. For instance, cultured meat has been reported to require more energy than pork (Ye, Zhou, Guan, & Sun, 2022) and emit greenhouse gases comparable to beef (Klein, 2023; Lynch & Pierrehumbert, 2019). The technology behind cultured meat production is still in its infancy, and its environmental impact is closely tied to the specific production system and scale (Treich, 2021). As of now, opinions on the environmental impact of cultured meat production remain divided and inconclusive. Therefore, the overall environmental impact of cultured meat needs further assessment, monitoring, and clarification (FAO, 2022).

Being cultured within a bioreactor under controlled artificial conditions, cultured meat is less reliant on environmental factors such as weather and geography and addresses some an ethical issue associated with traditional animal husbandry such as livestock slaughter (Stephens et al., 2018). Additionally, stem cells from farm animals are predicted to produce a million times more meat than corresponding farm animal carcasses and cultured meat can be fully consumed instead of wasting some inedible parts of traditional meat (Connor, 2013). Some authors claim that cultured meat has great potential to be a sustainable source of protein (Jairath, Mal, Gopinath, & Singh, 2021). Strict control in the production of cultured meat avoids the reproduction of pathogenic bacteria and thus reduces biological risks and the possibility of disease occurrence (Post, 2012). These benefits promote the popularity and development of cultured meat. However, some challenges persist. Although cultured meat can be easily produced on a small scale, it becomes expensive for large-scale production. Therefore, one challenge is the optimization of scale-up production, such as innovating alternatives to replace the rare fetal bovine serum and recapitulating the *in vivo* myogenic environment (Ye et al., 2022). Consumer acceptance presents an additional hurdle, as it remains relatively low in most countries. For instance, some consumers are concerned about the safety of cultured meat due to the lack of transparency in its production techniques. Others experience nauseous reactions when hearing about this concept and firmly refuse to take any risks in consuming cultured meat (Liu, Hocquette, Ellies-Oury, Chriki, & Hocquette, 2021). Various terminologies associated with cultured meat may also affect consumers' acceptance. Hence, an appropriate term is crucial to promote the development of this industry and an international harmonized name is imminent (Hallman et al., 2023).

It is most important to assess the safety of novel food before its introduction to guarantee consumer health. Some regulations and legislations have been enacted to assess the nutrition and risks associated with cultured meat. While the general production process of cultured meat has been extensively reported, key details such as relevant commercial production technologies and the composition of the final product remain unclear, presenting challenges for regulatory authorities. Furthermore, the rapid advancement of cultured meat research leads to frequent updates in regulations. However, the safety assessment of cultured meat is currently less structured due to the novel and complex production processes, as well as the unforeseen biological mechanisms and potential consequences associated with each product (Petetin, 2014). Consequently, the lack of a systematic and in-depth risk assessment is a significant barrier to the commercialization of cultured meat. This chapter comprehensively summarizes the possible risks that could be introduced during the production of cultured meat and proposes countermeasures. The overarching goal is to consolidate this information and develop systematic safety assessment of cultured meat products in the near future.

Key Technologies and Possible Risks that Could Be Introduced During Cultured Meat Production

Although some companies have successfully produced cultured meat products, the detailed production process remains as trade secrets un- known to the public. Lack of transparency remains a major challenge in the food production process, particularly in ensuring food safety and health. However, the critical steps involved in constructing complex structures of livestock muscle from a small number of cells have been extensively discussed. Generally, four key technical aspects are involved: cell collection, cell culture, biomass accumulation, and food processing. These steps may be omitted, repeated, or modified by different manufacturers, depending on their specific goals, such as producing disorganized muscle fibers or creating muscle flakes through overlapping layers of muscle fibers and fat (Li et al., 2022; Ong et al., 2021). The production of cultured meat involves complicated advanced technologies and multiple novel ingredient inputs. Consequently, potential hazard and risk characteristics may arise during cultured meat pro- duction which are described as follows.

Cell Collection

The collection of suitable cells is the basis of cultured meat production, where their ability to proliferate and differentiate is essential. Embryonic stem cells (so-called pluripotent stem cells), induced pluripotent stem cells, mesenchymal stem cells and muscle stem cells (so-called satellite cells) are often used as the first-choice seed or starter cells to produce muscle fiber or adipose tissue (Choi et al., 2021). It is speculated that stem cells from farm animals could produce over a million times more meat than the corresponding farm animal carcasses (Connor, 2013). These cells can be obtained in two ways. The first method involves isolating primary cells through tissue enzymatic dissociation, followed by filtration, differential centrifugation, and cell separation based on their physical characteristics (e.g., density), biological factors (e.g., adhesion time to the cell culture plate), or molecular traits (e.g., specific gene or protein expression) (Bodiou, Moutsatsou, & Post, 2020). This approach is technically challenging and costly to isolate sufficient desired cell type from heterogenous tissues (Choi et al., 2021). Another viable approach is the induction from somatic cells via genetic engineering, chemical approach or spontaneous mutations (Bhat, Morton, Mason, Bekhit, & Bhat, 2019). The acquisition of such seed cell line may involve unintentional harmful genetic alteration, and genetic drift may occur at any stages of subsequent cell growth, proliferation, differentiation and maturation (Stephens et al., 2018). To monitor and mitigate genetic drift, cells can be tested using genetic stability assays and chromosomal analyses, while freshly stored starter cells can be used for each batch (Ketelings, Kremers, & Boer, 2021; Sewalt, Shanahan, Gregg, La Marta, & Carrillo, 2016). Additionally, genome instability and genetic drift may lead to the production of undesirable substances in the final product at levels that pose food safety risks. The Singapore Food Agency has proposed effective strategies to identify potential undesirable substances through a systematic review of scientific literature, screening substances against relevant allergen and toxicology databases, and comparing the end-product cells to the starter cells using global untargeted and sequential targeted omics techniques (SFA, 2022). Microbiological contamination (bacteria, fungi and viruses) may occur at any time during the entire production process of cultured meat, which may originate from any items and operators involved in its production. Microbiological risk can be accessed based on the guidelines provided by FAO (FAO, 2021a). The health status (including viral or parasitic infections) of the source animal providing the cells should be thoroughly assessed to prevent the introduction of biohazards into the starter cells. For instance, prions are typically present in the brain, spinal cord, lymphoid tissue, tonsils, appendix, enteric nervous system, and blood of infected animals (Gough & Maddison, 2010). Therefore, these tissues should be avoided. Moreover, the source animal should be tested to ensure it is free from bovine spongiform encephalopathy, further minimizing the risk of prion contamination (Ong et al., 2021). It has been reported that prions can be removed from large volumes of culture medium supplemented with fetal bovine serum through hollow fiber anion-exchange membrane chromatography (Chou, Bailey, Avory, Tanimoto, & Burnouf, 2015). Standard operating procedures for sterilization and microbiological examination are also critical to minimize the risk of microbial foodborne diseases. Currently, the process of cell collection from animals can be aseptically controlled, which may help reduce or eliminate the need for antibiotics and mitigate their associated residual risks (Okocha, Olatoye, & Adedeji, 2018; Rottenberg, 2018).

Cell Culture

Stem cells are initially grown exponentially without differentiation in regular culture dishes or flasks containing culture medium, which generally contains various nutrients such as glucose, amino acids, fatty acids, vitamins and minerals, growth factors, hormones and fetal bovine serum (Tyagi & Mani, 2023). The media components influence both the growth efficiency of the cells and the organoleptic and nutritional characteristics of the final product, which must be optimized according to the cell species, type, and developmental stage (O'Neill et al., 2022). Fetal bovine serum, derived from bovine embryos, contains 200–400 different proteins and thousands of small molecule metabolites at indeterminate concentrations (Post et al., 2020). It provides essential nutrients such as hormonal and differentiation factors (e.g., growth hormone and insulin), transport proteins (e.g., transferrin and transcortin), and growth factors (e.g., epidermal, endothelial, fibroblast, and insulin-like growth factors) to promote cell growth (Hadi & Brightwell, 2021). However, its undefined composition carries unknown risks for the end products. Furthermore, its use in cultured meat pro- duction is unanimously opposed by scientific, regulatory and industry experts due to its unsustainable, expensive and unethical nature (Ketelings et al., 2021; Post et al., 2020). Therefore, the development of serum-free medium with chemically defined components is important for the commercial production of cultured meat. Currently, some serum-free media, such as AIM-V medium, Gibco medium, and Ultroser G medium, are commercially available for specific cell types like Caco-2 cells and myosatellite cells (Datar & Betti, 2010; Van der Valk et al., 2010). However, since various cells grow under specific conditions, a universal serum-free medium may not be achievable (Van der Valk et al., 2010), and some commercial media may perform less effectively compared to fetal bovine serum (Kolkmann, Post, Rutjens, Van, & Moutsatsou, 2020). Additionally, the details of the production of commercial media are generally not publicly disclosed, making it difficult to monitor any abnormal changes during cell proliferation, differentiation, and maturation, as well as to predict the risks associated with the final product. As cultured meat is a product of cellular agriculture, life cycle assessments must also be conducted to evaluate the environmental impact of the commercial media used in its production (Stephens et al., 2018). Additionally, as cell culture media components originating from animal may harbor viruses, prions and bacteria, their sources need to be properly characterized to avoid potential contamination (Ong et al., 2021). The defined animal-component-free medium (such as microorganism- or plant-based) has been developed for the routine culture of human embryonic stem cell and induced pluripotent stem cell at low cost, with simplified quality control and high safety (Chen et al., 2011). However, one must be mindful of the potential risks associated with allergens derived from plant-originated proteins (Datar & Betti, 2010). Encouragingly, in early 2022, a chemically defined, serum-free media formulation capable of supporting bovine satellite cell differentiation without the need for serum or transgene expression to produce cultured meat was reported (Messmer et al., 2022). Fat plays a significant role in meat palatability, and a simplified, defined serum-free medium with high-quality differentiation properties has been developed for cultivating adipose cells across species (Mitic, Cantoni, Borlin, Post, & Jackisch, 2023). Furthermore, a patent for a serum-free medium for culturing bovine progenitor cells has been published (Moutsatsou et al., 2023). These studies suggest that fetal bovine serum may eventually be completely replaced. However, these studies have been conducted in laboratory settings, which remain distant from industry application. The development of serum-free or animal-component-free media is expected to utilize existing materials approved for food production, implying that they must meet the corresponding specifications or criteria. If novel components are introduced, toxicological testing will be required to evaluate any residues in the final product (Ketelings et al., 2021). It is credited to Paracelsus who expressed the classic toxicology maxim "All things are poison, and nothing is without poison; the dose makes the poison." Although the addition of some components is permissible, their concentrations need to be properly controlled due to dose-dependent toxicity. For example, allergic and reproductive/developmental/carcinogenic toxicity can be introduced by high concentrations of wheat gluten and growth factors, respectively (Ong et al., 2021; Radosevi´c, Dukic, Andlar, Slivac, & Gaurina, 2016).

Biomass Accumulation

Economical large-scale production is essential for commercialization of cultured meat products. Hence, bioreactors that mimic *in vivo* environment are developed to achieve rapid cell proliferation. Prior to being added to bioreactor, the cells with naturally desired traits are selected or they may be modified by genetic engineering to optimize certain properties, such as rapid proliferation, extended life cycle, increased intra- cellular heme protein expression and optimal flavor profiles (Rubio et al., 2020; Wang, Chen, Yan, & Pei, 2019). However, genetic engineering may present regulatory challenges and issues with consumer acceptance, potentially rendering products unmarketable. For instance, several European countries, including France, Germany, and Greece, have banned the production and sale of genetically modified foods (Petetin, 2014). In addition to monitoring genetic drift, genetic stability, and potential tumorigenicity, assays should be conducted to assess the safety of genetically engineered cells and their products. The potential environmental impact of modified genes should also be considered, as dietary genes may integrate or transfer into human blood, tissue, and the gastrointestinal tract (Nawaz et al., 2019).

The formation of three-dimensional muscle tissue (tissue-engineering techniques) relies on the support of the scaffolds, which provide a temporary or permanent adherent surface for cell attachment, growth and easy harvesting. Furthermore, the scaffolds also infuse the muscle fibers with nutrients and stimulate them mechanically or electrically to approach the true muscle state (Bhat, Bhat, & Pathak, 2014). Some animal-, plant-, and marine-derived as well as biosynthetic materials could be suitable candidates for scaffold design due to their sustainable origins and ability to meet Food and Drug Administration (FDA) food additive requirements (Schneider, 2013). Notably, myoblasts tend to reside within animal-derived materials, such as collagen-based scaffolds, which recapitulate the natural physiological niche of cells. Not surprisingly, most successful cultured meat to date has been produced on scaffolds made of collagen (Bhat et al., 2019; Stephens et al., 2018). However, animal-derived materials are not sustainable and do not comply with animal welfare strategies. The expected scaffold for cultured meat production should not only provide a biocompatible and high surface area-to-volume ratio surface to support muscle and adipose cell growth but preferably comprise edible or degradable biological or synthetic materials (Li et al., 2022). Otherwise, cells growing on those scaffolds that are inedible or non-degradable need to be further separated from the scaffold before myogenic differentiation (Choi et al., 2021). The disadvantages of separating sheets of cells from scaffolds are damages of cells and the extracellular matrix, as well as the shrinkage and aggregation of the cell sheet without any support (Bhat et al., 2019; Canavan, Cheng, Graham, Ratner, & Castner, 2005). Shima and colleagues constructed three-dimensional (3D) contractile human skeletal muscle tissues with contractile properties from human skeletal myocyte cell lines using 3D culture technology (Datar & Betti, 2010). An edible soy protein-based scaffold was developed to create a 3D-engineered bovine muscle tissue (Ben-Arye et al., 2020). Additionally, another edible scaffold composed of gelatin and soymilk was successfully synthesized for the production of fat-containing cultured meat (Li et al., 2022). Despite these advancements, careful attention must be given to ensure that materials derived from known allergens are not inadvertently introduced (FAO, 2022).

Bioreactor systems supply optimal physical (temperature, humidity, pressure), chemical (pH, oxygen concentration) and biological (cell density, medium nutrition) environments to promote cell growth within scaffold. Currently, various bioreactors, such as stirred tank, airlift, and rocking/wave bioreactors, have been developed for cell culture. However, each of these systems has specific drawbacks that require optimization (Post et al., 2020). The conditions necessary for cell growth change throughout the stages of proliferation, differentiation, and maturation. These conditions should be adjusted accordingly by optimizing factors such as cell media composition, scaffolds, and bioreactor parameters. After the desired biomass is accumulated, the product is collected. For sustainable development, medium or other essential components need to be recovered for recycling, which will come true in the future.

Food Processing

After the cells have differentiated into myotubes and developed into myofibers, they are collected and may be mixed with other substances, such as hemoglobin and flavoring agents, to enhance the color, aroma, and taste of the final products (Simsa et al., 2019). Additionally, further processing is required to convert them into commodities, with the specific processing methods depending on the nature of the desired product. Currently, it is uncertain whether conventional food processing methods pose risks to the cultured meat products, such as the human or environmental hazards resulting from the unpredicted reaction of novel components in cultured meat by heating or microwave radiation.

Risk Assessment Strategies in Cultured Meat Production

The Premise of Safety-Standardized Production of Cultured Meat

Hazards and risks can be unintentionally introduced at any point during the production of cultured meat. As such, it is essential to adhere to standardized procedures and best practices to ensure safety. These include Good Laboratory Practices (EFSA, 2021), Good Manufacturing Practices (FDA, 2015), Good Cell Culture Practices (Bal-Price & Coecke, 2011), and the Code of Hygienic Practices (CAC, 2005; Hadi & Brightwell, 2021). Additionally, the implementation of effective hazard and risk management systems is critical. Such systems are now mandatory in many countries and primarily consist of Hazard Analysis and Critical Control Point (HACCP) and Hazard Analysis and Risk-Based Preventative Controls (HARPS). Briefly, the potential risks and hazards at each step of cultured meat production are pre-estimated and subsequently, the risks are decreased or prevented by alternating certain inputs or adjusting specific procedures.

Risk Assessment Strategies in Cultured Meat Production

Given the diverse and unknown hazards and risks that may arise during the production of cultured meat products, risk assessment strategies are described as follows.

Routine Biochemical Detection

Although it is uncertain whether all the existing standards of traditional meat products apply to cultured meat products, they provide effective references. Therefore, the common harmful microorganisms, viruses and mycoplasma in cultured meat can be monitored in similar ways as traditional meat. Unintended residues and harmful byproducts may form as a result of adding various ingredients and exposure to different materials and equipment during the production of cultured meat. To ensure safety, the production of cultured meat must adhere to the same maximum permissible levels of metals, natural toxins, agricultural or veterinary chemicals, and environmental contaminants that are established for traditional livestock products. These regulations have been outlined in countries such as Australia and Canada (Ong et al., 2021). Their risk assessment can be performed according to the guidelines provided by FAO (FAO, 2021b). Antibiotics can be avoided with a sterile environment, but they may still be used by some manufacturers, which can be detected, characterized and quantified in cultured meat by multiple bioassay methods. However, consumers need not worry too much about the residue of antibiotics because many currently authorized antibiotics used in agricultural animals also are authorized for humans (NRC, 1999). Beyond these authorized components, some novel materials in cultured meat may pose challenges for their safety assessment. The detection of novel components needs to be developed specifically based on conventional analytical methods (chromatographic, mass spectrometric, immunological methods, etc.) and existing bioassays. The detailed composition (e.g., macronutrients, micronutrients, bioactive compounds, toxins, allergens) of cultured meat products needs to be determined and quantified through laboratory assays (Rubio, Fish, Trimmer, & Kaplan, 2019) and compared to conventional meat products. Generally, novel products that are compositionally, nutritionally, and functionally equivalent to already accepted foods are theoretically as safe as their counterparts. However, substantial equivalence does not guarantee the absolute safety of the novel food; it merely provides a "reasonable certainty" that it is as harmless as conventional food, which may not convince all scientists (Ghisleri, Anado´n, Recuerda, & Peralta, 2009). The composition of cultured meat products is expectedly different from that of traditional meat products since there are under- lying compositional differences among species, breeds, genders or batches of conventional meat, in addition to the novel components that may be present in cultured meat (Constable et al., 2007). Consequently, even if the major component of cultured meat conforms to the range of traditional meat products specified by some agencies, further risk evaluation may still need to be performed on the novel components.

Allergenicity Testing

If a new protein is found in cultured meat products as compared with traditional meat products, it generally increases the product allergenicity because most food allergens are glycosylated proteins. Therefore, its amino acid sequence should be aligned and its structure should be compared with known allergenic proteins recorded in various databases using bioinformatics analysis. The databases relevant to allergenicity testing include food allergy research and resource program (FARRP) databases (https://farrp.unl.edu/resources/farrp-databases), WHO/IUIS allergen nomenclature database (http://allergen.org/), allergome database (http://www.allergome.org), protein family (Pfam) database (http://pfam.xfam.org/), AllFam database (http://www.meduniwien.ac.at/allfam/), structural database of allergenic proteins (SDAP) data- base (https://fermi.utmb.edu/) and UniProtKB database (https://www. uniprot.org/uniprot/) (Accessed September 26, 2022) (SFA, 2022). If homologous sequences or structural similarities to known allergens are observed, further allergenicity tests [*in vitro* digestive stability test (Brodkorb et al., 2019; Minekus et al., 2014), allergenicity assay using cell lines and rodents, human serum IgE test, skin prick testing, controlled food challenge study, etc.] are further considered. Clinical studies on allergenicity are considered unethical due to the severity of the clinical manifestations that might occur, which can only be con- ducted after obtaining adverse results from *in vitro* allergenicity tests. In addition, attention should be paid to the influence of the processing method on the putative allergenic potential during the consumption of cultured meat (Verhoeckx, Broekman, Knulst, & Houben, 2016). If the identified allergens are detected in the cultured meat product, the allergen information needs to be labeled clearly to guide consumers appropriately.

Toxicity Testing

All the materials, processes and relevant details involved in the production of cultured meat should be considered thoroughly for designing the appropriate toxicity tests. Toxicity testing primarily includes genotoxicity, cytotoxicity, and toxicity tolerance tests, which can be conducted based on the recent recommendations from the FDA regarding cultured chicken produced by UPSIDE Foods, or the requirements and tiered approach outlined in the European Food Safety Authority (EFSA) "Guidance for Submission for Food

Additive Evaluations." This approach aims to save resources and reduce animal usage (EFSA, 2012; FDA, 2022; FDA, 2023). The rationale behind the tiered approach is to start with less complex tests, assess whether these tests are sufficient for risk assessment, and select higher-tier tests if necessary. The results of studies conducted at higher tiers will, in principle, supersede those derived from lower-tier tests.

Genotoxicity refers to the damage to the DNA or chromosomes of mammalian cells, resulting in mutagenic or carcinogenic effects (Berlo, Clift, Albrecht, & Schins, 2012), which can be assessed by a series of *in vitro* and *in vivo* evaluation approaches. As described by EFSA Scientific Committee, the purpose of genotoxicity testing for risk assessment of substances in food is to identify substances that may cause genetic damage to humans, predict potential genotoxic carcinogens in the absence of carcinogenicity data and understand the mechanism of action of chemical carcinogens (EFSA, 2011a). A stepwise battery approach consisting of Ames bacterial mutagenicity test (OECD 471, 2020) and *in vitro* mammalian cell micronucleus test (OECD 487, 2016) is recommended because this combination can detect all genotoxic mechanisms (three genetic endpoints) relevant to mutagenesis (gene mutations, structural chromosome aberrations and aneuploidy) with minimum number of tests and at high sensitivity (EFSA, 2011a). If all *in vitro* tests are negative without uncertainty, contradiction, or ambiguity, the tested component is generally considered not to be genotoxic. Otherwise, one or more *in vivo* assays (such as the erythrocyte micronucleus test (OECD 474, 2016), comet assay (OECD 489, 2016), transgenic rodent assay (OECD 488, 2020), chromosomal aberration test (OECD 475, 2016), or rodent dominant lethal test (OECD 478, 2016)) should be conducted to further validate the potential genotoxicity. However, we emphasize avoiding *in vivo* (animal or human) experiments whenever possible (Ong et al., 2021). Cultured meat, as a whole food, is typically not tested for classical genotoxicity by itself, but fractions of it can be tested (Ververis et al., 2020). Cytotoxicity refers to abnormal cell growth, function, morphology, and ultimately disintegration and death caused by chemicals (Mukherjee, 2019). Relevant cytotoxicity assays (including dye exclusion, colorimetric assay, fluorometric assay, and luminometric assay) along with their advantages and disadvantages have been reviewed (Li et al., 2022). Additionally, after *in vitro* gastrointestinal digestion, the digested products may be subjected to *in vitro* gastrointestinal cell toxicity testing to verify their safety in the gut. Co-culturing of cells (Caco-2, HT29-MTX and Raji B) has further been suggested to simulate the complex gastro- intestinal environment for physiologically-relevant toxicity testing (Pradhan et al., 2020). It is important to emphasize that the *in vitro* tests mentioned above are focused on the ingredients of cultured meat, rather than the whole product itself, which cannot be fully dissolved in a given matrix without specialized treatment. However, under the appropriate conditions, whole novel foods can be tested in a manner similar to complex mixtures.

In some regions, risk assessment of novel food based on rodent model is mandatory. Therefore, there is an urgent need to develop, validate and apply reliable and representative alternative strategies for *in vivo* testing of cultured meat product that do not use animals. Currently, subchronic toxicity testing is generally performed for at least 90 days (OECD 408, 2018), which aims to identify any adverse effects following prolonged exposure through appropriate oral routes. Based on OECD 408 guidance, EFSA Scientific Committee further complemented the specific advice for conducting a repeated-dose 90-day oral toxicity study in rodents on whole food/feed (EFSA, 2011a). In animal feeding toxicity experiments, animals should be fed more test samples than the amount expected for human consumption. However, it is difficult for animals to consume large amounts of cultured meats due to their limited stomach volume. Additionally, as cultured meat product cannot be orally-gavaged, the dose is limited by preference. Furthermore, excessive intake of cultured meat product may lead to nutritional imbalance and metabolic disturbance (EFSA, 2011b). Ultimately, we may not be able to distinguish the adverse effects caused directly by cultured meat products from those due to nutritional imbalances induced by excessive consumption. Therefore, animal oral toxicity studies may not be applicable to whole cultured meat products but may be suitable for testing their major novel ingredients. Animal characteristics such as behavior, clinical signs, food and water consumption, body and organ weight, ophthalmology, clinical pathology, and histopathology are analyzed to reflect the multiple effects (e.g., immunotoxicity and neurotoxicity) of the tested components. In 90-day subchronic toxicity studies, if abnormalities in the reproductive system are observed, further *in vivo* reproductive and developmental toxicity testing such as prenatal developmental toxicity study (OECD 414, 2018), extended one-generation reproduction toxicity study (OECD 443, 2011) and two-generation reproductive toxicity study (OECD 416, 2001) may be required. If toxicities are found, tier 2 toxicity tests including chronic testing (12 months) (OECD 452, 2018), carcinogenicity testing (OECD 451, 2018) or combined chronic/carcinogenicity testing (OECD 453, 2018) are conducted. The integrated toxicity assessment for tiers 1 and 2 further determines the tier 3 toxicity tests. For example, the potential impact on the developing immune and nervous systems is also reflected in the extended one-generation reproduction toxicity study. If adverse effects are observed, tier 3 specialized toxicity assessments, such as neurotoxicity (OECD 424, 1997), developmental neurotoxicity (OECD 426, 2007), and immunotoxicity (WHO/IPCS, 2012), along with their modes of action, should be considered accordingly. Duplicate animal tests should be avoided whenever possible, and animal testing must comply with the relevant guidance documents, such as Regulation (EU) 2010/63 (EFSA, 2021) or similar legislation in non-EU countries (EFSA, 2005, 2011). In some cases, human studies may also need to be investigated further on a case-by-case basis. It is essential that all the data mentioned above be integrated using the weight-of-evidence approach to evaluate the safety of cultured meat products under the proposed conditions of use.

Kinetics Testing

Kinetics testing refers to absorption, distribution, metabolism and excretion (ADME) analysis, which can characterize how the nutrients of novel food are systemically absorbed, distributed throughout the body and eliminated via the liver and kidney (EFSA, 2016). Generally, absorption tests (including *in vitro*, *in vivo* and *ex vivo* absorption and bioavailability models) and *in vitro* gastrointestinal metabolism (e.g., the Ussing chamber and inverted sac models) are initially performed ac- cording to the tiered testing guideline described in the EFSA "Guidance for submission for food additive evaluations" (EFSA, 2012) and the advantages of tiered testing strategies have been described in detail elsewhere (Becker, Plunkett, Borzelleca, & Kaplan, 2007). Only compounds that are systemically absorbed, broken down, or metabolized into components that may be absorbed during gastrointestinal digestion require subsequent ADME (Absorption, Distribution, Metabolism, and Excretion) tests or higher-tier toxicological studies to generate more comprehensive data. ADME testing can be conducted through either single-dose studies (OECD 417, 2010) or repeated-dose studies, particularly when bioaccumulation is observed in single-dose ADME tests. Kinetic data, when combined with nutritional information, serve as references for further toxicity assessments. As cultured meat is considered a whole food, traditional toxicity studies may not be applicable. Instead, ADME analysis of individual components - such as impurities, by-products, residues, or chemical and microbiological contaminants - can be performed. However, selecting specific compounds for study presents challenges due to the complex composition of cultured meat (Ververis et al., 2020). Toxicologically relevant components in the whole food are generally considered to be the major components with known or proven biological or toxicological activity (EFSA, 2021). To replace, refine or reduce (3-R) the use of animals for ADME testing, preliminary alternative *in vitro* or *in silico* methods are preferred. Notably, physiologically-based toxicokinetic modeling (PBTK) modeling is an in-silico method for predicting *in vivo* kinetics of major component in cultured meat product by integrating human anatomical and physiological data, physicochemical data of the implicated component and the real-world consumption scenario of the cultured meat product. PBTK modeling has immense potential to contribute to the kinetics studies of cultured meat and other novel food in the future.

Test samples are generally considered safe for human consumption if they have similar components to the food that has been consumed safely or if the results of animal toxicology assays indicate that they are not significantly toxic. Post-marketing monitoring, such as tracking consumer feedback (especially regarding adverse effects), is crucial once a novel food is introduced to the market. This data can serve as clinical evidence for conducting a comprehensive safety and risk assessment of the food. For instance, France has implemented a nutritional vigilance system to monitor and analyze adverse effects following the consumption of specific food supplements or foods (Amat & Rihouey-Robini, 2014). At present, no significant risks to consumers have been identified for certain novel foods that have undergone hazard and risk management systems and passed extensive toxicological testing, according to industrial representatives and scientific experts.

Regulations or Legislations of Cultured Meat

Cultured meat is an unquestionable novel food. Consequently, each cultured meat product must be evaluated by the regulations or legislations of the corresponding jurisdictions before it can be introduced to the market to ensure food safety and human health. Currently, there is no firm definition of food safety, but European food law (known as the General Food Law) (Regulation (EU) No 178/2002) states that unsafe food is either harmful to health or unfit for human consumption, which depends on the usual conditions of consumption of the food and the information provided to consumers about the details of the food and its potential negative effects (EFSA, 2002). No regulation or legislation regarding food safety can ensure that food will never pose a risk to any consumer, but a legal framework can control the potential risks that occur in food consumption and establish reasonable certainty that the intended use will not cause harm (OECD, 1993). It is important to note that these regulations and legislations are primarily based on previous scientific advancements and, as such, are unlikely to be fully optimized in the short term. This presents certain limitations when evaluating the safety of rapidly advancing cultured meat technologies (de Boer & Bast, 2018). In the European Union (EU), many regulations related to the safety of novel foods have been in place for a significant period, and no substantial regulatory risks have been identified. Consequently, scientific, regulatory, and industry experts believe that current EU legislation is sufficient to ensure the safe introduction of cultured meat products to the market. As a result, cultured meat must undergo routine food safety assessments, such as microbiological examinations as outlined in Regulation (EC) No 2073/2005 and HACCP monitoring in Regulation (EC) No 852/2004. Meanwhile, cultured meat belongs to novel food (foods or ingredients that have not been used for human consumption to a significant degree in the EU before May 15, 1997) and it is particularly emphasized that "cultured meat" is not meat according to the EU regulation, it can only be authorized to enter the market by the EFSA after complying with the criteria of Regulation (EU) No 2015/2283 (also known as the Novel Food Regulation), which mentions unambiguous specific requirements and gives some guidance documents for the risky assessment analysis of novel food in detail to ensure safety within the different aspects, such as kinetics (ADME), nutritional com- ponents and toxicological and allergenic properties (EFSA, 2011a; 2011b, 2021). When the production of cultured meat involves the genetic modification of stem cells, it must undergo thorough procedures, specifically in accordance with (EC) No 1829/2003 and (EC) No 1830/2003, to minimize potential risks to health and the environment (Hadi & Brightwell, 2021; Ketelings et al., 2021). In 2019, the Singapore Food Agency issued comprehensive guidance (Requirements for the Safety Assessment of Novel Foods and Novel Food Ingredients) for the safety assessment of novel foods, which includes cultured meat. This guidance outlines specific requirements for cultured meat, such as cell lines, culturing media, scaffolding structures, and the genetic stability of cell cultures. It serves as a detailed reference for applicants conducting safety assessments of cultured meat and will be continually revised and updated (SFA, 2022). Novel food products aimed to market in Singapore must also comply with the relevant basic legislations, such as the Sale of Food Act, Wholesome Meat and

Fish Act and their subsidiary legislation (SFA, 2022). Encouragingly, cultured chicken meat has been sold in the Singapore market since 2020, which underscores the feasibility to control for unknown hazards or risks that could emerge during cultured chicken production and consumption. However, in other countries and regions, guidance, regulation, or legislation on the risk assessment of cultured meat has not been similarly established. In the United States, a regulatory and management agreement concerning the production of cultured meat was jointly established by the United States Department of Agriculture (USDA) and the FDA in 2019, while the Federal Meat Inspection Act and Poultry Product Inspection Act have been proposed as supporting legislation (Post et al., 2020). However, these regulations do not provide specific guidance to food businesses regarding the safety and risk assessment of novel foods, which creates challenges for bringing such products to market (Choudhury, Tseng, & Swartz, 2020).

In November 2022 and March 2023, UPSIDE Foods (the world's first cultured meat company) successfully completed the FDA's first and second premarket consultation on cultured chicken products, bringing cultured meat one step closer to the United States of America market (FDA, 2022, 2023). In addition, Australia-New Zealand Food Standard Code Standard 1.5.1 describes the pre-market assessment criteria for novel food intended for sale in Australia and New Zealand (Hadi & Brightwell, 2021). Different regions have varying approaches to regulating the introduction of novel foods, such as cultured meat, into local markets. As a result, some products that have been authorized in certain regions are restricted from entering others. To address this issue, it is essential for global experts to collaborate in developing regulatory agreements for cultured meat products. This would facilitate their approval and ensure their safe consumption across global markets.

Conclusion and Future Perspectives

With the development of cell culture and tissue engineering and the increasing demand for sustainable animal-based proteins, cultured meat has the potential to become a sustainable source of nutritional protein in the future due to its relative ethical and healthy properties. However, further inter-disciplinary research needs to be conducted before cultured meat become part of our food staples. Currently, some technical challenges in producing cultured meat products in large quantities economically still exist. To overcome these challenges, it is important to develop (i) alternative to fetal bovine serum or chemically defined serum-free media that is applicable at the industrial level, (ii) edible, sustainable and high-performance scaffold materials, (iii) *in vitro* myogenic environment that recapitulates real muscle formation, such as electrical or mechanical stimulation to give cultured meat the same taste and texture as the farming animal meat and (iv) recycling methods to maximize the use of all materials in cultured meat production. Additionally, consumer acceptance of cultured meat products constitutes an additional hurdle that needs to be suitably addressed and an international harmonized terminology of this product is necessary. As a novel food, cultured meat may introduce uncertain risks and hazards due to the new technologies involved in its production, especially since a comprehensive risk assessment system has not yet been fully established. Currently, there is no universal approach to testing cultured meat; therefore, effective risk and safety assessment strategies should be tailored to the specific hazards associated with its production. Moreover, the regulatory framework for evaluating cultured meat varies across countries. Moving forward, it is crucial for scientific, regulatory, and industry experts to collaborate in developing sound and harmonized regulations or legislation. This will enable effective risk assessment and ensure the safe global consumption of innovative cultured meat.

References

Amat, E., & Rihouey-Robini, L. (2014). New report on the French "nutrivigilance" and cultivation of permanent fish cell line CCO in serum-free medium and influence of protein hydrolysates on growth performance. *Cytotechnology,* 68, 115–121.

Becker, R. A., Plunkett, L. M., Borzelleca, J. F., & Kaplan, A. M. (2007). Tiered toxicity testing: Evaluation of toxicity-based decision triggers for human health hazard characterization. *Food and Chemical Toxicology*, 45, 2454–2469.

Ben-Arye, T., Shandalov, Y., Ben-Shaul, S., Landau, S., Zagury, Y., Ianovici, I., et al., (2020). Textured soy protein scaffolds enable the generation of three-dimensional bovine skeletal muscle tissue for cell-based meat. *Nature Food*, 1, 210–220.

Bhat, Z., Bhat, H., & Pathak, V. (2014). Chapter 79-prospects for *in vitro* cultured meat-a future harvest. *Principles of Tissue Engineering* (4th ed.). Academic Press.

Bodiou, V., Moutsatsou, P., & Post, M. J. (2020). Microcarriers for upscaling cultured meat production. *Frontiers in Nutrition,* 7, 10.

Bringing cultured meat to market: Technical, socio-political, and regulatory challenges in cellular agriculture. *Trends in Food Science & Technology*, 78, 155–166.

Canavan, H. E., Cheng, X., Graham, D. J., Ratner, B. D., & Castner, D. G. (2005). Cell sheet detachment affects the extracellular matrix: A surface science study comparing thermal liftoff, enzymatic, and mechanical methods. Journal of Biomedical Materials

Research Part A: *An Official Journal of the Society for Biomaterials,* 75, 1–13. the Japanese Society for Biomaterials, and the Australian Society for Biomaterials and the Korean Society for Biomaterials. Chemicals. Section.

Chen, G., Gulbranson, D. R., Hou, Z., Bolin, J. M., Ruotti, V., Probasco, M. D., et al., (2011). Chemically defined conditions for human iPSC derivation and culture. *Nature Methods*, 8, 424–429.

Chen, L., Guttieres, D., Koenigsberg, A., Barone, P. W., Sinskey, A. J., & Springs, S. L. (2022). Large-scale cultured meat production: Trends, challenges and promising biomanufacturing technologies. *Biomaterials,* 280, Article 121274.

Choi, K. H., Yoon, J. W., Kim, M., Lee, H. J., Jeong, J., Ryu, M., et al., (2021). Muscle stem cell isolation and *in vitro* culture for meat production: A methodological review. *Comprehensive Reviews in Food Science and Food Safety,* 20, 429–457.

Chou, M. L., Bailey, A., Avory, T., Tanimoto, J., & Burnouf, T. (2015). Removal of transmissible spongiform encephalopathy prion from large volumes of cell culture media supplemented with fetal bovine serum by using hollow fiber anion-exchange membrane chromatography. *PLoS One,* 10, Article e0122300.

Choudhury, D., Tseng, T. W., & Swartz, E. (2020). The business of cultured meat. *Trends in Biotechnology*, 38, 573–577.

Chriki, S., Ellies-Oury, M. P., & Hocquette, J. (2022). Is "cultured meat" a viable alternative to slaughtering animals and a good comprise between animal welfare and human expectations? *Animal Frontiers,* 12(1), 35–42.

History of safe use as applied to the safety assessment of novel foods and foods derived from genetically modified organisms. *Food and Chemical Toxicology,* 45, 2513–2525.

INFOGEST static *in vitro* simulation of gastrointestinal food digestion. *Nature Protocols,* 14, 991–1014.

Jairath, G., Mal, G., Gopinath, D., & Singh, B. (2021). A holistic approach to access the viability of cultured meat: A review. *Trends in Food Science & Technology*, 110, 700–710.

Ketelings, L., Kremers, S., & Boer, A. (2021). The barriers and drivers of a safe market introduction of cultured meat: A qualitative study. *Food Control,* 130, Article 108299.

Kumar, A., Sood, A., & Han, S. S. (2023). Technological and structural aspects of scaffold manufacturing for cultured meat: Recent advances, challenges, and opportunities. *Critical Reviews in Food Science and Nutrition*, 63, 585–612.

Li, C. H., Yang, I. H., Ke, C. J., Chi, C. Y., Matahum, J., Kuan, C. Y., et al., (2022). The production of fat-containing cultured meat by stacking aligned muscle layers and adipose layers formed from gelatin-soymilk scaffold. *Frontiers in Bioengineering and Biotechnology,* 10, Article 875069.

Newscientist. Kolkmann, A. M., Post, M. J., Rutjens, M. A. M., Van, A. L. M., & Moutsatsou, P. (2020). Serum-free media for the growth of primary bovine myoblasts. *Cytotechnology,* 72, 111–120.

O'Neill, E., Ansel, J., Kwong, G., Plastino, M., Nelson, J., Baar, K., et al., (2022). Spent media analysis suggests cultivated meat media will require species and cell type optimization. *Npj Science of Food,* 6, 46.

OECD 408. (2018). Repeated dose 90-day oral toxicity study in rodents, 4. *OECD Guidelines for the Testing of Chemicals.* Section.

Petetin, L. (2014). Frankenburgers, risks and approval. *European Journal of Risk Regulation,* 5, 168–186.

Physical, chemical, and toxicological characterization of fibrillated forms of cellulose using an *in vitro* gastrointestinal digestion and co-culture model. *Toxicology Research,* 9, 290–301.

Post, M. J. (2012). Cultured meat from stem cells: Challenges and prospects. *Meat Science*, 92, 297–301.

Post, M. J., Levenberg, S., Kaplan, D. L., Genovese, N., Fu, J., Bryant, C. J., et al., (2020). Scientific, sustainability and regulatory challenges of cultured meat. *Nature Food,* 1, 403–415.

Rottenberg, C. (2018). *The use of cell culture technology to develop products derived from livestock and poultry*. Day One Presentation: USDA/FDA Joint Public Meeting.

Part 2. Mycoprotein

Chapter 1

Fungi in the Food Chain

Abstract

The rapid development of human civilizations has created significant pressure to develop new food products with improved nutritional value and reduced environmental impact. Filamentous fungi, a group of microorganisms historically used in various foods, have recently garnered attention in both research and commercial sectors for their potential in innovative applications, particularly as alternative proteins. Advances in the cultivation and processing of filamentous fungi have created new opportunities to control the textures, flavors, and nutritional properties of fungi-based foods. This chapter highlights these technological advances, discusses the current state of fungi-based food products, and explores their future potential, with a focus on sustainable practices such as circular bioprocessing for utilizing agricultural and food processing byproducts.

Keywords: fungal biomass, mycoprotein, fungi-based food products, sustainable food production

Introduction

Fungi have been cultivated and utilized for the production of food, food ingredients (e.g., citric acid, vinegar), feed, enzymes, pharmaceuticals, nutraceuticals, and more for generations. The edible portions of fungal biomass have popularly been consumed in the form of mushrooms or as a component of foodstuffs, for instance in huitlacoche, a delicacy comprised of Ustilago maydis fungus galls and also in Roquefort and Camembert cheeses (Gilbert, 1957; Valverde, 1995). Edible mushrooms are foraged and/or cultivated worldwide and are considered to be an important component of healthy human diets (USDA, 2021). The mycelia of certain fungal species are known to contain vitamins, dietary fiber, essential amino acids, and healthy lipids. A balanced intake of these components has been linked to various health benefits, such as reduced cholesterol levels (Derbyshire, 2021; Hashempour, 2020; Colosimo, 2020). However, despite these health benefits, many individuals fail to consume the recommended amounts of these nutrients. For example, the USDA estimates that at least 70% of the U.S. population over the age of one consumes fewer mushrooms and other vegetables than recommended.

Interest in the utilization of fungal cells and materials as alternative protein sources to livestock-derived proteins has been present since before the mid twentieth century (Gilbert, 1957). The most common rationales for the use of fungi or other microorganisms as alternative proteins are due to increased resource use efficiencies (Souza, 2019; Smetana, 2015), complete amino acid profiles or other nutritional benefits (Souza, 2019), and decreased ethical concerns (Thavamani, 2020). Certain species of filamentous fungi have higher protein contents than most mushrooms and therefore are good candidates as alternative protein sources (Souza, 2018). Other species may be valued for their favorable flavors, textures, healthy lipid contents, or probiotic properties, among others. Recently, there has been growing interest in producing filamentous fungi in industrial and commercial fermentations for food applications, with some new products successfully reaching the market. Specifically, recent technological advances (e.g., food 3D printing) have enabled modifications in the textures, colors, shapes, and sizes of food products. These modern technologies and innovations open up exciting new opportunities for the utilization of filamentous fungi in a wide range of food products, in addition to alternative proteins.

This chapter reviews the recent advances and challenges for sustainable production of fungi as a future food source. Strategies for creating fungal products with various morphologies, textures, and shapes are outlined and the state of knowledge on the nutritional characteristics of these products is compiled. In addition, we detail how increasingly circular production methods are used for fungal cultivations to provide environmental benefits such as increased resource efficiency and reduced greenhouse gas emissions.

Fungal Biomass as Food

Mycoprotein

The most common mycoprotein product on the market, called QuornTM, has been produced since 1985 following 15 years of research and development (Wiebe, 2009). There are many available products ranging from alternative chicken patties to sausages and burgers. These products are produced using a pressure cycle fermenter operated in continuous aerobic mode with the filamentous fungi *Fusarium venenatum*, which produces mycoprotein products between 60% and 90% protein on a dry basis (Batt, 2014). The branched nature of the fungal mycelium approximates the organization of muscle fibers and aids in the development of a meat-like texture (Finnigan, 2011). Recently, initiatives have emerged to harness fungi for the production of various products, including steak and bacon, made from fungal mycelium. While these ventures are still in their early stages, they are characterized by their efforts to replicate structures and materials that appeal to consumer preferences.

Beverages, Baked Goods, and Novel Food Products

In addition to uses as an alternative to animal protein products, fungal mycelium can also be used as an alternative to a wide array of other food products. For example, researchers have recently developed a fungal cultivation strategy to produce fungi-based beverages (Pflueger-Peters, 2020). The resulting product is low in sugar, high in protein and fiber, and rich in mono- and polyunsaturated fats. This innovation demonstrates the potential of expanding edible mycelium into new categories of food products, beyond just alternative proteins.

Fungal biomass can also be used as innovative ingredients due to its special nutrition and physical properties. For instance, Stoffel et al., (Stoffel, 2021) used *Pleurotus albidus* mycoprotein flour to replace wheat flour to produce cookies. The authors reported that using mycoprotein flour significantly increased the nutritional value of the cookies as measured by the contents of protein, dietary fiber, and phenolic compounds. The addition of mycoprotein also increased the hardness and altered the color of the cookies.

Fungal mycelium can also be utilized as a carrier medium for other microorganisms due to its ability to attach to the cell walls of other microorganisms that may provide additional nutritional benefits or textures, flavors, and/or colors to final food products. For example, several species of fungi are known to be capable of flocculating microalgal cells from suspension in liquid cultures (Gultom, 2013; Zhou, 2012; Muradov, 2015; Wrede, 2014). Depending on the specific strains of algae and fungi used, the resulting pelletized biomass may exhibit higher levels of protein or certain fatty acids in the final product compared to either component alone (Wrede, 2014). In addition, due to their small size, microalgae require energy-intensive processes to harvest and incorporate into food products, especially those that require a dewatered or dry microalgae product (Milledge, 2012; Draaisma, 2013). Fungal-assisted harvesting of microalgae cells may provide mutual benefits by facilitating the harvest into macroscopic fungal pellets and producing a novel food or food ingredient (Pflueger-Peters, 2020). The authors of this article have developed fungi-based products using *A. awamori*, *A. oryzae*, and other fungi, along with microalgae. The resulting microbial biomass contains 30-60% protein and is rich in beneficial lipids, dietary fibers, and vitamins, making it both healthy and easily digestible.

Fungi-Based Food Manufacturing

When fungi are utilized in food manufacturing processes, special care must be taken to connect the various characteristics of the product to the desired market. For food applications, regional marketplaces may have historic or cultural expectations relating to appearance, taste, and texture of certain products, especially those that are designed to mimic traditional foods (e.g., alternative proteins).

Control of Food Properties

Flavor

The control of flavor is one of the most important characteristics of a final food product. For fungi-based food products, the choice of microorganism is perhaps the most important aspect in controlling the flavor of the resulting product. For example, mycelia derived from fungi typically cultivated as mushrooms retain similar flavor characteristics, even when grown in submerged culture (Gilbert, 1960). Depending on the desired product, this can be either a desirable or undesirable trait, particularly for strongly flavored fungal species. Less flavorful strains, on the other hand, can be enhanced with supplemental flavoring materials and are often valued more for their textural contributions to the food product. In addition to the strain of fungi used in these products the choice of medium is also a critical component that can play a role in the delivery of certain flavors to the fungal biomass for instance fungi grown in potato dextrose broth may contain distinct flavors of potatoes while those grown utilizing a fruit juice will retain certain flavor components of the juice in the retained biomass. Spices, herbs, and extracts can be added to the fermentation broth or directly to the final mycelium product.

Texture

The texture of fungi derived food products can be controlled in several ways depending on the growth mode of the fungal microorganisms. Firstly, texture can be modified from the addition of certain chemicals such as albumin protein, as used in the production of Quorn™ mycoprotein (Finnigan, 2011). The production of gelled albumin and fiber bundles from the combined effects of albumin addition, steaming, and freezing are thought to impart "fibrosity" to the Quorn™ material (Batt, 2014). Another factor influencing product texture is fungal morphology. The size of pellets grown in submerged cultures is influenced by mass transfer and mechanical effects, which can impact the resulting density and porosity. For example, excessively large fungal pellets may have an interior with very low oxygen levels, leading to anaerobic conditions. These conditions may trigger autolysis within the pellet's core, resulting in a hollow texture with low biomass density (Espinosa-Ortiz, 2016). Alternatively, when pellets are grown with very small diameters, the entire cross-section of the pellet can be more fully occupied by fungal biomass due to enhanced oxygen transfer throughout the pellet's width. Other methods of modifying the texture of fungal-derived food products may occur post-cultivation and involve processes such as pressing, extrusion, and other mechanical treatments.

Texture is normally measured using a texture profile analysis (TPA) methodology (Breene, 1975). This methodology utilizes a specialized piece of equipment that measures the force intensity and deformation of a sample during a, usually, two cycle compression using an appropriate probe. Texture includes many aspects of the force and compressibility profile of a material and may be expressed using parameters such as hardness, fractureability, cohesiveness, springiness, gumminess chewiness, and resilience. Hardness is defined as the maximum force of the first compression while gumminess measures how well the product withstands a second deformation. Resilience is measured by how well the product regains its original height after compression.

Color

The color of fungal-based food products can be modified either during or after the cultivation of the fungal pellet, depending on the intended application.

Fungal cells possess the ability to adsorb specific molecules from liquid media, enabling pigments present in the cultivation medium to be absorbed by the fungi and incorporated into the mycelial material. For example, fungi have demonstrated effectiveness in removing dyes from industrial wastewater and have been explored for their potential to treat water before it is disposed of or released into the environment (Espinosa-Ortiz, 2016). Similarly to flavor compounds that can be delivered to fungal biomass, colors and pigments are highly dependent on the characteristics of the fungus being cultivated and may be impacted by the fungus' own pigments or the enzymes and other products created by the organism(s) during cultivation.

Nutritional Profile

The nutritional profile of fungal derived food products can also be modified. Similarly, to the addition of flavor and color compounds to the fungal biomass during cultivation, molecules of nutritional significance can also potentially be selectively attached to the fungal surface. In addition, the biological activity of the resulting biomass, products can also be modified by making use of the innate ability of certain strains of fungi to harvest other microorganisms from the cultivation media. This characteristic can, in effect, create products with not only superior nutritional performance but also potentially biological performance by mediating delivery of active microbes to the digestive system.

3D Printing of Food

3D printing of food is an emerging area of research and commercial interest. Advantages of 3D printing over traditional methods of producing final food products include the ability to accurately control the shape and composition of final food materials throughout the final product. Additionally, 3D printing holds potential as a valuable tool for creating functional foods enriched with bioactive compounds from fruits and vegetables, which are directly incorporated into the printed products. This method is particularly advantageous when the bioactive compounds are sensitive to traditional food processing techniques (Tomašević, 2021). Food 3D printing can take various mechanical forms, with the most common methods including extrusion for direct printing, inkjet printing, selective sintering with hot air, laser jetting, and binder jetting (Dankar, 2018; Sun, 2015). A wide range of foods, such as chocolate, mashed potatoes, pizza, peanut butter, and snacks, have already been 3D printed (Derossi, 2018). Moreover, commercial companies are investigating the use of extrusion-based 3D printers to develop plant-based steak and pork products (Scott, 2021). In one study, Derossi et al., (2018) utilized a 3D food printer to create fruit-based snacks that were enriched with energy, calcium, iron, and vitamin D. They found that flow rate of material through the 3D printer was related to the amount of deposited material as well as the density and porosity of the final product. A wide variety of material and process parameters are yet to be fully explored in scientific literature for wide classes of 3D printed foods. Dankar and colleagues (Dankar, 2018) reviewed the material and process parameters affecting 3D printing of foods, which included critical material parameters such as: the feed material texture, viscosity, color, and gelation temperature as well as process parameters such as: the printing speed, feed pressure, nozzle size, and the distance between the nozzle and the printing bed, among others.

3D printing has recently been demonstrated with fungal biomass for application in construction and packaging materials (Dankar, 2018). This study utilized a biomass substrate derived from waste agricultural materials which was inoculated with fungi and then applied through a 3D printer to create a product of desired shape, size, and biological activity. Interestingly, the study demonstrated that biological activity of the fungal material was preserved following 3D printing, which may have implications for food product formulations desiring "probiotic" or "biologically active" labeling. The printing of fungal biomass was not without challenges with the authors noting problems such as nozzle clogging, poor print quality, biological contamination, and destruction of fungal biological activity. Accordingly, there are many knowledge gaps yet to be explored in the 3D printing of fungal foods. Areas of interest include the optimization of environmental conditions for primary and secondary colonization of the fungi, mixing conditions, operational parameters of 3D printing, as well as investigations into drying or further processing steps that may follow the printing phase. 3D printed fungal food products are in their infancy, but future applications may focus on improving or mimicking characteristics of existing food products utilizing the precision application of 3D printers.

Fungi-Based Food Products on the Market

A variety of fungi-based products are available on the market or are under development. Fungi based food products have emerged from mushroom cultivations to other processed foods such as alternative proteins that mimic livestock meat, beverage products such as coffees and teas, dairy-free cheese and milk, and several others. In addition to their use in food products, fungi are increasingly being explored for a variety of new applications, including health and beauty products, textiles, biodegradable packaging, pet food, and environmental remediation, as well as in agriculture, among others. At the time of publication, the websites of many of these ventures were mapped interactively on an online platform (Cabane, 2021).

Environmental Impact of Fungi-Based Foods

The environmental impacts of cultivating fungi-based food products have been quantified in several studies utilizing life cycle assessment (LCA) methodologies, which have been well reviewed by Souza Filho et al., (Souza, 2019). Briefly, mycoprotein generally performs better than animal-derived protein across impact categories like global warming potential, land use efficiency, and energy use but the results are sensitive to the choice of functional unit (e.g., on a per kg of product basis versus on a per kg of protein basis). The environmental impact of fungal cultivations can be improved by utilizing agricultural and food processing byproducts, as discussed earlier (Smetana, 2018). Unfortunately, LCA studies have thus far only focused on mycoprotein applications. As more fungi-based food products are developed, additional environmental studies should be conducted to either compare their performance to an existing product (such as mycoprotein) or to quantify their impacts relative to similar foods (such as new food innovations).

Summary and Research Needs

The use of filamentous fungi in future foods is a promising and active area of research and commercial venture. The use of fungi as alternative protein sources (mycoprotein) to animal proteins has increased in breadth in recent years to include foods with different or more challenging textures to emulate (e.g., steak and bacon). The application of 3D printing in food has recently gained significant attention, offering potential to enhance and modify the texture and nutritional properties of both existing and novel food products. While much of the focus from academia and industry has been on mycoprotein applications, there has also been growing interest in the use of fungi across a variety of food categories, including beverages and baked goods. These innovations aim to either replicate and improve existing food products or to create entirely new product categories. However, the application of fungi in these new products and classes is in its scientific infancy.

Future research topics may include the investigation of fungal biomass as alternative materials to a wide variety of traditional food products in a comparable manner as they have been applied as alternative proteins. The unique structure, life cycle, and nutritional characteristics of filamentous fungi species endow them with special abilities to emulate materials and deliver nutritional or environmental benefits compared to the traditional source. For example, certain high-sugar desserts or snack products could potentially be entirely or partially replaced with filamentous fungi, offering similar flavors and textures while providing enhanced nutritional value. However, for each new product concept, extensive research and development are required to understand the associated unit operations for production. Specifically, when considering the 3D printing of fungi-based foods, critical aspects such as material characteristics, operational parameters, species-specific dependencies, and downstream processes remain largely unexplored in the scientific literature. In addition, environmental and nutritional studies should accompany future food product development and quantify the benefits, costs, and risks of the products to the environment and human health.

Conclusion

This chapter has provided a foundational understanding of fungi's role in the food chain, with a specific focus on mycoprotein as a promising meat alternative. The potential of mycoprotein to address global protein shortages, improve sustainability, and provide a nutritious, versatile food source is increasingly recognized. By utilizing fungi-based systems, we can reduce the environmental footprint associated with traditional meat production, offering a more sustainable pathway to meet the growing demand for protein.

While mycoprotein production offers substantial benefits, challenges remain in scaling up production methods, ensuring product quality, and addressing consumer acceptance. Further research into fermentation processes, nutritional optimization, and the economic feasibility of mycoprotein-based products will be crucial in unlocking its full potential.

In conclusion, mycoprotein stands at the intersection of innovation and sustainability, offering an exciting opportunity to reshape the future of food production. As research progresses and technologies evolve, mycoprotein could play a key role in diversifying our food sources and contributing to more sustainable and ethical food systems worldwide.

References

Aldridge, S., J. Parascandola, J. L. Sturchio, The discovery and development of Penicillin 1928-1945, *Am. Chem. Soc. R. Soc. Chem.* (1999).

Arranz-Otaegui, A., L. G. Carretero, M. N. Ramsey, et al., Archaeobotanical evidence reveals the origins of bread 14,400 years ago in northeastern Jordan, *Proc. Natl. Acad. Sci.* U.S.A. 115 (2018) 7925-7930. https://doi.org/10.1073/pnas.1801071115.

Batt, C. A., M. Lou Tortorello, Encyclopedia of food microbiology, Second Edition, 2014. O.F. Cabane, The fungi industry landscape, *KindEarth.Tech.* (2021). http://fungi.ketmaps.com/.

Bhardwaj, A., J. Vasselli, M. Lucht, et al., 3D printing of biomass-fungi composite material: a preliminary study, *Manuf. Lett.* 24 (2020) 96-99. https://doi.org/10.1016/j.mfglet.2020.04.005.

Breene, W. M., Application of texture profile analysis to instrumental food texture evaluation, *J. Texture Stud.* 6 (1975) 53-82. https://doi.org/10.1111/j.1745-4603.1975.tb01118.x.

Carlsen, M., A. B. Spohr, J. Nielsen, et al., Morphology and physiology of an α-amylase producing strain of Aspergillus oryzae during batch cultivations, *Biotechnol. Bioeng.* 49 (1996) 266-276.

Christias, C., C. Couvaraki, S. G. Georgopoulos, et al., Protein content and amino acid composition of certain fungi evaluated for microbial protein production, *Appl. Microbiol.* 29 (1975). https://doi.org/10.1128/aem.29.2.250-254.1975.

Cobb, M., Oswald Avery, DNA, and the transformation of biology, *Curr. Biol.* 24 (2014) 55-60. https://doi.org/10.1016/j.cub.2013.11.060.

Colin, V. L., M. D. Baigorí, L. M. Pera, Tailoring fungal morphology of Aspergillus niger MYA 135 by altering the hyphal morphology and the conidia adhesion capacity: biotechnological applications, *AMB Express.* 3 (2013) 27. https://doi.org/10.1186/2191-0855-3-27.

Cui, Y. Q., R. G. J. M. van der Lans, M. L. F. Giuseppin, et al., Influence of fermentation conditions and scale on the submerged fermentation of Aspergillus awamori, *Enzyme Microb. Technol.* 23 (1998) 157-167. https://doi.org/10.1016/S0141-0229(98)00041-6.

Dankar, I., A. Haddarah, F. E. L. Omar, et al., 3D printing technology: the new era for food customization and elaboration, *Trends Food Sci. Technol.* 75 (2018) 231-242. https://doi.org/10.1016/j.tifs.2018.03.018.

Derbyshire, E. J., J. Delange, Fungal protein – what is it and what is the health evidence? A systematic review focusing on mycoprotein, *Front. Sustain. Food Syst.* 5 (2021). https://doi.org/10.3389/fsufs.2021.581682.

Derossi, A., R. Caporizzi, D. Azzollini, et al., Application of 3D printing for customized food. A case on the development of a fruit-based snack for children, *J. Food Eng.* 220 (2018) 65-75. https://doi.org/10.1016/j.jfoodeng.2017.05.015.

Draaisma, R. B., R. H. Wijffels, P. M. Slegers, et al., Food commodities from microalgae, *Curr. Opin. Biotechnol.* 24 (2013) 169-177. https://doi.org/10.1016/j.copbio.2012.09.012.

Dunlop, M. V., S. P. Kilroe, J. L. Bowtell, et al., Mycoprotein represents a bioavailable and insulinotropic non-animal-derived dietary protein source: a dose-response study, *Br. J. Nutr.* 118 (2017) 673-685. https://doi.org/10.1017/S0007114517002409.

Duygu Ozsoy, H., E. B. Arikan, C. Cinkir, et al., Fungal oil production from oleaginous fungi Mucor circinelloides and Aspergillus oryzae cultivated on sugar beet pulp, *APJES.* 3 (2015).

Espinosa-Ortiz, E. J., E. R. Rene, K. Pakshirajan, et al., Fungal pelleted reactors in wastewater treatment: applications and perspectives, *Chem. Eng. J.* 283 (2016) 553-571. https://doi.org/10.1016/j.cej.2015.07.068.

Finnigan, T. J. A., Mycoprotein : origins, production and properties, in: *Handb. Food Proteins,* 2011. https://doi.org/10.1533/9780857093639.335.

Franklin, R. E., R. G. Gosling, Molecular configuration in sodium thymonucleate, *Nature* 171 (1953) 740–741. https://doi.org/10.1038/171740a0.

Gema, H., A. Kavadia, D. Dimou, et al., Production of γ-linolenic acid by Cunninghamella echinulata cultivated on glucose and orange peel, *Appl. Microbiol. Biotechnol.* 58 (2002) 303-307. https://doi.org/10.1007/s00253-https://doi.org/10.1016/B978-012373944-5.00347-3.

Gilbert, F. A., R. F. Robinson, Food from fungi, *Econ. Bot.* 11 (1957) 126-145. https://doi.org/10.1007/BF02985303.

Gilbert, F. A., The submerged culture of Morchella, *Mycologia* 52 (1960) 201-209. https://doi.org/10.2307/3756006.

Hashempour-Baltork, F., K. Khosravi-Darani, H. Hosseini, et al., Mycoproteins as safe meat substitutes, *J. Clean. Prod.* 253 (2020) 119958. https://doi.org/10.1016/j.jclepro.2020.119958.

Jin, B., X.Q. Yan, Q. Yu, et al., A comprehensive pilot plant system for fungal biomass protein production and wastewater reclamation, *Adv. Environ. Res.* 6 (2002) 179-189. https://doi.org/10.1016/S1093-0191(01)00049-1.

Lahue, C., A. A. Madden, R. R. Dunn, et al., History and domestication of Saccharomyces cerevisiae in bread baking, *Front. Genet.* 11 (2020). https://doi.org/10.3389/fgene.2020.584718.

Liu, Y. S., J. Y. Wu, Effects of Tween 80 and pH on mycelial pellets and exopolysaccharide production in liquid culture of a medicinal fungus, *J. Ind. Microbiol. Biotechnol.* 39 (2012) 623-628.

Maheshwari, R., G. Bharadwaj, M. K. Bhat, Thermophilic fungi: their physiology and enzymes, *Microbiol. Mol. Biol. Rev.* 64 (2000). https://doi.org/10.1128/mmbr.64.3.461-488.2000.

Manan, M. A., C. Webb, Design aspects of solid state fermentation as applied to microbial bioprocessing, *J. Appl. Biotechnol. Bioeng.* 4 (2017) 511-532. https://doi.org/10.15406/jabb.2017.04.00094.

Metz, B., N. W. F. Kossen, The growth of molds in the form of pellets–a literature review, *Biotechnol. Bioeng.* 19 (1977) 781-799. https://doi.org/10.1002/bit.260190602.

Meyer, V., E. Y. Basenko, J. P. Benz, et al., Growing a circular economy with fungal biotechnology: a white paper, *Fungal Biol. Biotechnol.* 7 (2020) 1-23. https://doi.org/10.1186/s40694-020-00095-z.

Monteyne, A. J., M. O. C. Coelho, C. Porter, et al., Mycoprotein ingestion stimulates protein synthesis rates to a greater extent than milk protein in rested and exercised skeletal muscle of healthy young men: a randomized controlled trial, *Am. J. Clin. Nutr.* 112 (2020) 318-333. https://doi.org/10.1093/ajcn/nqaa092.

Nair, R. B., P. R. Lennartsson, M. J. Taherzadeh, Mycelial pellet formation by edible ascomycete filamentous fungi, Neurospora intermedia, *AMB Express.* 6 (2016) 31. https://doi.org/10.1186/s13568-016-0203-2.

Nigam, P. S., A. Singh, Single cell protein: mycelial fungi, in: Encycl. *Food Microbiol. Second Ed.,* 2014. https://doi.org/10.1016/B978-0-12-384730- 0.00311-6.

Pflueger-Peters, N., C. Ditterich, UC *Davis engineers fight food insecurity through sustainable agriculture,* (2020). https://bae.ucdavis.edu/news/uc-davis-engineers-fight-food-insecurity-through-sustainable-agriculture.

Ruxton, C. H. S., B. McMillan, The impact of mycoprotein on blood cholesterol levels: a pilot study, *Br. Food J.* 112 (2010) 1092-1101. https://doi.org/10.1108/00070701011080221.

Sajad Hashemi, S., K. Karimi, M. J. Taherzadeh, Integrated process for protein, pigments, and biogas production from baker's yeast wastewater using filamentous fungi, *Bioresour. Technol.* 337 (2021) 125356. https://doi.org/10.1016/j.biortech.2021.125356.

Saleh, A. A., A. Ohtsuka, M. Yamamoto, et al., Aspergillus awamori feeding modifies lipid metabolism in rats, *Biomed Res. Int.* 2013 (2013) 594393. https://doi.org/10.1155/2013/594393.

Samuel, D., Investigation of ancient Egyptian baking and brewing methods by correlative microscopy, *Science* 273 (1996) 488-490. https://doi.org/10.1126/science.273.5274.488.

Satari, B., K. Karimi, M. J. Taherzadeh, et al., Co-production of fungal biomass derived constituents and ethanol from citruswastes free sugars without auxiliary nutrients in airlift bioreactor, *Int. J. Mol. Sci.* 17 (2016) 302. https://doi.org/10.3390/ijms17030302.

Skory, C. D., S. N. Freer, R. J. Bothast, Screening for ethanol- producing filamentous fungi, *Biotechnol. Lett.* 19 (1997) 203-206. https://doi.org/10.1023/A:1018337003433.

Souza Filho, P. F., D. Andersson, J.A. Ferreira, et al., Mycoprotein : environmental impact and health aspects, *World J. Microbiol. Biotechnol.* 35 (2019) 147. https://doi.org/10.1007/s11274-019-2723-9.

Souza Filho, P. F., R. B. Nair, D. Andersson, et al., Vegan-mycoprotein concentrate from pea-processing industry byproduct using edible filamentous fungi, *Fungal Biol. Biotechnol.* 5 (2018). https://doi.org/10.1186/s40694- 018-0050-9.

Stoffel, F., W. de O. Santana, R. C. Fontana, et al., Use of Pleurotus albidus mycoprotein flour to produce cookies: evaluation of nutritional enrichment and biological activity,

Innov. *Food Sci. Emerg. Technol.* 68 (2021) 102642. https://doi.org/10.1016/j.ifset.2021.102642.

Thavamani, A., T. J. Sferra, S. Sankararaman, Meet the meat alternatives: the value of alternative protein sources, *Curr. Nutr. Rep.* 9 (2020) 346-355. https://doi.org/10.1007/s13668-020-00341-1.

Tomašević, I., P. Putnik, F. Valjak, et al., 3D printing as novel tool for fruit- based functional food production, *Curr. Opin. Food Sci.* 41 (2021) 138-145. https://doi.org/10.1016/j.cofs.2021.03.015.

Turnbull, W. H., A. R. Leeds, D. G. Edwards, Mycoprotein reduces blood lipids in free-living subjects, *Am. J. Clin. Nutr.* 55 (1992) 415-419. https://doi.org/10.1093/ajcn/55.2.415.

USDA, *Dietary Guidelines for Americans,* 2020-2025. 9th Edition, 2020. DietaryGuidelines.gov (accessed August 26, 2021).

Valverde, M. E., O. Paredes-Lapez, F. Guevara-Lara, et al., Huitlacoche (Ustilago maydis) as a food source-biology, composition, and production, *Crit. Rev. Food Sci. Nutr.* 35 (1995) 191-229. https://doi.org/10.1080/10408399509527699.

Wiebe, M. G., QuornTM myco-protein - overview of a successful fungal product, *Mycologist* 18 (2004) 17-20. https://doi.org/10.1017/S0269915X04001089.

Wrede, D., M. Taha, A. F. Miranda, et al., Co-cultivation of fungal and microalgal cells as an efficient system for harvesting microalgal cells, lipid production and wastewater treatment, *PLoS One* 9 (2014) e113497. https://doi.org/10.1371/journal.pone.0113497.

Zhang, L., Y. Lou, M. A. I. Schutyser, 3D printing of cereal-based food structures containing probiotics, *Food Struct.* 18 (2018) 14-22. https://doi.org/10.1016/j.foostr.2018.10.002.

Zhou, W., Y. Cheng, Y. Li, et al., Novel fungal pelletization-assisted technology for algae harvesting and wastewater treatment, *Appl. Biochem. Biotechnol.* 167 (2012) 214-228. https://doi.org/10.1007/s12010-012- 9667-y.

Zhou, Y., J. Du, G. T. Tsao, Mycelial pellet formation by Rhizopus oryzae ATCC 20344, *Appl. Biochem. Biotechnol.* 84 (2000) 779-789. https://doi.org/10.1385/abab:84-86:1-9:779.

Zhu, Y., Immobilized cell fermentation for production of chemicals and fuels, in: Bioprocess. Value-Added Prod. from Renew. *Resour.* 2007. https://doi.org/10.1016/B978-044452114-9/50015-3.

Žnidaršič, P., R. Komel, A. Pavko, Influence of some environmental factors on Rhizopus nigricans submerged growth in the form of pellets, *World J. Microbiol. Biotechnol.* 16 (2000) 589. https://doi.org/10.1023/A:1008967519157.

Chapter 2

Mycoprotein: A Promising Source of Meat Alternative

Abstract

The growing global population has increased the demand for food, underscoring the limitations of animal-based protein production, such as high costs, time, energy use, and environmental impact. Mycoproteins, derived from microbial sources, offer a sustainable alternative to meat by utilizing agro-industrial waste for production and addressing environmental concerns. Mycoproteins provide essential amino acids, carbohydrates, vitamins, and carotenes, and can be produced cost-effectively, independent of climate or land constraints. This chapter reviews the sensory attributes, consumer acceptance, and formulation of mycoprotein-based meat substitutes. It also examines environmental, economic, and health aspects, including impacts on cholesterol, satiety, glycemic response, and foodborne pathogens, along with ethical and halal considerations.

Keywords: mycoprotein, meat alternatives, sustainable protein, fungi-based foods

Introduction

In the last six decades, the global population has been increased up to 250% with an increase from 2.6 to 7 billion individuals (Gabriel et al., 2014). If population growth continues with the current rate, individuals may reach 9 billion by 2042 and can result in challenges to provide foods (Upadhyaya et al., 2016). Based on estimations by world health organization (WHO), starvation, malnutrition and relative diseases in developing countries are the major factors of death in 12,000,000 humans every year (Gabriel et al., 2014). Nearly one billion people worldwide cannot afford foods that provide adequate energy and protein (Godfray et al., 2010). Serious clinical issues, such as muscle weakness, growth failure, and a weakened immune system, arise from a lack of sufficient protein sources. In contrast, the high consumption of meat products by Western populations (86.7 kg of meat with bones per capita annually) and some developing countries is considered one of the most pressing global issues (Stehfest et al., 2009; Bonny et al., 2015). In addition to health problems for the consumers, further consumption of meat can seriously affect climate change (Tilman and Clark, 2014). The introduction of suitable substitutes for animal proteins, particularly meats, that incur lower costs and consume fewer resources, is of significant importance in the food industry (Aiking, 2011; Gerbens-Leenes et al., 2010). Artificial meats can be categorized into modified meats (derived from genetically modified organisms or GMOs), cultured meats (produced through *in vitro* cultures of cells or tissues), and meat substitutes sourced from plants and single-cell proteins (SCP) (Maga & Murray, 2010; Van Der Spiegel et al., 2013). SCP refers to proteins of microbial origin that are produced from pure or mixed cultures of bacteria, yeast, fungi, and microalgae (Upadhyaya et al., 2016). One of the various sources of SCP is filamentous fungus. The produced proteins are called mycoproteins such as those from *Fusarium venenatum* currently used in food products with the commercial name of Quorn™ (Gour, 2015). Mycoproteins are described as "Generally Recognized as Safe" (FDA, 2002).

In general, mycoproteins include 13 g of fats, 45 g of proteins, 10 g of carbohydrates, 25 g of fibers and several vitamins and minerals in 100 g of dry matter (Finnigan et al., 2017). Studies on human volunteers have shown that the biological value of proteins in mycoproteins is comparable to that of milk proteins. Additionally, toxicology studies have confirmed that mycoproteins do not negatively affect the normal growth of humans or animals. Both short-term and long-term consumption of mycoproteins do not raise significant health concerns (Finnigan et al., 2017). The Protein Digestibility Corrected Amino Acid (PDCAAS) score for mycoproteins is nearly 1.0, indicating a high-quality protein source. The fiber content in mycoproteins does not have any undesirable effects on mineral absorption. Although case reports have documented instances of intolerance to mycoproteins, the prevalence of such intolerance is lower than that observed with soy or eggs (Finnigan et al., 2017). Growing concerns about food insecurity and the lack of healthy food options, driven by an increasing global population and environmental challenges, have spurred further research into the production of Single-Cell Protein (SCP) as a meat substitute. Previous reviews have examined various methods and microorganisms used for SCP production, as well as the optimal processes for achieving higher yields and improving the nutritional properties of mycoproteins. There is a little information on various aspects of the meat substitution by mycoproteins. Therefore, the aim of the current chapter was to review sensory attributes, consumer acceptance, application, formulation and environmental, economic marketing, health and safety aspects of mycoproteins, as well as their effects on satiety, foodborne pathogens, allergy and waste use. Moreover, ethical and halal concerns of the mycoproteins were reviewed.

Application and Formulation of Mycoproteins in Meat Products

Fungi have been used in human foods for a long time. Quorn is the greatest example of a well-known mycoprotein product, which is commercialized through 15 countries worldwide (Wiebe, 2004). Mycoproteins are principally appropriate to imitate the taste and consistency of meats. This promises appropriation of the myco-proteins as substitutes for conventional animal-based products. Currently, mycoproteins produced by Quorn are estimated as 25000-ton DM/year, with a global market value of nearly 214 million Euro, which is suggested to grow by 20% annually in future (Matassa et al., 2016). The production processes of Quorn minces and pieces are based on continuous batch method. For the production of both forms, mycoproteins are treated with a small quantities of egg albumen and water. In Quorn mince, malt extract is added to the mixture, while in Quorn pieces, a natural flavoring is incorporated, contributing to the overall taste. The mixing process occurs at low shear, with a batch size of nearly two tons. The mixing of these two components can be challenging, as egg albumen is a powder with complex hydration properties, and mycoprotein is a viscoelastic substance. However, the simplicity of the process reduces this difficulty. Additionally, a small quantity of calcium is added to improve interhyphal crosslinking and enhance the meat-like texture of the final product. The mixed mass is then transferred to the forming equipment to create shaped blocks using pressure. Next, the temperature is raised to 85-90°C with live steam, which denatures the proteins. The blocks are then cooled to -18°C for 30 minutes. Freezing is a crucial step in forming the desirable meaty texture, achieved through the controlled growth of ice crystals. These crystals can efficiently bind filamentous hyphae to form fibrous bundles with a final texture that can be described as meat-like (Knight et al., 2001). Stability of this process and final products is linked to the strength of mycoproteins and egg networks with the gelled proteins and entangled fibers (Finnigan et al., 2017).

Environmental Effects

One of the solutions to solve the current environmental problems is mycoprotein, which serves as a novel healthy protein with low environmental effects. It is estimated that the world population increases by 30% by 2050, compared to the current time, and food production needs to increase by 70% (FAO, 2009). This has led to a nearly 200 million-ton increase in meat production and a nearly 1 billion-ton rise in cereal harvesting (FAO, 2009). Over the past few decades, one common solution has been the use of fertilizers, pesticides, and antibiotics (as growth promoters) in agriculture (Hashempour-Baltork et al., 2019). However, the widespread use of these chemicals has disrupted ecological balance and biodiversity. Additionally, antibiotic-resistant bacteria have contributed to the resurgence of epidemic diseases worldwide (Lymbery, 2014). Livestock production also has significant negative effects on greenhouse gas emissions, contributing to climate change, polluting freshwater resources, and threatening natural biodiversity (Tilman and Clark, 2014; Gerber et al., 2013).

Carbon Footprints

Nearly 29% of global greenhouse gas emissions are linked to agriculture and food production, with livestock production accounting for nearly half of this total (Bailey et al., 2014). The carbon footprint of mycoproteins is at least ten times lower than that of beef (Dairy UK, 2010) and four times lower than that of chicken (Macleod et al., 2013). Therefore, mycoprotein products can play a considerable role in decrease of challenges in food production systems and its greenhouse gas effects.

Water Footprints

Human activities linked to agriculture and food production are responsible for nearly 92% of all freshwater uses (Gerbens-Leenes et al., 2013). Reports have indicated that more than 15,000 l or 15,500 m^3/ton of water are needed to produce 1 kg of beef (Dibb and Fitzpatrick, 2014), instead of 3918 m^3/ton of water needed to produce 1 kg of chicken meat (Hoekstra and Chapagain, 2006). Moreover, various volumes of water are used in production of meats based on the animal species and location and method of breeding, (Finnigan et al., 2017). Thus, production of mycoproteins can play a significant role in providing energy and protein water efficiently.

Land Use

Recently, livestock production uses more than 70% of the agricultural lands. A study was carried out to compare landscape requirements for mycoprotein production with those for beef and chicken meat production. In general, the land required for the production of beef and chicken meat is at least ten and two times larger, respectively, than the land needed for the production of mycoproteins (Trust, 2014; Williams et al., 2006). In mycoprotein production, a carbohydrate extracted from agro-industrial waste serves as the carbon source. As a result, mycoprotein production significantly contributes to improving the overall protein balance (Finnigan et al., 2017).

Reactive Nitrogen Species (RNS)

Other important properties include the nutrients chain recycling, mainly RNS (NH^+, NO^- and NO^-), in ecosphere by fixing gaseous N_2 (Haber-Bosch process) (Rockstro€m et al., 2009). According to Van Grinsven et al., (Van Grinsven et al., 2013), if the cost- benefit of the agricultural products ranges 20-80 billion Euros annually, the detriment cost (from human health and environment aspects) of N pollution by soil-plant ranges 35-230 billion Euros annually. The microbial protein production can reduce the high nitrogen of agriculture systems (linked to nitrogen fertilizer) and hence lead to a significant reduction in adverse effects on ecosystems, eutrophic resources and nitrous oxide (N_2O) emissions (Matassa et al., 2015). By establishment of official legislations in microbial protein production (Øverland et al., 2010), further opportunities are created in nutrient and carbon source recoveries as parts of waste recycling schemes (Matassa et al., 2016). To allow an extensive adoption of the microbial protein biotechnology, these problems must be solved.

Economic Aspects and Market Affordability

Microorganisms are usually known as good sources of various value-added products that are rich in proteins for use in animal and human diets (Revuelta, 2016). In 2016, Matassa et al., (2016) outlined the average quantities of proteins produced by bacteria, microalgae, and fungi, along with other potential value-added products from these microorganisms. Fungi, in particular, are significant producers of microbial proteins and play an important role in the saccharification of fibers from corn. After a biological pre-treatment of corn fibers with fungi, the resulting residues contain glucose, arabinose, and xylose (Matassa et al., 2016; Leathers, 2003). These carbohydrates can then be utilized for the production of microbial proteins. In 2015, Bonny et al., (2015) assessed the market acceptability of various types of artificial meats compared to traditional meats. Market acceptability encompasses factors such as potential for mass production, costs, and regulatory considerations. According to their findings, mycoproteins were deemed favorable across all parameters when compared to traditional meats. Moreover, they demonstrated that mycoproteins held greater potential for meat replacement than other artificial meats and protein alternatives, including insect proteins, cultured meats, and genetically modified meats. Nowadays, several synthetic meats produced from mycoproteins or proteins with plant origins are available in the market; however, these synthetic meats constitute a small part of the market (1-2%) (Hoek et al., 2004). According to Beer (2015), the global market value of mycoproteins (Quorn) is 0.214 billion Euros annually. These products are now well known to consumers and classified in groups with properties similar to processed meats (Hoek et al., 2011a). However, there are some negative reports regarding the taste and texture of plant-originated proteins (Hoek et al., 2011b). As a result, these meat alternatives are not considered true substitutes by consumers who are not vegetarians or vegans (Hoek et al., 2004, 2011a).

Health and Safety

Effects on Total Cholesterol, Low-Density Lipoprotein (LDL) Cholesterol and High-Density Lipoprotein (HDL) Cholesterol

Clinical trials carried out with mycoproteins have demonstrated that these proteins are capable of decreasing blood total and LDL cholesterols as well as increasing blood HDL cholesterol (Homma et al., 1995; Nakamura et al., 1994). Mycoproteins are rich sources of proteins, high in fibers and low in total fats and saturated fatty acids (SFA). It has been known for many years that people consuming such proteins are at a lower risk of chronic diseases such as cardiovascular diseases (CVD) and strokes (Denny and Buttriss, 2007).

Effects on Satiety

Previous studies by Turnbull et al., (1993) and Burley et al., (1993) have verified that diets containing mycoproteins can positively affect appetite regulation. Recent studies have further validated this hypothesis (Bottin, 2013). In a study conducted by Williamson et al., the effects of mycoprotein, chicken meat, and tofu consumption prior to meals on satiety were examined (Williamson et al., 2006). The results indicated that mycoprotein significantly reduced satiety, in contrast to chicken meat, which had a lesser effect. Production of short-chain fatty acids (SCFA) from the gut digestion of mycoprotein fibers can by linked to appetite regulation due to send of satiation signals from the colon to the brain (Finnigan et al., 2017; Farhadi et al., 2013; Ahmadi et al., 2017).

Effects on Glycemic Responses

Incorporation of mycoproteins in common meals can result in control of glycemic responses (Turnbull and Ward, 1995). Exact mechanisms of the mycoprotein effects on insulin concentrations and blood glucose have not been described; however, it can be linked to its high content of fibers. Soluble fibers improve glycemic responses by slowing diffusion of glucose through the small intestinal wall (Edwards et al., 1988). Furthermore, fiber reduces the absorption rate of glucose by slowing the passage of food through the gastrointestinal tract (GIT) (Leclere et al., 1994). Mycoproteins contain low levels of carbohydrates; therefore, improvements in glycemic responses cannot be solely attributed to the delayed breakdown and absorption of carbohydrates.

Foodborne Pathogens

In USA, millions of infections annually originate from foodborne pathogens such as *Campylobacter* Spp., *Salmonella* Spp. and *Escherichia coli*. Meat products are responsible for transfer of nearly 22% of these pathogens to humans (Foodnet, 2014; Bakhtiary et al., 2016). Vaccination of animals or use of other approaches against infectious diseases is not only expensive and time-consuming, it is hard to accomplish under some special conditions (Mahmoudzadeh et al., 2016; Rezaei et al., 2012). Production of mycoproteins may improve health and safety by decreasing possibility of foodborne infections or contaminants. Indeed, myco- proteins decrease human-animal interaction risks by decreasing risk of zoonoses and infectious diseases (Bonny et al., 2015). Despite the significant nutritional value of microbial proteins for humans and animals, several challenges hinder their global acceptance. These challenges include: 1) Microbial proteins contain higher nucleic acid (NA) concentrations compared to plant and animal-based proteins. The utilization of proteins with high NA content (18–25 g/100 g of protein dry weight) increases blood uric acid levels, potentially leading to urolithiasis (Nasseri et al., 2011); 2) Sensory properties, such as color and flavor, may be altered, especially in microalgae and yeasts. Additionally, the consumption of foreign proteins can cause skin and gastrointestinal reactions, including nausea and queasiness (Upadhyaya et al., 2016); and 3) Filamentous fungi grow more slowly than yeasts and bacteria, and the risk of contamination is higher. Furthermore, the safety of selected strains (e.g., the production of mycotoxins) must be verified (Upadhyaya et al., 2016). These issues may impact the global acceptance of single-cell proteins (SCP) (Upadhyaya et al., 2016). As a result, regulations have been established in several countries to ensure the safety of such foods (Bagchi, 2006).

Allergenic Effects

Mycoproteins may cause several gastrointestinal reactions and sometimes life-threatening allergies. In 2018, Jacobson and Deporter (2018) studied adverse reactions linked to mycoprotein containing foods (Quorn brand). Analysis of 1,752 reactions showed that Quorn products caused emergence of allergic and gastrointestinal signs. A total of 312 individuals experienced allergic reactions, including anaphylaxis and urticaria, within 4 hours of consumption. However, 13 individuals showed no reactions after their first consumption. Furthermore, 188 individuals experienced repeated reactions after consuming Quorn multiple times. Gastrointestinal symptoms, such as vomiting and diarrhea, affected 1,692 people within 8 hours of consuming Quorn products, with 66.6% of cases occurring between 46- and 180-minutes post-consumption. Symptoms ranged from mild nausea to severe vomiting, which required medical attention. While the definition of food safety may vary across different regions, international standards have been established by the Joint Food and Drug Administration and the World Health Organization Expert Committee on Food Additives (WHO, 2017). Regulations often differ based on the type of product. Although single-cell proteins (SCPs) are primarily intended for use as food or animal feed (providing nutrition), they may also be marketed as food additives (e.g., for coloring purposes). Therefore, effective screening strategies are essential before the sale of new food products or additives (Smedley, 2013). Potential allergies, RNA contents and possible microbial toxin productions are key concerns of mycoprotein consumption. However, high NA contents in microbial proteins can be decreased by up to 71% using chemical treatments (sodium hydroxide) and thermal shocks (Santin et al., 2003). Mycoprotein containing foods must generally be screened for the presence or absence of toxic compounds before using as human foods (Wiebe, 2004). The *F. venenatum* and *Yarrowia lipolytica* are the best-known mycoprotein producers, which do not produce mycotoxins under defined culture conditions (Groenewald et al., 2014). The primary toxicity tests for mycoproteins were carried out within 16 years to receive approvals for trade from UK to other countries (Solomons, 1986).

Mycotoxin Analysis

Toxicology tests on the Quorn Strain A 3/5 have shown that this strain produces no toxic metabolites (Solomons, 1986). However, it is important to note that many Fusaria species produce significant quantities of mycotoxins, including deoxynivalenols and zearalenones (Marasas et al., 1984). Deoxynivalenols can inhibit protein synthesis, while zearalenones are estrogenic metabolites that pose potential risks to plants, animals, and humans (Desjardins et al., 1993; Marasas et al., 1984). O'Donnell et al., (1998) investigated two Quorn strains and three commercial Quorn products (pieces, mince, and quarter pounders) for the production of trichothecene mycotoxins. Phylogenetic analysis of DNA sequences was performed using polymerase chain reaction (PCR) on nuclear 28S ribosomal DNA, and the analysis of mycotoxins was conducted using electron ionization (EI) spectra and retention times, which were compared to Sigma standards. Estimation of mycotoxins in Quorn strain has shown that four types of trichothecenes are formed at low levels. Furthermore, no trichothecenes were seen in analysis of the Quorn products sold for the human consumption (O'donnell et al., 1998). No mycotoxins in three commercial Quorn mycoprotein products (at 0.5-ppm limit of detection) revealed that toxins either were not produced during cultivation or were removed during further processing. Nevertheless, caused by the potential of Quorn Strain NRRL 25416 *F. venenatum* to produce four types of trichothecenes, it seemed necessary to careful monitoring of all commercial mycoprotein batches to ensure the safety of mycoprotein products.

Waste Use

Use of various wastes as substrates for the SCP production is satisfying from cost and sustainability perspectives but may be challenging from the safety prospect. Therefore, sources of the feedstock must carefully be selected. For example, Quorn is produced from glucose (hydrolyzed starch) using approved media and processes, which meet GLP standards (Wiebe, 2004). In addition to safety challenges linked to the use of waste -derived substrates for the SCP production, the public approval of waste-derived foods is an essential factor while using SCPs in human diets (Ritala et al., 2017).

Ethics and Halal Concerns

The market for meat analogs includes both vegetarians and non-vegetarians who wish to reduce their meat consumption for health reasons or ethical beliefs. Additionally, it caters to individuals following specific religious dietary regulations, such as Halal, Kashrut, and Buddhist dietary practices (Asgar et al., 2010). Furthermore, with the appearance of bovine spongiform encephalopathy (BSE) or mad cow disease in the 1980s, direct use of plant proteins in foods has increased over the years, mainly in Europe (Morrison et al., 1999). Religious beliefs and lifestyles may prevent consumers from consuming meats and meat products. For example, pork is prohibited in Islam, and there are specific regulations for the slaughter of meat animals. The global trade of halal foods is estimated at nearly 80 billion US dollars per year, accounting for 12% of total trade in agri-food products (Anonymous, 2007). This is predicted to rise to 20% by 2025 since Muslims constitute 30% of the world population by 2025 (Anonymous, 2007). The development of non-meat protein substitutes is particularly beneficial for halal foods (Asgar et al., 2010). As mentioned earlier, meat protein alternatives also cater to other ethnic groups, including Jews and Hindus (Lusas, 1996). Notably, mycoprotein products have received certification from the Halal Food Authority, with the halal certification logos displayed on their packaging (Foods, 2009).

Conclusion and Future Prospects

Increased world populations and demands for nutritional foods require increases in meat and cereal productions. In recent decades, use of fertilizers, pesticides and antibiotics (as growth promoter) in meat production has replied to these increased demands. However, these chemicals may lead to unbalances in nature and emergence of antibiotic-resistant bacteria. Livestock adversely affect green- house gas emissions and climate changes, polluting freshwater resources, degrading lands, enhancing RNS and hence threatening natural biodiversity. As previously described, the market of meat analogs comprises vegetarians and non-vegetarians, who are interested to lessen their meat consumption due to health, ethical or religious reasons. Thus, the emergence of nonmeat protein substitutes with high qualities in texture and flavor can be used in international trades. In general, mycoproteins demonstrate significant contributions as nutrient supplements and meat substitutes in foods.

Mycoproteins, as substitute nutritional supplements, have the potential to address food shortage issues in rapidly growing populations, particularly in developing countries like Iran. Recent studies indicate that the biological value and essential amino acid content of mycoproteins are comparable to that of meats, while the vitamin B content in mycoproteins is notably higher than that in meat. From an environmental perspective, mycoprotein production requires significantly less water and land than traditional meat production, with a reduction of approximately 20 and 23 times, respectively. This makes mycoproteins a sustainable and eco-friendly alternative to meat, offering both nutritional benefits and environmental advantages. Furthermore, foodborne diseases and blood cholesterol levels can be limited using mycoproteins. Interestingly, mycoproteins received a higher sensory acceptance, compared to that other meat alternatives such as Tofu, Goodbite chicken style and Tivall stir-fry pieces did. In toxicology of mycoprotein formulations, no detections of mycotoxins in commercial Quorn mycoprotein products have revealed that toxins are not produced during cultivation or are removed during further processing. Furthermore, mycoprotein products have been certified by the Halal Food Authority for consumption by Muslims. However, despite their highlighted advantages, mycoproteins have not gained widespread popularity worldwide, primarily due to the lack of proper production technologies. Additionally, high nucleic acid (NA) content and potential allergic effects are other factors limiting their global use as food. Therefore, further studies are required to develop novel methods to address these issues in mycoprotein production. However, limitations such as incomplete data hinder the progress of such studies. For example, the number of studies on allergic and toxic effects, as well as the optimization of mycoprotein production, is very limited. No published articles were found on the mutagenicity potential of fungi for the induction of mycotoxins in safe species. Furthermore, several articles lack sufficient statistical quality for meaningful comparisons. Nevertheless, this chapter remains valuable and has contributed to a better understanding of the characteristics of mycoprotein products. It provides essential information to raise awareness about mycoprotein products and highlights the need for the development and validation of toxicological tests and appropriate assessments to reduce allergic manifestations.

Conclusion

This chapter has thoroughly explored mycoprotein as a promising meat alternative, highlighting its nutritional benefits, environmental advantages, and potential applications in the food industry. As a high-quality protein source derived from fungi, mycoprotein offers a sustainable and scalable solution to meet the growing global demand for protein. Its versatility in food products, from meat analogues to baked goods and beverages, makes it a strong contender in the future of alternative proteins.

While mycoprotein presents numerous benefits, several challenges remain, particularly in scaling production, optimizing flavor profiles, and ensuring cost-effectiveness. Additionally, the environmental impact of large-scale mycoprotein production, though considerably lower than traditional meat, must be continually assessed to ensure its sustainability in the long term.

In conclusion, mycoprotein represents a vital component of the broader movement toward sustainable food systems. As the industry continues to innovate in production methods and consumer acceptance, mycoprotein has the potential to play a central role in reducing the environmental footprint of protein production while addressing global food security challenges. Further research and technological advancements will be essential to realize its full potential.

References

Ahmadi, N., Khosravi-Darani, K., Mohammad Mortazavian, A., Mashayekh, S. M., 2017. Effects of process variables on fed-batch production of propionic acid. *J. Food Process.* Preserv. 41, e12853.

Aiking, H., 2011. Future protein supply. *Trends Food Sci. Technol.* 22, 112e120.

Aiking, H., De Boer, J., 2006. Background, Aims and Scope. *Sustainable Protein Production And Consumption: Pigs Or Peas*? Springer.

Anonymous, 2007. *Global halal food market.* Available from: http://ats.agr.gc.ca/africa/4352_e.htm#_Toc171306542. (Accessed 18 February 2010).

Asgar, M., Fazilah, A., Huda, N., Bhat, R., Karim, A., 2010. Nonmeat protein alternatives as meat extenders and meat analogs. *Compr. Rev. Food Sci. Food Saf.* 9, 513e529.

Bagchi, D., 2006. Nutraceuticals and functional foods regulations in the United States and around the world. *Toxicology* 221, 1.

Beer, 2015. *No success model for new protein ingredients* [WWW Document]. http://www.foodnavigator.com/Market-Trends/No-success-model-for-new-proteiningredients.

Bekatorou, A., Psarianos, C., Koutinas, A. A., 2006. Production of food grade yeasts. *Food Technol. Biotechnol.* 44.

Benelux, G. P. S., 2009. GfK Jaargids 2009. Inzichten en marktontwikkelingen in de Benelux. Dongen. *GfK Panel Services Benelux bv,* The Netherlands.

Bonny, S. P., Gardner, G. E., Pethick, D. W., Hocquette, J. F., 2015. What is artificial meat and what does it mean for the future of the meat industry? *Journal of Inte- grative Agriculture* 14, 255e263.

Burley, V., Paul, A., Blundell, J., 1993. Influence of a high-fiber food (myco-protein) on appetite: effects on satiation (within meals) and satiety (following meals). *Eur. J. Clin. Nutr.* 47, 409-409.

Chandrani-Wijeyaratne, S., Tayathilake, A., 2000. Characteristics of two yeast strain (Candida tropicalis) isolated from Caryotaurens (Khitul) toddy for single cell protein production. *J. Natl. Sci. Found. Sri Lanka* 28, 79e86.

Dairy Uk, D.C., *The Carbon Trust,* 2010. Guidelines for the Carbon Footprinting of Dairy Products in the UK.

Denny Aisbitt, B., Lunn, J., 2008. Mycoprotein and health. *Nutr. Bull.* 33, 298e310.

Denny, Buttriss, J., 2007. Plant foods and health: focus on plant bioactives. *Synthesis report* 4, 1-64.

Desjardins, A. E., Hohn, T. M., Mccormick, S. P., 1993. Trichothecene biosynthesis in Fusarium species: chemistry, genetics, and significance. *Microbiol. Mol. Biol. Rev.* 57, 595e604.

Dibb, S., Fitzpatrick, I., 2014. Let's Talk about Meat: Changing Dietary Behaviour for the 21st Century. *Eating Better.*

Edwards, C., Johnson, I., Read, N., 1988. Do viscous polysaccharides slow absorption by inhibiting diffusion or convection? *Eur. J. Clin. Nutr.* 42, 307e312.

Ejiofor, A. O., Chisti, Y., Moo-Young, M., 1996. Culture of Saccharomyces cerevisiae. Am. J. Food Technol. 68, 100e134.

Elzerman, J. E., Hoek, A. C., Van Boekel, M. A., Luning, P.A., 2011. Consumer acceptance and appropriateness of meat substitutes in a meal context. *Food Qual. Prefer.* 22, 233e240.

FAO, 2009. How to Feed the World in 2050. High-Level Expert Forum, Rome. Farhadi, S., Khosravi-Darani, K., Mashayekh, M., Mortazavian, A. M., Mohammadi, A., Shahraz, F., 2013. Production of propionic acid in a fermented dairy beverage. *International Journal of Dairy Technology* 66, 127e134.

Finnigan, T., Needham, L., Abbott, C., 2017. Mycoprotein : a Healthy New Protein with a Low Environmental Impact. *Sustainable Protein Sources.* Elsevier.

Foodnet, F. D., *A S, N.,* 2014. http://www.cdc.gov/foodnet/reports/annual-reports-2014.html.

Foods, M., 2009. *Frequently asked questions.* Available from: http://www.quorn.co.uk/FAQs/.

Gabriel, A., Victor, N., Du Preez James, C., 2014. Cactus pear biomass, a potential lignocellulose raw material for single cell protein production (SCP): a review. *Int J Curr Microbiol App Sci* 3, 171e197.

Gerbens-Leenes, P., Mekonnen, M., Hoekstra, A.Y., 2013. The water footprint of poultry, pork and beef: a comparative study in different countries and pro- duction systems. *Water Resour. Ind.* 1, 25e36.

Gerbens-Leenes, P., Nonhebel, S., Krol, M. S., 2010. Food consumption patterns and economic growth. Increasing affluence and the use of natural resources. *Appetite* 55, 597e608.

Gerber, P. J., Steinfeld, H., Henderson, B., Mottet, A., Opio, C., Dijkman, J., Falcucci, A., Tempio, G., 2013. Tackling Climate Change through Livestock: a Global Assessment of Emissions and Mitigation Opportunities. *Food and Agriculture Organization of the United Nations* (FAO).

Godfray, H. C. J., Beddington, J. R., Crute, I. R., Haddad, L., Lawrence, D., Muir, J. F., Pretty, J., Robinson, S., Thomas, S. M., Toulmin, C., 2010. Food security: the challenge of feeding 9 billion people. *Science* 327, 812e818.

Gour, S., M. N., Singh, A., Pradeep, B., 2015. Single cell protein production: a review. *Int. J. Curr. Microbiol. App. Sci.* 4 (9), 251e262.

Groenewald, M., Boekhout, T., Neuveglise, C., Gaillardin, C., Van Dijck, P.W., Wyss, M., 2014. Yarrowia lipolytica: safety assessment of an oleaginous yeast with a great industrial potential. *Crit. Rev. Microbiol.* 40, 187e206.

Hashempour-Baltork, F., Hosseini, H., Shojaee-Aliabadi, S., Torbati, M., Mirza Alizadeh, A., Alizadeh, M., 2019. Drug resistance and the prevention strategies in food borne bacteria: an update review. *Adv. Pharmaceut. Bull., inpress* 9 (3), 335e347. https://doi.org/10.15171/apb.2019.041, 31592430.

Hoek, A.C., Luning, P.A., Stafleu, A., DE Graaf, C., 2004. Food-related lifestyle and health attitudes of Dutch vegetarians, non-vegetarian consumers of meat substitutes, and meat consumers. *Appetite* 42, 265e272.

Hoek, A. C., Luning, P. A., Weijzen, P., Engels, W., Kok, F. J., DE Graaf, C., 2011a. Replacement of meat by meat substitutes. A survey on person-and product- related factors in consumer acceptance. *Appetite* 56, 662e673.

Hoek, A. C., Van Boekel, M. A., Voordouw, J., Luning, P.A., 2011b. Identification of new food alternatives: how do consumers categorize meat and meat substitutes? *Food Qual. Prefer.* 22, 371e383.

Hoekstra, A. Y., Chapagain, A. K., 2006. Water Footprints of Nations: Water Use by People as a Function of Their Consumption Pattern. *Integrated Assessment of Water Resources and Global Change*. Springer.

Homma, Y., Nakamura, H., Kumagai, Y., Ryuzo, A., Saito, Y., Ishikawa, T., Takada, K., Yamagami, H., Kikuchi, H., Inadera, H., 1995. Effects of eight-week ingestion of mycoprotein on plasma levels of lipids and Apo (Lipo) proteins. *Prog. Med.* 15, 183e195.

Hoseyni, S., Khosravi-Darani, M. M. K., 2010. Production and rheological evaluation of mycoprotein produced from fusarium venenatum ATCC 20334 by surface culture method. *Seed* 72, 48.

Hosseini, S., Khosravi-Darani, K., 2010. Response surface methodology for myco- protein production by Fusarium venenatum ATCC 20334. *J. Bioprocess. Biotech.* 1, 1e5.

Hosseini, S., Khosravi-Darani, K., Mohammadifar, M., Nikoopour, H., 2009. Production of mycoprotein by Fusarium venenatum growth on modified vogel me- dium. *Asian J. Chem.* 21, 4017.

Jacobson, M. F., Deporter, J., 2018. Self-reported adverse reactions associated with mycoprotein (Quorn-brand) containing foods. *Ann. Allergy Asthma Immunol.* 120, 626e630.

Jonker, D., 1995. Determination of the Net Protein Utilisation, Digestibility, and Biological Value of Three Samples of Freeze Dried Mycoprotein in Rats. *The Hague, The Netherlands. TNO Nutrition and Food Research Institute* (Report V95, 134).

Knight, N., Roberts, G., Shelton, D., 2001. The thermal stability of Quorn™ pieces. *Int. J. Food Sci. Technol.* 36, 47e52.

Leathers, T. D., 2003. Bioconversions of maize residues to value-added coproducts using yeast-like fungi. *FEMS Yeast Res.* 3, 133e140.

Leclere, C. J., Champ, M., Boillot, J., Guille, G., Lecannu, G., Molis, C., Bornet, F., Krempf, M., Delort-Laval, J., Galmiche, J., 1994. Role of viscous guar gums in lowering the glycemic response after a solid meal. *Am. J. Clin. Nutr.* 59, 914e921.

Lusas, E., 1996. Modern texturized soy proteins: preparation and uses. *Food Technol.* 50, 132e135.

Macleod, M., Gerber, P., Mottet, A., Tempio, G., Falcucci, A., Opio, C., Vellinga, T., Henderson, B., Steinfeld, H., 2013. Greenhouse Gas Emissions from Pig and Chicken Supply ChainseA Global Life Cycle Assessment. *Food and Agriculture Organization of the United Nations (FAO),* Rome, p. 171.

Maga, E., Murray, J., 2010. Welfare applications of genetically engineered animals for use in agriculture. *J. Anim. Sci.* 88, 1588e1591.

Mahmoudzadeh, M., Hosseini, H., Nasrollahzadeh, J., Mousavi Khaneghah, A., Rismanchi, M., Djalma Chaves, R., Shahraz, F., Azizkhani, M., Mahmoudzadeh, L., Haslberger, A., 2016. *Antibacterial Activity of Carum Copticum Essential Oil against Escherichia Coli O157:H7 in Meat: Stx Genes Expression.*

Marasas, W. F. O., Nelson, P. E., Toussoun, T., 1984. *Toxigenic Fusarium Species. Identity and Mycotoxicology.* Pennsylvania State University.

Matassa, S., Batstone, D. J., HüLsen, T., Schnoor, J., Verstraete, W., 2015. *Can Direct Conversion of Used Nitrogen to New Feed and Protein Help Feed the World?* ACS Publications.

Matassa, S., Boon, N., Pikaar, I., Verstraete, W., 2016. Microbial protein: future sustainable food supply route with low environmental footprint. *Microb. Bio- technol.* 9, 568e575.

Meiselman, H. L., Johnson, J., Reeve, W., Crouch, J., 2000. Demonstrations of the influence of the eating environment on food acceptance. *Appetite* 35, 231e237.

Morrison, N., Clark, R., Chen, Y., Talashek, T., Sworn, G., 1999. Gelatin Alternatives for the Food Industry. *Physical Chemistry and Industrial Application of Gellan Gum.* Springer.

Nakamura, H., Ishikawa, T., Akanuma, M., Nishiwaki, M., Yamashita, T., Tomiyasu, K., Yoshida, H., Nishio, E., Hosoai, K., Shiga, H., 1994. Effect of mycoprotein intake on serum lipids of healthy subjects. *Prog. Med.* 14, 1972e1976.

Nasseri, A., Rasoul-Amini, S., Morowvat, M., Ghasemi, Y., 2011. Single cell protein: production and process. *Am. J. Food Technol.* 6, 103e116.

O'donnell, K., Cigelnik, E., Casper, H., 1998. Molecular phylogenetic, morphological, and mycotoxin data support reidentification of the quorn mycoprotein fungus asFusarium venenatum. *Fungal Genet. Biol.* 23, 57e67.

Øverland, M., Tauson, A. H., Shearer, K., Skrede, A., 2010. Evaluation of methane-utilising bacteria products as feed ingredients for monogastric animals. *Arch. Anim. Nutr.* 64, 171e189.

Ravindra, P., 2000. Value-added food: single cell protein. *Biotechnol. Adv.* 18, 459e479.

Reihani, S. F. S., Khosravi-Darani, K., 2018. Influencing factors on single cell protein production by submerged fermentation: a review. *Electron. J. Biotechnol.* 37, 34e40. https://doi.org/10.1016/j.ejbt.2018.11.005.

Reihani, S. F. S., Khosravi-Darani, K., 2019. Mycoprotein production from date waste using Fusarium venenatum in a submerged culture. *Appl. Food Biotechnol.* 5, 243e352.

Rezaei, M., Hosseini, H., Safari, R., Yaghoubzade, Z., 2012. Inhibitory effect of nisin on Listeria monocytogenes inoculated into surimi and minced meat. *J. Fasa Univ. Med. Sci.* 1, 221e226.

Ritala, A., Ha€kkinen, S. T., Toivari, M., Wiebe, M. G., 2017. Single cell proteindstate-of-the-art, industrial landscape and patents 2001e2016. Front. Microbiol. 8, 2009. Rockstro€m, J., Steffen, W., Noone, K., Persson, Å., Chapin Iii, F. S., Lambin, E. F., Lenton, T. M., Scheffer, M., Folke, C., Schellnhuber, H.J., 2009. A safe operating space for humanity. *Nature* 461, 472.

Rudravaram, R., Chandel, A. K., Rao, L., Hui, Y., Ravindra, P., 2009. Bio (Single Cell) protein: issues of production, toxins and commercialisation status. In: Ashworth, G.S., Azevedo, P. (Eds.), *Agricultural Wastes.* Hauppage), New York, NY, pp. 129e153.

Santin, E., Paulillo, A. C., Maiorka, A., Nakaghi, L. S. O., Macari, M., Silva, A., Alessi, A. C., 2003. Evaluation of the efficacy of Saccharomyces cerevisiae cell wall to ameliorate the toxic effects of aflatoxin in broilers. *Int. J. Poult. Sci.* 2, 341e344.

Smedley, K. O., 2013. Comparison of Regulatory Management of Authorized In- gredients, Approval Processes, and Risk-Assessment Procedures for Feed Ingredients. *International Feed Industry Federation*, 2013.

Solomons, G., 1986. Microbial proteins and regulatory clearance for RHM myco- protein. *Microbial Biomass Proteins* 19e26.

Souza Filho, P. F., Nair, R. B., Andersson, D., Lennartsson, P.R., Taherzadeh, M.J., 2018. Vegan-mycoprotein concentrate from pea-processing industry byproduct using edible filamentous fungi. *Fungal Biol. Biotechnol.* 5, 5.

Stehfest, E., Bouwman, L., Van Vuuren, D. P., Den Elzen, M. G., Eickhout, B., Kabat, P., 2009. Climate benefits of changing diet. *Clim. Change* 95, 83e102.

Suman, G., Nupur, M., Anuradha, S., Pradeep, B., 2015. Single cell protein production: a review. *Int. J. Curr. Microbiol. Appl. Sci.* 4, 251e262.

Tilman, D., Clark, M., 2014. Global diets link environmental sustainability and hu- man health. *Nature* 515, 518.

Trust, C., 2014. Quorn, Beef and Chicken Footprints. *Internal Company Report.*

Turnbull, W. H., Walton, J., Leeds, A.R., 1993. Acute effects of mycoprotein on sub-sequent energy intake and appetite variables. *Am. J. Clin. Nutr.* 58, 507e512.

Turnbull, W. H., Ward, T., 1995. Mycoprotein reduces glycemia and insulinemia when taken with an oral-glucose-tolerance test. Am. J. Clin. Nutr. 61, 135e140. Udall, J.N., Lo, C. W., Young, V. R., Scrimshaw, N. S., 1984. The tolerance and nutritional value of two microfungal foods in human subjects. *Am. J. Clin. Nutr.* 40,285-292.

Ugalde, U., Castrillo, J., 2002. Single cell proteins from fungi and yeasts. *Appl. $ycol. Biotechnol.* 2, 123e149.

Upadhyaya, S., Tiwari, S., Arora, N., Singh, D., 2016. Microbial Protein: a Valuable Component for Future Food Security. *Microbes and Environmental Management.* Studium Press, New Delhi.

US FDA, 2002. Marlow Foods. *GRAS Notification for Mycoprotein (submitted as: US FDA.* Marlow Foods Ltd, Stokesley (UK). GRN 091).

Van Der Spiegel, M., Noordam, M., Van Der Fels-Klerx, H., 2013. Safety of novel protein sources (insects, microalgae, seaweed, duckweed, and rapeseed) and legislative aspects for their application in food and feed production. *Compr. Rev. Food Sci. Food Saf.* 12, 662e678.

Van Grinsven, H. J., Holland, M., Jacobsen, B. H., Klimont, Z., Sutton, M. A., Jaap Willems, W., 2013. Costs and benefits of nitrogen for Europe and implications for mitigation. *Environ. Sci. Technol.* 47, 3571e3579.

Verbeke, W., Pe´rez-Cueto, F. J., DE Barcellos, M. D., Krystallis, A., Grunert, K. G., 2010. European citizen and consumer attitudes and preferences regarding beef and pork. *Meat Sci.* 84, 284c292.

Waite, R. I., Beveridge, M., Brummett, R., Castine, S., Chaiyawannakarn, N., Kaushik, S., Mungkung, R., Nawapakpilai, S., Phillips, M., 2014. In: *Improving productivity and environmental performance of aquaculture.* WorldFish Pub- lisher, pp. 1e60.

Wiebe, M. G., 2004. QuornTM Myco-protein - overview of a successful fungal product. *Mycologist* 18, 17e20.

Williams, A., Audsley, E., Sandars, D., 2006. *Determining the environmental burdens and resource use in the production of agricultural and horticultural commodities: defra project report IS0205.* Zu finden in: http://randd.defra.gov.uk/Default.aspx.

Williamson, D., Geiselman, P., Lovejoy, J., Greenway, F., Volaufova, J., Martin, C., Arnett, C., Ortego, L., 2006. Effects of consuming mycoprotein, tofu or chicken upon subsequent eating behaviour, hunger and safety. *Appetite* 46, 41-48.

Chapter 3

The Critical Role of Mycoprotein in Supporting Muscle and Metabolic Health

Abstract

The growing global population has led to an increased demand for dietary protein to support health and adaptation. Although animal-derived proteins are well-established nutritionally, sustainability challenges have sparked growing interest in alternative protein sources. Mycoprotein, a protein-rich, high-fiber food produced through fungal fermentation, offers a sustainable option. Research shows mycoprotein can reduce cholesterol, improve postprandial glycemic control, and enhance satiety, potentially due to its fiber content. Its amino acid profile and bioavailability make it a promising protein source for supporting muscle protein metabolism, offering benefits for both athletes and healthy aging. This chapter examines the current evidence on mycoprotein's metabolic effects and identifies key research questions for further exploration.

Keywords: mycoprotein, skeletal muscle adaptation, cardio-metabolic health, nutritional benefits

Introduction

Developing a nutritionally sustainable future is an urgent contemporary issue. The world's population is projected to increase from ~7.3 billion to > 9 billion by 2050. This is coupled with global trends concerning rises in urbanization, social mobility, and wealth creation, all factors expected to exacerbate global food demand. As a result, current and future generations are required to view developments in the understanding of human nutrition through the lens of mounting challenges associated with the sustainability of increased production.

When considering global dietary protein production requirements, demographic demands are also compounded by accumulating scientific data to support protein consumption at levels greater than currently accepted RDAs in various populations. For example, evidence indicates that modest increases in dietary protein intake above the current Recommended Dietary Allowances (RDAs) support muscle mass maintenance in older adults, promote and retain muscular adaptations in athletes, and contribute to successful weight management. Therefore, it is evident that the global demand for dietary protein production is an urgent societal issue that continues to gain momentum.

Crucially, the majority of data supporting the refinement of dietary protein requirements has been obtained from studies examining the *in vivo* metabolic handling and/or adaptive responses to animal-derived protein ingestion. The carbon, water, and land use footprints of animal-derived protein production is 8-80, 50-150, and 30-220 times greater, respectively, than those of many plant-based proteins (variation dependent on protein source and methods used to quantify). Furthermore, vegan, vegetarian, and flexitarian diets are becoming increasingly popular. As a result, research into non-animal-derived protein sources is relevant to an expanding demographic, and the impact of such evidence will grow correspondingly. It is, therefore, vital that the scientific community begin to examine the metabolic handling and nutritional value of alternative, non-animal derived sustainable protein sources.

Mycoprotein

Mycoprotein is a whole food source produced by continuous flow fermentation of the filamentous fungus *Fusarium venenatum* (for a detailed description of the production processes please see reference 14). The resulting product is a high-protein, high-fiber, and relatively low-energy complete food source that is textured (through freezing) and flavored into a variety of products sold under the trade name Quorn (Marlow Foods, Stokesley, North Yorkshire, UK). Notably, the sustainability advantages of mycoprotein production make it an appealing alternative protein source, helping to address environmental concerns linked to the growing demand for dietary protein.

Following its development in the 1960s, initial human experimentation during the 1970s established the basic feasibility, tolerability, and metabolic impact of mycoprotein consumption, prior to it being available for general sale in 1985. However, the now fully established commercial viability, environmental advantages, and alternative potential applications of mycoprotein for metabolic health, skeletal muscle maintenance, and reconditioning have recently reignited research interest in this novel food source.

Mycoprotein, Dietary Fiber and Cardio-Metabolic Health

An initial human investigation aimed at establishing tolerability of mycoprotein made an interesting ancillary observation. Human volunteers who consumed 20 g mycoprotein (dry weight) per day for 30 days (consumed as supplemental cookies) showed a ~7% decrease in blood cholesterol concentrations, which replicated earlier findings in animals. Early research primarily focused on the potential health impact of mycoprotein, particularly regarding its dietary fiber content. This emphasis was driven by a substantial body of epidemiological studies that consistently demonstrated a link between higher fiber intake - typically from fruits, vegetables, and cereals - and beneficial health outcomes. These studies have shown that increased fiber consumption is associated with lower blood cholesterol levels, improved blood lipid profiles, and a reduced incidence of myocardial infarction and coronary heart disease. Such findings have been confirmed during intervention studies where increasing dietary fiber consumption has been reported to improve peripheral insulin sensitivity, and lower blood cholesterol concentrations and glycated hemoglobin (HbA1c) in both healthy individuals and patients with type-2 diabetes.

Follow-up studies focusing on mycoprotein consumption and cardio-metabolic health confirmed and extended these effects on blood lipid profiles. Turnbull and colleagues conducted a 3-week dietary intervention study in which participants consumed 191 g of mycoprotein-containing products (equivalent to approximately 40 g of dry mycoprotein) per day. This intervention was part of a fully controlled, laboratory-supervised diet designed to maintain energy balance in individuals with mildly elevated blood cholesterol levels. The tightly controlled study found that the mycoprotein intervention led to a significant reduction in total blood cholesterol (from 5.54 to 4.81 mmol/L, a 13% decrease) and low-density lipoprotein (LDL) cholesterol (from 4.16 to 3.78 mmol/L, a 9% decrease). Additionally, high-density lipoprotein (HDL) cholesterol levels increased from 0.58 to 0.65 mmol/L, representing a 12% increase. These findings were particularly notable as the control group exhibited opposing responses, with no significant changes observed in their cholesterol levels. Given that the energy, macronutrient, lipid composition, and cholesterol content of the diets were similar across groups, it was assumed that fiber content was the causative component.

The increased dietary fiber content could have exerted its cholesterol-lowering effect by altering LDL cholesterol synthesis/degradation, enhancing cholesterol clearance in peripheral tissues, and/or increasing the binding of fiber to neutral sterols, cholesterol, or bile acids in the intestine, thereby reducing cholesterol entering the circulating pool. However, it is important to note that the beneficial effects of higher fiber diets on circulating cholesterol concentrations do not always extend to improvements in the specific lipid sub-fractions of LDL and HDL. It is thus interesting to ponder whether the type, rather than simply the amount, of dietary fiber contained within mycoprotein may, at least in part, explain the beneficial effects of mycoprotein consumption on circulating cholesterol.

Dietary fibers contained within mycoprotein predominantly comprise 2/3 β-glucan and 1/3 chitin, which form a fibrous insoluble matrix that is relatively rare in more traditional food sources. Consistent with the significance of fiber type, follow-up research by Turnbull and colleagues reported similar effects of mycoprotein consumption under free-living conditions. Specifically, they observed a 0.95 mmol/L (16%) reduction in total cholesterol and a 0.34 mmol/L (21%) reduction in LDL cholesterol concentrations, despite maintaining consistent overall energy, macronutrient, and fiber content (approximately 6 g) across groups. Recent *in vitro* studies have delved deeper into the mechanisms behind these effects, beginning to reveal potential pathways through which the specific fiber profile of mycoprotein may influence gut microbiota, thereby contributing to its cholesterol-lowering effects in humans.

Proteins and dietary fibers that enter the large intestine become available for fermentation by the gut microbiota. The fermentation of dietary fibers primarily produces short-chain fatty acids (SCFAs), including acetate, propionate, and butyrate, in a molar ratio of approximately 60:20:20. In contrast, the fermentation of protein-derived amino acids generates phenols, amines, ammonia, branched-chain fatty acids, and additional SCFAs. The gut microbiota prioritizes fiber fermentation over protein fermentation, meaning that when fiber fermentation is active, the fate of dietary protein-derived amino acids shifts toward bacterial cell biomass rather than metabolism. As a result, the transition from fiber to protein fermentation has been shown to significantly affect the composition of the gut microbiota. SCFA production, and propionate in particular, has been shown to reduce hepatic cholesterol synthesis via inhibition of β-hydroxy-β-methylglutaryl coenzyme A (HMG-CoA) reductase (the rate-limiting enzyme within cholesterol synthesis) and suppress adipose tissue lipolysis. In human studies using inulin propionate ester, which delivers propionate directly to the large intestine, propionate has been demonstrated to reduce LDL cholesterol and improve liver function and insulin sensitivity. However, the current evidence around propionate is inconsistent, with another study suggesting that its consumption leads to insulin resistance and compensatory hyperinsulinemia. Using *in vitro* colonic models, mycoprotein and its purified dietary fiber have been shown to be fermentable, producing SCFAs. Both mycoprotein and purified mycoprotein dietary fiber have been shown to increase the production of propionate and butyrate, while decreasing acetate levels. Additionally, the elevated colonic production of propionate has been found to inhibit the incorporation of plasma acetate into cholesterol. These findings suggest that the distinctive fiber profile present in mycoprotein warrants further *in vivo* research to explore its digestive and metabolic properties.

The beneficial metabolic effects of mycoprotein consumption have also been shown to extend to acute postprandial glycemic control. It was reported that 20 g mycoprotein (dry weight) consumed during an oral glucose tolerance test resulted in reduced post-prandial glycaemia and insulinemia compared with an isonitrogenous, isoenergetic control condition (soy and skimmed milk) in healthy, young adults. In a recent study, reduced post-prandial insulinemia, but not glycaemia, was also shown with mycoprotein consumption (around 40 g dry weight) compared with an energy- and macronutrient-matched chicken meal in overweight adults. Once again, the underlying mechanism is likely related to both the amount (4 and 7 g, respectively) and the type of fiber present in mycoprotein in these two studies. Viscous polysaccharides, such as those found in mycoprotein, can help reduce postprandial glycemia and insulinemia. Additionally, previous research has shown that 5 g of β-glucan can influence glycemia and insulinemia when consumed alongside a high-carbohydrate meal.

Though the chitin-glucan matrix is insoluble and not viscous, chitin is likely to undergo alkaline deacetylation to produce the viscous polysaccharide chitosan at some stage of the gastrointestinal tract. This may, in turn, increase resistance to the flow induced by gastrointestinal motility, thereby reducing small intestine contact time, which leads to slower gastric emptying and, consequently, altered nutrient absorption. Regardless of the mechanism, it is crucial for health-related applications to note that no data are currently available regarding whether these acute effects on post-prandial glycaemia result in significant changes in insulin sensitivity or habitual glycemic control when mycoprotein is incorporated into the daily diet.

Mycoprotein and Skeletal Muscle Adaptation

Adequate dietary protein intake is required for skeletal muscle mass maintenance and reconditioning. Skeletal muscle mass and protein quality are regulated through dynamic fluctuations in muscle protein synthesis and breakdown rates. During the overnight fasted state, muscle protein breakdown exceeds protein synthesis, resulting in net muscle protein loss. However, protein ingestion temporarily increases muscle protein synthesis rates (for 2-5 hours), primarily due to elevated plasma levels of essential amino acids, with leucine being especially important in this process. Protein ingestion also stimulates pancreatic insulin secretion, which inhibits muscle protein breakdown, contributing to net muscle protein accretion ("the anabolic response") in the post-prandial state, and offsetting fasted protein losses. It is these diurnal oscillations in muscle protein balance that ultimately allow individuals to maintain muscle mass.

Individuals performing structured and prolonged physical activity will elicit skeletal muscle adaptive responses, such as increased muscle mass, muscle quality, contractile function, and/or muscle oxidative capacity. Performing physical activity stimulates muscle protein synthesis rates, and to a lesser extent muscle protein breakdown rates, improving muscle protein balance for up to 48 h. The accumulation of exercise-induced periods of muscle protein accretion plays a crucial role in driving skeletal muscle reconditioning. Following resistance training, this process primarily involves the synthesis of myofibrillar proteins, which support adaptations related to strength and muscle mass. In contrast, endurance exercise primarily stimulates the synthesis of mitochondrial proteins, enhancing oxidative capacity. Consuming dietary protein in close temporal proximity to physical activity has been shown to further enhance the muscle protein synthetic response, surpassing the effects of either stimulus alone. Consequently, strategically increasing dietary protein intake during prolonged training can significantly augment the skeletal muscle adaptive response to exercise.

Since post-prandial muscle protein breakdown rates seem to be maximally inhibited with only mild elevations in circulating insulin, the anabolic potential of post-exercise dietary protein ingestion is believed to depend on its ability to stimulate muscle protein synthesis rates. Animal-derived proteins typically exhibit high bioavailability, leading to rapid and/or sustained post-prandial aminoacidaemia and/or leucinaemia following ingestion. Consequently, animal-derived dietary protein sources have been shown to be superior to plant-based protein sources in their ability to stimulate muscle protein synthesis rates in humans. However, to date, wheat and soy (both relatively low in leucine and/or essential amino acids) are the only non-animal-derived protein sources to be evaluated for their anabolic potential.

Mycoprotein is rich in essential amino acids (~41% of total protein) and relatively high in leucine (~6% of total protein) and possesses a high PDCAAS score (0.99; an indirect indication of a protein's digestibility). The *in vivo* amino acid bioavailability of mycoprotein was recently investigated in comparison to isolated milk protein. Milk protein was selected as the control comparator since this contains a high essential amino acid (~49% of total protein) and leucine content (~11% of total protein), a PDCAAS score of 1.0, and is consequently typically thought of as a near gold standard protein source with respect to its potency for stimulating muscle protein synthesis rates and optimizing training adaptations. The findings indicated that, in healthy young men, the bioavailability of essential amino acids and leucine in the hours following the ingestion of protein-matched boluses of milk protein and mycoprotein was comparable, although the response was slower and more sustained with mycoprotein ingestion.

It is of note that to protein-match these conditions, approximately double the mass (and energy) of mycoprotein was consumed due to its "whole food" nature. It was also found that the amino acid bioavailability of mycoprotein increases in a dose-response fashion until between 60 and 80 g of mycoprotein (i.e., 27-36 g of protein; 2.1-2.9 g leucine) is consumed. It seems likely that mycoprotein ingestion would stimulate a robust and, in larger quantities, an optimal muscle protein synthetic response, making it a potential alternative protein source to support muscle tissue reconditioning during prolonged training - questions that remain to be addressed. However, the magnitude of this response, when compared with other protein sources, would likely depend on whether the overall systemic availability of (essential) amino acids or the speed at which they become available is the more crucial regulatory factor.

An interesting additional consideration is that mycoprotein represents a whole food source rather than an isolated protein. The latter has generally been employed in studies addressing post-prandial muscle protein synthetic responses. While co-ingestion of carbohydrates or fats with isolated protein does not appear to modulate the post-prandial muscle protein synthesis response, emerging evidence suggests that protein consumed from whole food sources may offer an anabolic advantage. The mechanisms behind this effect remain unclear, with potential contributions from differences in energy content, macro- and micronutrient profiles, or the specific food matrix of the protein source. Nevertheless, evaluating the anabolic response to whole food sources is becoming an increasingly important area of research, essential for translating laboratory findings into practical dietary protein recommendations.

Conclusions and Future Directions

Developing sustainable dietary protein sources is a pressing socio-economic and environmental concern, and there is an obvious need to develop a robust evidence base to inform the use of such alternative sources. There is evidence that the incorporation of modest amounts of mycoprotein into the diet positively influences certain circulating lipid sub-fractions, and acute mycoprotein ingestion attenuates postprandial glycaemia and/or insulinemia. These data are striking as they occur in the face of energy-balanced conditions (i.e., are not an artefact of lower overall energy intake and/or consequent weight loss).

These responses may be influenced by the distinct digestive and metabolic properties of the chitin and β-glucan fibers present in mycoprotein, although a comprehensive *in vivo* mechanistic understanding has yet to be established. The effect of mycoprotein incorporation into the diet on circulating cholesterol levels remains unknown, and a complete characterization of the lipid sub-fraction responses is still lacking. Additionally, it is unclear whether changes in acute postprandial glycemic control result in improved insulin sensitivity and/or long-term glycemic control when mycoprotein is included in the daily diet.

Mycoprotein is a source of nutrients that can effectively induce satiety as evidenced by a reduced *ad libitum* energy intake, suggesting it may be a useful tool within weight management. This is especially true when considered alongside its potential as a high-quality protein source and as a modulator of postprandial glycaemia. As such, research into the ability of mycoprotein to modulate habitual glycemic control, caloric intake, and weight management is clearly warranted.

Emerging data suggest that mycoprotein is a bioavailable and insulinotropic protein source, which would likely be effective in stimulating muscle protein anabolism. As a result, the consumption of mycoprotein as a dietary protein source to enhance muscle protein synthesis and promote muscle adaptation or maintenance in various populations (e.g., athletes, older adults) represents a promising area for future research.

Key Points

- Environmental concerns over increased dietary protein production require the development of robust investigation into the nutritional value of alternative, sustainably produced dietary protein sources.
- Mycoprotein is a sustainably produced fungal-derived dietary protein source that has been shown to improve blood lipid profiles and acute post-prandial glucose control, and provides a potent satiety effect.
- Mycoprotein has a favorable amino acid composition and bioavailability when considering its potential to stimulate muscle protein synthesis rates.
- Future work should assess the anabolic potential of mycoprotein in various situations (e.g., resting, exercise) and populations (e.g., athletes, older adults).

References

Allès, B., Baudry J, Méjean C, Touvier M, Péneau S, Hercberg S, Kesse-Guyot E. Comparison of sociodemographic and nutritional characteristics between self-reported vegetarians, vegans, and meat-eaters from the Nutrinet-Sante study. *Nutrients* 2017; 9:1023.

Anderson, J., Tietyen-Clark J. Dietary fiber: hyperlipidemia, hypertension, and coronary heart disease. *Am J Gastroenterol* 1986; 81:907-919.

Bauer, J., Biolo G, Cederholm T, Cesari M, Cruz-Jentoft A, Morley J, Phillips S, Sieber C, Stehle P, Teta D, Visvanathan R, Volpi E, Boirie Y. Evidence-based recommendations for optimal dietary protein intake in older people: a position paper from the PROT-AGE Study Group. *J Am Med Dir Assoc* 2013; 14:542-559.

Beals, J., Mackenzie R, van Vliet S, Skinner S, Pagni B, Niemiro G, Ulanov A, Li Z, Dilger A, Paluska S, De Lisio M, Burd N. Protein-Rich Food Ingestion Stimulates Mitochondrial Protein Synthesis in Sedentary Young Adults of Different BMIs. *J Clin Endocrinol Metab* 2017; 102:3415-3424.

Beasley, J., Wertheim B, LaCroix A, Prentice R, Neuhouser M, Tinker L, Kritchevsky S, Shikany J, Eaton C, Chen Z, Thomson C. Biomarker-calibrated protein intake and physical function in the Women's Health Initiative. *J Am Geriatr Soc* 2013; 61:1863-1871.

Bertenshaw, E. J., Lluch A, Yeomans MR. Perceived thickness and creaminess modulates the short-term satiating effects of high-protein drinks. *British Journal of Nutrition* 2013; 110:578-586.

Blackburn, N., Holgate A, Read N. Does guar gum improve post-prandial hyperglycaemia in humans by reducing small intestinal contact area? *Br J Nutr* 1984; 52:197-204.

Bottin, J., Swann J, Cropp E, Chambers E, Ford H, Ghatei M, Frost G. Mycoprotein reduces energy intake and postprandial insulin release without altering glucagon-like peptide-1 and peptide tyrosine-tyrosine concentrations in healthy overweight and obese adults: a randomised-controlled trial. *Br J Nutr* 2016; 116:360-374.

Burd, N., Beals J, Martinez I, Salvador A, Skinner S. Food-First Approach to Enhance the Regulation of Post-exercise Skeletal Muscle Protein Synthesis and Remodeling. *Sports Med* 2019; 49:59-68.

Burd, N., Gorissen S, van Vliet S, Snijders T, van Loon L. Differences in postprandial protein handling after beef compared with milk ingestion during postexercise recovery: a randomized controlled trial. *Am J Clin Nutr* 2015; 102:828-836.

Burley, V., Paul A, Blundell J. Influence of a high-fiber food (myco-protein) on appetite: effects on satiation (within meals) and satiety (following meals). *European journal of clinical nutrition* 1993; 47:409-409.

Cermak, N., Res P, de Groot L, Saris W, van Loon L. Protein supplementation augments the adaptive response of skeletal muscle to resistance-type exercise training: a meta-analysis. *Am J Clin Nutr* 2012; 96:1454-1464.

Chambers, E. S., Byrne CS, Morrison DJ, Murphy KG, Preston T, Tedford C, Garcia- Perez I, Fountana S, Serrano-Contreras JI, Holmes E. Dietary supplementation with inulin-propionate ester or inulin improves insulin sensitivity in adults with overweight and obesity with distinct effects on the gut microbiota, plasma metabolome and systemic inflammatory responses: a randomised cross-over trial. *Gut* 2019:gutjnl- 2019-318424.

Chambers, E. S., Viardot A, Psichas A, Morrison DJ, Murphy KG, Zac-Varghese SE, MacDougall K, Preston T, Tedford C, Finlayson GS. Effects of targeted delivery of propionate to the human colon on appetite regulation, body weight maintenance and adiposity in overweight adults. *Gut* 2015; 64:1744-1754.

Chen, W., Anderson J, Jennings D. Propionate may mediate the hypocholesterolemic effects of certain soluble plant fibers in cholesterol-fed rats. *Proc Soc Exp Biol Med* 1984; 175:215-218.

Cummings, J., Pomare E, Branch W, Naylor C, Macfarlane G. Short chain fatty acids in human large intestine, portal, hepatic and venous blood. *Gut* 1987; 28:1221-1227.

Cuthbertson, D., Smith K, Babraj J, Leese G, Waddell T, Atherton P, Wackerhage H, Taylor P, Rennie M. Anabolic signaling deficits underlie amino acid resistance of wasting, aging muscle. *FASEB J* 2005; 19:422-424.

David, L. A., Maurice CF, Carmody RN, Gootenberg DB, Button JE, Wolfe BE, Ling AV, Devlin AS, Varma Y, Fischbach MA. Diet rapidly and reproducibly alters the human gut microbiome. *Nature* 2014; 505:559.

Dunlop, M., Kilroe S, Bowtell J, Finnigan T, Salmon D, Wall B. Mycoprotein represents a bioavailable and insulinotropic non-animal derived dietary protein source: a dose-response study *Br J Nutr* 2017; 11:1-13.

Finnigan, T. Mycoprotein: a healthy new protein with a low environmental impact. Sustainable protein sources (Ed Nadathur, SR, Wanasundra, J & Scanlin, L 2012; Elsevier Publishing:305-326 Carbon Trust. *Quorn, beef and chicken footprints*. 2014.

Finnigan, T. Mycoprotein; origin, production and properties. *Handbook of food proteins* (Ed Phillips, GO and William, PA) 2012; Woodhead publishing.

Fukagawa, N., Anderson J, Hageman G, Young V, Minaker K. High-carbohydrate, high-fiber diets increase peripheral insulin sensitivity in healthy young and old adults. *Am J Clin Nutr* 1990; 52:524-528.

Ge, H., Li X, Weiszmann J, Wang P, Baribault H, Chen J, Tian H, Li Y. Activation of G protein-coupled receptor 43 in adipocytes leads to inhibition of lipolysis and suppression of plasma free fatty acids. *Endocrinology* 2008; 149:4519-4526.

Gorissen, S. H., Crombag JJ, Senden JM, Waterval WH, Bierau J, Verdijk LB, van Loon LJ. Protein content and amino acid composition of commercially available plant-based protein isolates. *Amino acids* 2018; 50:1685-1695.

Gorissen, S., Burd N, HM H, Gijsen A, Groen B, van Loon L. Carbohydrate co- ingestion delays dietary protein digestion and absorption but does not modulate postprandial muscle protein accretion. *J Clin Endocrinol Metab* 2014; 99:2250-2258.

Gorissen, S., Horstman A, Franssen R, Crombag J, Langer H, Bierau J, Respondek F, van Loon L. Ingestion of Wheat Protein Increases *In Vivo* Muscle Protein Synthesis Rates in Healthy Older Men in a Randomized Trial. *J Nutr* 2016; 146:1651-1659.

Greenhaff, P., Karagounis L, Peirce N, Simpson E, Hazell M, Layfield R, Wackerhage H, Smith K, Atherton P, Selby A, Rennie M. Disassociation between the effects of amino acids and insulin on signaling, ubiquitin ligases, and protein turnover in human muscle. *Am J Physiol Endocrinol Metab* 2008; 295:595-604.

Hall, K. D., Bemis T, Brychta R, Chen KY, Courville A, Crayner EJ, Goodwin S, Guo J, Howard L, Knuth ND. Calorie for calorie, dietary fat restriction results in more body fat loss than carbohydrate restriction in people with obesity. *Cell metabolism* 2015; 22:427-436.

Harris, H. C., Edwards CA, Morrison DJ. Short Chain Fatty Acid Production from Mycoprotein and Mycoprotein Fiber in an *In Vitro* Fermentation Model. *Nutrients* 2019; 11:800.

Hartman, J., Tang J, Wilkinson S, Tarnopolsky M, Lawrence R, Fullerton A, Phillips S. Consumption of fat-free fluid milk after resistance exercise promotes greater lean mass accretion than does consumption of soy or carbohydrate in young, novice, male weightlifters. *Am J Clin Nutr* 2007; 86:373-381.

Hochstenbach-Waelen, A., Veldhorst M, Nieuwenhuizen A, Westerterp-Plantenga M, Westerterp K. Comparison of 2 diets with either 25% or 10% of energy as casein on energy expenditure, substrate balance, and appetite profile. *Am J Clin Nutr* 2009; 89:831-838.

Hochstenbach-Waelen, A., Westerterp-Plantenga M, Veldhorst M, Westerterp K. Single-protein casein and gelatin diets affect energy expenditure similarly but substrate balance and appetite differently in adults. *J Nutr* 2009; 139:2285-2292.

Hogenkamp, P. S., Stafleu A, Mars M, Brunstrom JM, de Graaf C. Texture, not flavor, determines expected satiation of dairy products. *Appetite* 2011; 57:635-641.

Homma, Y., Nakamura H, Kumagai Y, Ryuzo A, Saito Y, Ishikawa T, Takada K, Yamagami H, Kikuchi H, Inadera H. Effects of eight week ingestion of mycoprotein on plasma levels of lipids and Apo (Lipo) proteins. *Progress in Medicine* 1995; 15:183-195.

Isanejad, M., Mursu J, Sirola J, Kröger H, Rikkonen T, Tuppurainen M, Erkkilä A. Association of protein intake with the change of lean mass among elderly women: The

Osteoporosis Risk Factor and Prevention - Fracture Prevention Study (OSTPRE-FPS). *J Nutr Sci* 2015; 4:e41. doi: 10.1017/jns.2015.31. eCollection 2015.

Ishikawa, T. The effect of mycoprotein intake (12 and 24g/day) over four weeks on serum cholesterol levels. *Progress in Medicine* 1995; 15:61-74.

Jenkins, D., Kendall C, McKeown-Eyssen G, Josse R, Silverberg J, Booth G, Vidgen E, Josse A, Nguyen T, Corrigan S, Banach M, Ares S, Mitchell S, Emam A, Augustin L, Parker T, Leiter L. Effect of a low-glycemic index or a high-cereal fiber diet on type 2 diabetes: a randomized trial. *JAMA* 2008; 300:2742-2753.

Jenkins, D., Wolever T, Leeds A, Gassull M, Haisman P, Dilawari J, Goff D, Metz G, Alberti K. Dietary fibers, fiber analogues, and glucose tolerance: importance of viscosity. *Br Med J* 1978; 1:1392-1394.

Karl, J. P., Roberts SB. Energy density, energy intake, and body weight regulation in adults. *Advances in nutrition* 2014; 5:835-850.

Keim, N. L., Stern JS, Havel PJ. Relation between circulating leptin concentrations and appetite during a prolonged, moderate energy deficit in women. *The American journal of clinical nutrition 1*998; 68:794-801.

Kirby, R., Anderson J, Sieling B, Rees E, Chen W, Miller R, Kay R. Oat-bran intake selectively lowers serum low-density lipoprotein cholesterol concentrations of hypercholesterolemic men. *Am J Clin Nutr* 1981; 34:824-829.

Larsen, T., Dalskov S, van Baak M, Jebb S, Papadaki A, Pfeiffer A, Martinez J, Handjieva-Darlenska T, Kunešová M, Pihlsgård M, Stender S, Holst C, Saris W, Astrup A. Diets with high or low protein content and glycemic index for weight-loss maintenance. *N Engl J Med* 2010; 363:2102-2113.

Louard, R., Fryburg D, Gelfand R, Barrett E. Insulin sensitivity of protein and glucose metabolism in human forearm skeletal muscle. *J Clin Invest* 1992; 90:2348-2354.

Ludwig, D., Pereira M, Kroenke C, Hilner J, Van Horn L, Slattery M, Jacobs D. Dietary fiber, weight gain, and cardiovascular disease risk factors in young adults. *JAMA* 1999; 282:1539-1546.

Macfarlane, G., Gibson G, Cummings J. Comparison of fermentation reactions in different regions of the human colon. *Journal of Applied Bacteriology* 1992; 72:57-64.

McCrickerd, K., Chambers L, Brunstrom JM, Yeomans MR. Subtle changes in the flavor and texture of a drink enhance expectations of satiety. *Flavor* 2012; 1:20.

Mettler, S., Mitchell N, Tipton K. Increased protein intake reduces lean body mass loss during weight loss in athlete. *Med Sci Sports Exerc* 2010; 42:326-337.

Mitchell, C., McGregor R, D'Souza R, Thorstensen E, Markworth J, Fanning A, Poppitt S, Cameron-Smith D. Consumption of Milk Protein or Whey Protein Results in a Similar Increase in Muscle Protein Synthesis in Middle Aged Men. *Nutrients* 2015; 7:8685-8699.

Moore, D., Churchward-Venne T, Witard O, Breen L, Burd N, Tipton K, Phillips S. Protein Ingestion to Stimulate Myofibrillar Protein Synthesis Requires Greater Relative Protein Intakes in Healthy Older Versus Younger Men. *J Gerontol A Biol Sci Med Sci* 2014; 70:57-62.

Moore, D., Tang J, Burd N, Rerecich T, Tarnopolsky M, Phillips S. Differential stimulation of myofibrillar and sarcoplasmic protein synthesis with protein ingestion at rest and after resistance exercise. *J Physiol* 2009; 587:897-904.

Morton, R., Murphy K, McKellar S, Schoenfeld B, Henselmans M, Helms E, Aragon A, Devries M, Banfield L, Krieger J, Phillips S. A systematic review, meta-analysis and meta-regression of the effect of protein supplementation on resistance training-induced gains in muscle mass and strength in healthy adults. *Br J Sports Med* 2018; 52:376-384.

Nadathur, S., Wanasundara J, Scanlin L. Sustainable protein sources. *Elsevier press* 2016.

Nakamura, H, Ishikawa T, Akanuma M, Nishiwaki M, Yamashita T, Tomiyasu K, Yoshida H, Nishio E, Hosoai K, Shiga H. Effect of mycoprotein intake on serum lipids of healthy subjects. *Progress in Medicine* 1994; 14:1972-1976.

Nazare, J. A., Sylvie Normand, Angeliki Oste Triantafyllou, Aude Brac de la Perrière, Michel Desage, Martine Laville. Modulation of the postprandial phase by beta-glucan in overweight subjects: effects on glucoseand insulin kinetics. *Mol Nutr Food Res* 2009; 53:777-787.

Nazare, J., Normand S, Oste Triantafyllou A, Brac de la Perrière A, Desage M, Laville Owen D, Munday K, Taylor T, Turner M. Effects of wheat bran and a mould (Fusarium) on cholesterol excretion in rats. *Proc Nutr Soc* 1976; 35:38-39.

Owen, D., Munday K, Taylor T, Turner M. Hypocholesterolaemic action of wheat bran and a mould (Fusarium) in rats and hamsters. *Proc Nutr Soc* 1975; 34:16-17.

Owen, D., Munday K, Taylor T, Turner M. The possible nature of the hypocholesterolaemic action of a mould (Fusarium) [proceedings]. *Proc Nutr Soc* 1976; 35:127-128.

Paddon-Jones, D., Campbell W, Jacques P, Kritchevsky S, Moore L, Rodriguez N, van Loon L. Protein and healthy aging. *Am J Clin Nutr* 2015; pii: ajcn084061.

Pennings, B., Boirie Y, Senden J, Gijsen A, Kuipers H, van Loon L. Whey protein stimulates postprandial muscle protein accretion more effectively than do casein and casein hydrolysate in older men. *Am J Clin Nutr* 2011; 93:997-1005.

Pennings, B., Groen B, de Lange A, Gijsen A, Zorenc A, Senden J, van Loon L. Amino acid absorption and subsequent muscle protein accretion following graded intakes of whey protein in elderly men *Am J Physiol Endocrinol Metab* 2012; 302:E992-999.

Pennings, B., Koopman R, Beelen M, Senden J, Saris W, van Loon L. Exercising before protein intake allows for greater use of dietary protein-derived amino acids for de novo muscle protein synthesis in both young and elderly men. *Am J Clin Nutr* 2011; 93:322-331.

Phillips, S. Dietary protein requirements and adaptive advantages in athletes. *Br J Nutr* 2012; 108:158-167.

Phillips, S. Nutrition in the elderly: a recommendation for more (evenly distributed) protein? *Am J Clin Nutr* 2017; 106:12-13.

Phillips, S., Chevalier S, Leidy H. Protein "requirements" beyond the RDA: implications for optimizing health. *Appl Physiol Nutr Metab* 2016; 41:565-572.

Phillips, S., Tipton K, Aarsland A, Wolf S, Wolfe R. Mixed muscle protein synthesis and breakdown after resistance exercise in humans. *Am J Physiol* 1997; 273:99-107.

Ranganathan, J., Vennard D, Waite R, Dumas P, Lipinski B, Searchinger T. Shifting diets for a sustainable food future; working paper, installment 11 of 'Creating a sustainable food future.' *World resources institute* 2016; Washington, DC.

Ravussin, E., Lillioja S, Knowler WC, Christin L, Freymond D, Abbott WG, Boyce V, Howard BV, Bogardus C. Reduced rate of energy expenditure as a risk factor for body-weight gain. *New England Journal of Medicine* 1988; 318:467-472.

Rennie, M., Edwards R, Halliday D, Matthews D, Wolman S, Millward D. Muscle protein synthesis measured by stable isotope techniques in man: the effects of feeding and fasting. *Clin Sci (Lond)* 1982; 63:519-523.

Rieu, I., Balage M, Sornet C, Giraudet C, Pujos E, Grizard J, Mosoni L, Dardevet D. Leucine supplementation improves muscle protein synthesis in elderly men independently of hyperaminoacidaemia. *J Physiol* 2006; 575:305-315.

Rimm, E., Ascherio A, Giovannucci E, Spiegelman D, Stampfer M, Willett W. Vegetable, fruit, and cereal fiber intake and risk of coronary heart disease among men. *JAMA* 1996; 275:447-451.

Ruxton, C., McMillan B. The impact of mycoprotein on blood cholesterol levels: a pilot study. *British Food Journal* 2010; 112:1092-1101.

Staples, A., Burd N, West D, Currie K, Atherton P, Moore D, Rennie M, Macdonald M, Baker S, Phillips S. Carbohydrate does not augment exercise-induced protein accretion versus protein alone. *Med Sci Sports Exerc* 2011; 43:1154-1161.

Stefan, H. M. Gorissen, Nicholas A Burd, Irene Fleur Kramer, Janneau van Kranenburg, Annemie P Gijsen, Olav Rooyackers, Luc J C van Loon Co-ingesting milk fat with micellar casein does not affect postprandial protein handling in healthy older men. *Clin Nutr* 2017; 36:429-437.

Tang, J., Moore D, Kujbida G, Tarnopolsky M, Phillips S. Ingestion of whey hydrolysate, casein, or soy protein isolate: effects on mixed muscle protein synthesis at rest and following resistance exercise in young men. *J Appl Physiol* 2009; 107:987-992.

Tipton, K., Elliott T, Cree M, Aarsland A, Sanford A, Wolfe R. Stimulation of net muscle protein synthesis by whey protein ingestion before and after exercise. *Am J Physiol* 2007; 292:71-76.

Tipton, K., Gurkin B, Matin S, Wolfe R. Nonessential amino acids are not necessary to stimulate net muscle protein synthesis in healthy volunteers. *J Nutr Biochem* 1999; 10:89-95.

Tirosh, A., Calay ES, Tuncman G, Claiborn KC, Inouye KE, Eguchi K, Alcala M, Rathaus M, Hollander KS, Ron I. The short-chain fatty acid propionate increases glucagon and FABP4 production, impairing insulin action in mice and humans. *Science translational medicine* 2019; 11:eaav0120.

Trommelen, J., Betz M, van Loon L. The Muscle Protein Synthetic Response to Meal Ingestion Following Resistance-Type Exercise. *Sports Med* 2019; 49:185-197.

Turnbull, W., Leeds A, Edwards D. Mycoprotein reduces blood lipids in free-living subjects. *Am J Clin Nutr* 1992; 55:415-419.

Turnbull, W., Leeds A, Edwards G. Effect of mycoprotein on blood lipids. *Am J Clin Nutr* 1990; 52:646-650.

Turnbull, W., Walton J, Leeds A. Acute effects of mycoprotein on subsequent energy intake and appetite variables. *Am J Clin Nutr* 1993; 58:507-512.

Turnbull, W., Ward T. Mycoprotein reduces glycemia and insulinemia when taken with an oral-glucose-tolerance test. *Am J Clin Nutr* 1995; 61:135-140.

Udall, J., Clifford W, Young V, Scrimshaw N. The tolerance and nutritional value of two microfungal foods in human subjects. *Am J Clin Nutr* 1984; 40:285-292.

US Department of Agriculture Agricultural Research Service - Nutrient Data Laboratory. *USDA National Nutrient Database for Standard Reference*. Release 28. Version Current: September 2015;/nea/bhnrc/ndl.

van Vliet, S., Burd N, van Loon L. The Skeletal Muscle Anabolic Response to Plant- versus Animal-Based Protein Consumption. *J Nutr* 2015; 145:1981-1991.

van Vliet, S., Shy E, Abou Sawan S, Beals J, West D, Skinner S, Ulanov A, Li Z, Paluska S, Parsons C, Moore D, Burd N. Consumption of whole eggs promotes greater stimulation of postexercise muscle protein synthesis than consumption of isonitrogenous amounts of egg whites in young men. *Am J Clin Nutr* 2017; 106:1401-1412.

Veldhorst, M., Nieuwenhuizen A, Hochstenbach-Waelen A, Westerterp K, Engelen M, Brummer R, Deutz N, Westerterp-Plantenga M. A breakfast with alpha-lactalbumin, gelatin, or gelatin + TRP lowers energy intake at lunch compared with a breakfast with casein, soy, whey, or whey-GMP. *Clin Nutr* 2009; 28:147-155.

Volpi, E., Mittendorfer B, Rasmussen B, Wolfe R. The response of muscle protein anabolism to combined hyperaminoacidemia and glucose-induced hyperinsulinemia is impaired in the elderly. *J Clin Endocrinol Metab* 2000; 85:4481-4490.

Volpi, E., Mittendorfer B, Wolf S, Wolfe R. Oral amino acids stimulate muscle protein anabolism in the elderly despite higher first-pass splanchnic extraction. *Am J Physiol* 1999; 277:513-520.

Volpi, E., Sheffield-Moore M, Rasmussen B, Wolfe R. Basal muscle amino acid kinetics and protein synthesis in healthy young and older men. *JAMA* 2001; 286:1206-1212.

Wall, B., Burd N, Franssen R, Gorissen S, Snijders T, Senden J, Gijsen A, van Loon L. Presleep protein ingestion does not compromise the muscle protein synthetic response to protein ingested the following morning. *Am J Physiol Endocrinol Metab* 2016; 311:964-973.

Wall, B., Cermak N, van Loon L. Dietary protein considerations to support active aging. *Sports Med* 2014; 44.

Wall, B., Gorissen S, Pennings B, Koopman R, Groen B, Verdijk L, van Loon L. Aging Is Accompanied by a Blunted Muscle Protein Synthetic Response to Protein Ingestion. *PLOS one* 2016; 10:e0140903.

Wall, B., Hamer H, de Lange A, Kiskini A, Groen B, Senden J, Gijsen A, Verdijk L, van Loon L. Leucine co-ingestion improves post-prandial muscle protein accretion in elderly men. *Clin Nutr* 2013; 32:412-419.

Wall, B., Morton J, van Loon L. Strategies to maintain skeletal muscle mass in the injured athlete: Nutritional considerations and exercise mimetics. *Eur J Sport Sci* 2014; 15:53-62.

Weigle, D., Breen P, Matthys C, Callahan H, Meeuws K, Burden V, Purnell J. A high-protein diet induces sustained reductions in appetite, ad libitum caloric intake, and body weight despite compensatory changes in diurnal plasma leptin and ghrelin concentrations. *Am J Clin Nutr* 2005; 82:41-48.

Weiss, E. P., Jordan RC, Frese EM, Albert SG, Villareal DT. Effects of weight loss on lean mass, strength, bone, and aerobic capacity. *Medicine and science in sports and exercise* 2017; 49:206.

West, D., Burd N, Coffey V, Baker S, Burke L, Hawley J, Moore D, Stellingwerff T, Phillips S. Rapid aminoacidemia enhances myofibrillar protein synthesis and anabolic intramuscular signaling responses after resistance exercise. *Am J Clin Nutr* 2011; 94:795-803.

Westerterp-Plantenga, M., Lemmens S, Westerterp K. Dietary protein - its role in satiety, energetics, weight loss and health. *Br J Nutr* 2012; 108.

Westerterp-Plantenga, M., Nieuwenhuizen A, Tomé D, Soenen S, Westerterp K. Dietary protein, weight loss, and weight maintenance. *Annu Rev Nutr* 2009; 29:21-41.

Wilkes, E., Selby A, Atherton P, Patel R, Rankin D, Smith K, Rennie M. Blunting of insulin inhibition of proteolysis in legs of older subjects may contribute to age-related sarcopenia. *Am J Clin Nutr* 2009; 90:1343-1350.

Wilkinson, S., Phillips S, Atherton P, Patel R, Yarasheski K, Tarnopolsky M, Rennie M. Differential effects of resistance and endurance exercise in the fed state on signalling molecule phosphorylation and protein synthesis in human muscle. *J Physiol* 2008; 586:3701-3717.

Williamson, D., Geiselman P, Lovejoy J, Greenway F, Volaufova J, Martin C, Arnett C, Ortego L. Effects of consuming mycoprotein, tofu or chicken upon subsequent eating behaviour, hunger and safety. *Appetite* 2006; 46:41-48.

Wolever, T., Spadafora PJ, Cunnane SC, Pencharz PB. Propionate inhibits incorporation of colonic [1, 2-13C] acetate into plasma lipids in humans. *The American journal of clinical nutrition* 1995; 61:1241-1247.

Wolk, A., Manson J, Stampfer M, Colditz G, Hu F, Speizer F, Hennekens C, Willett W. Long-term intake of dietary fiber and decreased risk of coronary heart disease among women. *JAMA* 1999; 281:1998-2004.

World Health Organization. Obesity: preventing and managing the global epidemic. *World Health Organization;*2000. 9241208945.

Chapter 4

The Main Manufacturing Methods of Mycoprotein

Abstract

To meet global protein demands sustainably, mycoprotein, derived from filamentous fungi, is a viable meat alternative with nutritional and health benefits. It acts as a prebiotic, antioxidant, and regulator of blood cholesterol and glucose, supporting muscle protein development. Produced through submerged, solid-state, or surface fermentation, submerged fermentation offers higher yields and better nutritional content. Approved as GRAS by the FDA, species like *Fusarium venenatum* are commonly used. Mycoprotein has a smaller carbon footprint than beef and chicken, making it eco-friendly. Despite challenges like lower protein content and potential side effects, it holds promise as a meat substitute, especially for vegetarians.

Keywords: mycoprotein production, fermentation techniques, solid-state fermentation (SSF), nutritional assessment

History of Mycoprotein

The timeline for the commercialization of fungal protein is illustrated in Figure 1. In the 1960s, concerns arose globally among practitioners and officials about the potential for future protein shortages due to projected population growth. In response, food researchers sought to develop a palatable and cost-effective source of microbial protein. In 1967, *Fusarium venenatum*, a fungus discovered in a garden in Marlow, Buckinghamshire, was identified as a promising candidate for fungal protein production (Denny, 2008). Mycoprotein is the nucleic acid-depleted biomass that is cultivated from the mycelium (cells) of *F. venenatum* in continuous fermenters such as chemostats or turbidostats. Considerable research has been dedicated to assessing the safety of mycoprotein derived from *F. venenatum* for human consumption. It became clear that mycoprotein could be consumed without harming human or murine models when the nucleic acid content of the cells was reduced to safe levels (Asgar, 2010). The ATCC PTA-2684 strain was considered safe as it could not produce toxic microbial metabolites because the growth conditions (28 to 30°C; pH - 6.0) used for manufacture were inefficient. *F. venenatum* A3/5 was approved for trade as food in 1984 by the Ministry of Agriculture, Fisheries and Food in the United Kingdom (Asgar, 2010). A joint venture of two companies in the United Kingdom, namely RHM (Rank Hovis McDougall) and ICI, was solely responsible for the commercialization of mycoprotein.

Methods of Production

Mycoprotein is produced by fermenting agro-industrial waste through microbial fermentation processes. Agro-industrial wastes like seaweed waste (Landeta, 2021), date waste (Reihani, 2018), soy waste (Liang, 2023), pineapple peel waste (Umesh, 2023), and pea process byproducts can be used to produce mycoprotein biomass. There are two primary fermentation processes for the mass production of mycoprotein: solid-state fermentation (SSF) and submerged fermentation (SmF). Additionally, the surface culture method can be used to produce mycoprotein on a laboratory scale.

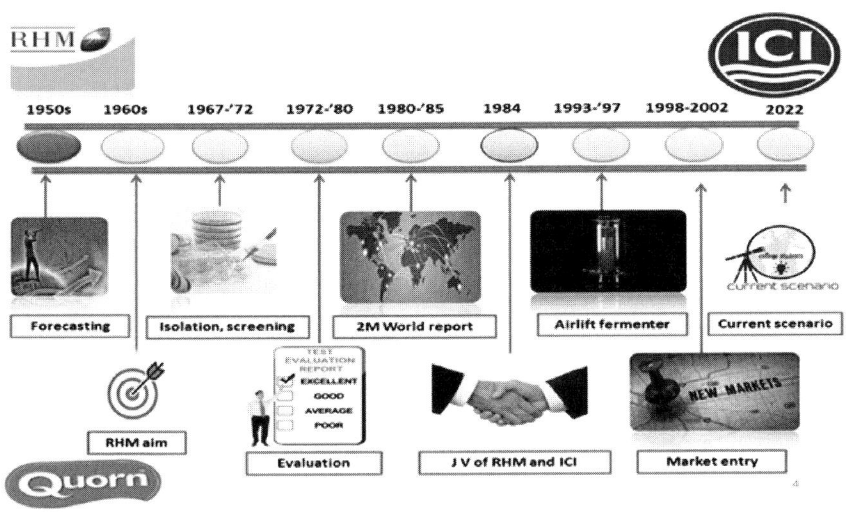

Figure 1. A historical overview of the development and commercialization of mycoprotein.

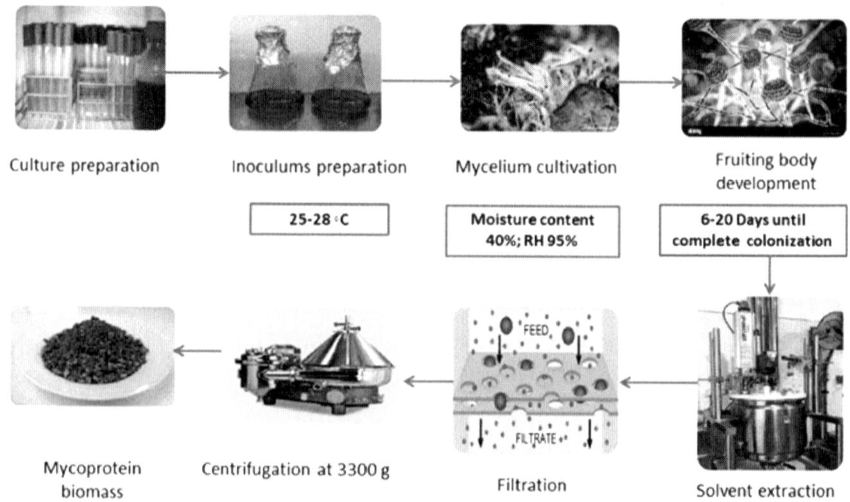

Figure 2. Process of mycoprotein preparation using solid-state fermentation.

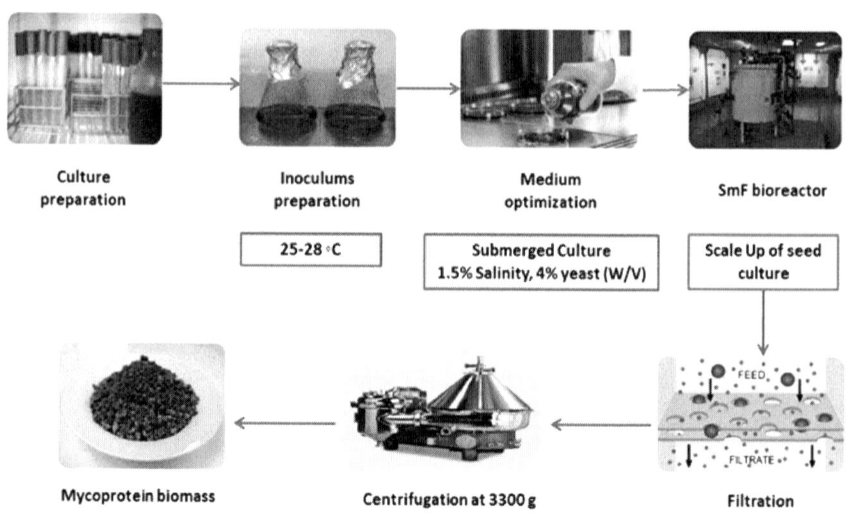

Figure 3. Process of mycoprotein production using submerged fermentation.

Solid-State Fermentation (SSF)

SSF is gaining increasing attention in the food and pharmaceutical industries due to its fermentation process, which is conducted on insoluble substrates with minimal use of liquid medium. The resulting mycoprotein exhibits a higher PDCAAS (protein digestibility-corrected amino acid score) than chicken and beef proteins, compensating for the deficiency of essential amino acids in plant-based protein sources (Kim, 2012). Stoffel et al., developed edible mycoproteins using SSF, employing moisture-rich brewers' spent grains and grape waste on a potato dextrose agar (PDA) medium with a common salt solution (pH 6.5). The mixture was then incubated at 28°C for 15 days. *Pleurotus albidus* cultivation generated more mycelial biomass (250 g/2 kg) than *Agaricus blazei* (212 g/2 kg) and *Auricularia fuscosuccinea* (56 g/2 kg) biomasses, respectively. Furthermore, as compared to the control and other strains, *Pleurotus albidus* biomass had greater protein content (23% w/w), total amino acids (8% w/w), and dietary fibers (34% w/w) (Ahmad, 2022).

The SSF method for mycoprotein synthesis consists of several successive phases, each contributing to the final protein-rich output (Figure 2). Preparing the solid substrate requires the careful blending of various raw materials (Cebran, 2023). This mixture serves the dual purpose of creating a nutrient-rich environment suitable for fungal growth while maintaining the structural integrity of the substrate. Commonly used elements in this substrate formulation include wheat, barley, oats, and other cereal grains. Following that, the chosen fungus strain, usually of the Fusarium genus, is injected into the substrate. Inoculation plays a significant role in the development of fungal spores or mycelium. Subsequent incubation is carried out in a regulated environment in a bioreactor (Schweiggert, 2020). Critical parameters such as temperature, humidity, and aeration are carefully managed throughout this phase to optimize fungal growth and subsequent protein synthesis. As the fungal biomass grows, it takes advantage of the available nutrients in the substrate, triggering a sequence of metabolic processes. This metabolic process for producing fungal mycelium produces mycelial biomass and mycoprotein. The synthesized mycoprotein is notable for its high concentration of critical amino acids and protein components.

The fungal biomass, primarily mycoprotein, is extracted from the solid substrate at an opportune juncture, specified by the end of the incubation time. This extraction method requires the careful separation of mycoprotein from both the solid substrate and any remaining fungus spores (Cebran, 2023). For this, after subsequent cultivation of fungal mycelium, it is dissolved in an organic solvent like hexane. Then, filtration takes place, followed by subsequent centrifugation to obtain pure mycoprotein biomass. The extracted mycoprotein undergoes several processing steps to enhance its sensory properties, nutritional profile, and overall quality. Texturization, flavor enhancement, and shaping are employed to mimic various meat-based products. Ultimately, the processed mycoprotein becomes a versatile component suitable for incorporation into a wide range of culinary items, including meat replacements (Okeudo, 2023).

Submerged Fermentation (SmF)

Submerged fermentation (SmF) is the process by which microorganisms grow (anaerobic/partially anaerobic) and decompose substrates (i.e., carbohydrates) in the presence of lots of free water (liquid medium) (Nevalainen, 2020). Despite the popularity of SmF in bioprocessing, SSF is becoming the center of attention because of higher productivity, less water consumption, and less contamination (Javourez, 2022). Submerged microbial cultivation involves growing microorganisms in a liquid medium enriched with carbon, nitrogen, and essential micronutrients. This method has become more prevalent than traditional fruiting body cultivation for the production of bioactive mushroom compounds in the industry, primarily due to the ease and efficiency with which growing conditions can be controlled. Additionally, the use of air-lift bioreactors in submerged culture systems provides an opportunity for scaling up industrial applications, further enhancing its potential for large-scale production (Bakratsas, 2023).

This process (Figure 3) is initiated by cultivating a pure culture of the designated fungal strain, typically *Fusarium* species, in a liquid growth medium to generate an inoculum. The next step involves the preparation of a nutrient-rich fermentation medium, containing carbon and nitrogen sources, vitamins, minerals, and other essential growth factors (Zhao, 2023). Once the medium is ready, the inoculum is introduced, serving as the source of fungal biomass and protein synthesis. The fermentation process, conducted in controlled bioreactors or fermentation vessels, requires precise regulation of environmental parameters, including temperature, pH, oxygen levels, and agitation (Dudekula, 2020). These conditions are meticulously modulated to optimize fungal proliferation and mycoprotein production. During the metabolic progression, the fungal cells actively metabolize the nutrients in the submerged medium, ultimately culminating in the biosynthesis of biomass enriched with mycoprotein. This proteinaceous entity is characterized by a notable content of essential amino acids, rendering it nutritionally significant (Deng, 2023). Notably, the submerged fermentation process is carefully monitored to assess parameters such as biomass concentration, mycoprotein yield, pH stability, and oxygen levels. Once the desired mycoprotein accumulation is achieved, the fermentation is terminated, and the fungal biomass is separated from the liquid medium through methods like centrifugation or filtration. The harvested biomass then undergoes further processing to improve its textural and sensory qualities (Hashempour, 2023). The resulting mycoprotein is then harnessed as a versatile ingredient in diverse food applications, effectively substituting conventional meat sources. This comprehensive process underscores the potential of submerged fermentation as a proficient route for mycoprotein production, albeit with considerations of energy demand due to the requisite liquid medium and agitation (Smetana, 2023).

The Quorn (submerged) fermenters, the largest continuous flow culture systems in biotechnology, use air-lift technology, distinct from conventional stirred-tank bioreactors. This design benefits mycelial production with low shear, fostering lengthy hyphal development vital for replicating meat-like textures. Air-lift fermenters require less energy than stirred tanks, efficiently mixing with a smaller energy footprint. These fermenters introduce sterile air bubbles to achieve oxygen saturation and medium mixing, contrasting mechanical impellers. Oxygen transfer primarily occurs near the base of the riser due to hydrostatic pressure and the turbulence from pumped air, optimizing gaseous-liquid oxygen transfer. Air and culture mix in the riser, creating an environment where air constitutes up to 50% of the volume, gradually decreasing as the riser ascends. In the upper riser, the low pressure releases unused gases. Oxygen-depleted culture is reoxygenated in the downcomer to prevent cell death. The cycle ends as the reoxygenated culture returns to the riser (Nevalainen, 2020).

Surface Culture Method (SCM)

Like the SSF and SmF, mycoprotein can also be cultivated using the SCM. However, establishing this system on a bigger scale is expensive and not economical, like SSF and SmF. In a study by Prakash et al., they made *Fusarium venenatum* grow well, producing 4.9 g of biomass per liter in just 3 days. Optimal mycoprotein production requires 20 g L^{-1} date juice, 4.48 g L^{-1} nitrogen, and 12.97% seed size, yielding 46 to 48% protein in dried cells. Heat treatment at 65°C for 30 min reduces single-stranded nucleic acid levels for safe food-grade use (Prakash, 2015). This method involves cultivating fungal strains on solid substrates, such as pea processing byproducts (PpB) (Souza,

2018). Controlled environmental conditions facilitate fungal growth and mycoprotein production. In this process (Figure 4), inoculation is performed using the spreading method, with fungal biomass growing on the surface of the matrix. Subsequently, filtration is conducted, followed by centrifugation to isolate the mycoprotein biomass.

Nutritional Status

Due to its relatively high protein value, increased fiber and reduced saturated fatty acid content, mycoprotein is essential for a healthy diet. Mycoprotein may provide several nutritional advantages, including improving satiety and controlling blood sugar and cholesterol levels, according to experimental studies (Ahmad, 2022). Mycoprotein must be of high protein quality if it is to be used in foods that are the main course of a meal. All of the necessary amino acids are present in mycoprotein. The ideal protein bioavailability -assessed amino acid level is 0.996, indicating that the protein was produced using the best ileostomy techniques and is of high quality (Edwards, 2010). Mycoprotein contains 6 g of fiber per 100 g, which classifies it as "rich in fiber" according to European Commission requirements (Derbyshire, 2019). The natural dietary fiber in mycoprotein consists of 12% soluble fiber and 88% insoluble fiber, with a small amount of chitin and a predominant presence of glucan, forming a "fibrous chitin–

glucan matrix" in the small intestine. Both linear and branched α-glucans from grains and yeast have been identified to actively participate in physiological processes related to fat metabolization and boost the immune system. Mycoprotein's energy composition is approximately one-third fat, as it contains minimal readily available carbohydrates. It primarily consists of less than 1.5 g of saturated long- and short-chain fatty acids per 100 g of solid. Additionally, this fungal protein is rich in mono- and polyunsaturated fatty acids (Finnigan, 2017). Mycoprotein contains several water-soluble B vitamins, including pyridoxine (0.1 mg), folate (114 μg), and cobalamin (0.72 μg) (Saeed, 2023). According to nutrient claims made by the European Commission, fungal protein is a potent source of various mineral components, including zinc, phosphorus, calcium, iron, potassium, and many more (Saeed, 2023).

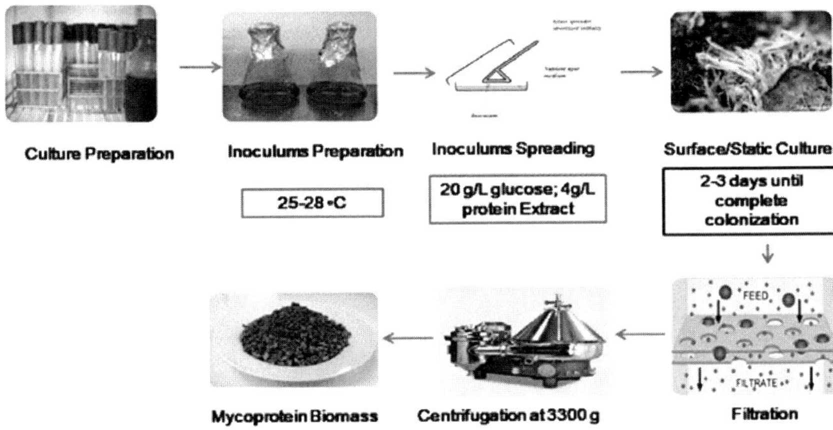

Figure 4. Process of mycoprotein preparation using the surface culture technique.

Safety Assessment

As approved by the US Food and Drug Administration in 2001, mycoprotein is considered safe (GRAS) for food consumption, excluding poultry and meat products. Despite mycoprotein being a potential allergen, the company named Quorn® (largest producer), Marlow Foods, UK is selling fungi-based meat alternatives operating in 20 different countries. Mycoprotein foods aren't advised for children under three because of the high energy demands for quickly growing newborns, relatively low energy value, and increased fiber content (Derbyshire, 2019).

Future Prospects

Compared to animal-based proteins, mycoprotein requires fewer resources, such as water, and generates lower greenhouse gas emissions, including carbon dioxide and nitrogen dioxide, than conventional animal protein production (Warner, 2019). Plant-based foods are generally less harmful to the environment. As a result, although certain principles are not universally adopted across all countries, they are gaining popularity and are increasingly being integrated into the Food-Based Dietary Guidelines (FBDG) (Stipanuk, 2018). Vegan diets have been linked to lower total cholesterol levels compared to meat, fish, and vegetarian diets, potentially due to dietary modifications

(Tong, 2019). There are still some misconceptions, however, with some nutritionists mistaking fungal protein for "plant-based" protein, which is inaccurate as this is mainly obtained from fungal sources (Van Loo, 2020). Fungal biotechnological approaches can also transform potential organic substances into nutrient-dense food proteins, which could help address some of the most critical global challenges (Meyer, 2020). The need for further research into how biodiversity can be harnessed as a source of sustainable and healthful protein has never been greater, despite the long history of humans utilizing biodiversity. The importance of protein diversity in modern diets is underscored by the rapid emergence of novel protein sources, primarily from plants and fungi, despite the long-standing dominance of wheat and soy proteins in the protein industry (Derbyshire, 2021). Given these factors, it is time to consider producing dietary protein from fungi, which contain mycelium mycoprotein and have been shown to positively impact both consumer health and the environment (Cherta-Murillo, 2022).

Conclusion

Replacing meat with meat substitutes might positively affect both societal and personal arenas. A unique alternative protein source called mycoprotein has a high level of commercial acceptance. Eating it might be great for people who don't eat meat. Due to its low fat, calories, fibers, and protein levels, it's a fantastic resource of dietary peptides and proteins. Numerous studies have shown that ingesting mycoprotein has various positive health effects, such as maintaining muscle, lowering low-density lipoprotein cholesterol, and controlling blood glucose and insulin levels. The high cost of protein sources limits their commercial viability in developing economies, and the raw materials required for their production often have a greater ecological impact compared to plant-based alternatives. To address these challenges, it is essential to prioritize innovative production techniques, such as submerged and solid-state fermentation, as well as the use of agricultural and food waste as substrates for protein production.

References

Ahmad, M. I., S. Farooq, Y. Alhamoud, C. Li and H. Zhang, A Review on Mycoprotein: History, Nutritional Composition, Production Methods, and Health Benefits, Trends *Food Sci. Technol.*, 2022, 121, 14–29, doi: 10.1016/j.tifs.2022.01.027.

Asgar, M., A. Fazilah, N. Huda, R. Bhat and A. Karim, Nonmeat Protein Alternatives as Meat Extenders and Meat Analogues, *Compr. Rev. Food Sci. Food Saf.*, 2010, 9(5), 513-529, doi: 10.1111/j.1541-4337.2010.00124.x.

Cebrian M., and J. Ibarruri, Filamentous fungi processing by solid-state fermentation, in *Current Developments in Biotechnology and Bioengineering*, Elsevier, 2023, pp. 251–292.

Cherta-Murillo, A., and G. Frost, The association of mycoprotein-based food consumption with diet quality, energy intake and non-communicable diseases' risk in the UK adult population using the National Diet and Nutrition Survey (NDNS) years 2008/2009–2016/2017: a cross- sectional study, *Br. J. Nutr.*, 2022, 127(11), 1685–1694, doi: 10.1017/S000711452100218X.

Coelho, M. O., A. J. Monteyne, M. V. Dunlop, H. C. Harris, D. J. Morrison, F. B. Stephens and B. T. Wall, Mycoprotein as a possible alternative source of dietary protein to support muscle and metabolic health, *Nutr. Rev.*, 2020, 78(6), 486–497, doi: 10. 1093/nutrit/nuz077.

Coelho, M., A. Monteyne, M. Dirks, T. Finnigan, F. Stephens and B. Wall, Daily mycoprotein consumption for 1 week does not affect insulin sensitivity or glycaemic control but modulates the plasma lipidome in healthy adults: a randomised controlled trial, *Br. J. Nutr.*, 2021, 125(2), 147–160, doi: 10.1017/S0007114520002524.

Dudekula, U. T., K. Doriya and S. K. Devarai, A critical review on submerged production of mushroom and their bioactive metabolites, *3 Biotech*, 2020, 10, 1–12.

Dunlop, M., S. Kilroe, J. Bowtell, T. Finnigan, D. Salmon and B. Wall, Mycoprotein represents a bioavailable and insulinotropic non-animal-derived dietary protein source: a dose–response study, *Br. J. Nutr.*, 2017, 118(9), 673–685, doi: 10.1017/S0007114517002409.

Finnigan, T., L. Needham and C. Abbott, Mycoprotein : a healthy new protein with a low environmental impact, in Sustainable protein sources, *Academic Press,* 2017, pp. 305–325.

Javourez, U., E. A. Rosero Delgado and L. Hamelin, Upgrading agrifood co-products via solid fermentation yields environmental benefits under specific conditions only, *Nat. Food,* 2022, 3(11), 911–920.

Meyer, V., E. Y. Basenko, J. P. Benz, G. H. Braus, M. X. Caddick, M. Csukai, R. P. De Vries, D. Endy, J. C. Frisvad and N. Gunde-Cimerman, Growing a Circular Economy with Fungal Biotechnology: A White Paper, *Fungal Biol. Biotechnol.,* 2020, 7(1), 1–23.

Okeudo-Cogan, M. C., B. S. Murray, R. Ettelaie, S. D. Connell, S. J. Radford, S. Micklethwaite and A. Sarkar, Understanding the microstructure of a functional meat analogue: demystifying interactions between fungal hyphae and egg white protein, *Food Hydrocolloids,* 2023, 140, 108606.

Schweiggert-Weisz, U., P. Eisner, S. Bader-Mittermaier and R. Osen, Food proteins from plants and fungi, *Curr. Opin. Food Sci.*, 2020, 32, 156–162.

Smetana, S., D. Ristic, D. Pleissner, H. L. Tuomisto, O. Parniakov and V. Heinz, Meat substitutes: resource demands and environmental footprints, *Resour., Conserv. Recycl.*, 2023, 190, 106831.

Thomas, A. B., T. D. Shetane and R. G. Singha, Employing Central Composite Design for Evaluation of Biomass Production by Fusarium venenatum: *In Vivo* Antioxidant and Antihyperlipidemic Properties, *Appl. Biochem. Biotechnol.*, 2017, 183, 91–109, doi: 10.1007/s12010-017-2432-5.

Tong, T. Y., P. N. Appleby, K. E. Bradbury, A. Perez-Cornago, R. C. Travis, R. Clarke and T. J. Key, Risks of Ischaemic Heart Disease and Stroke in Meat Eaters, Fish Eaters, and Vegetarians over 18 Years of Follow-up: Results from the Prospective EPIC-Oxford Study, *BMJ,* 2019, 366, doi: 10.1136/bmj.l4897.

Umesh, M., S. Suresh, A. S. Santosh, S. Prasad, A. Chinnathambi, S. Al Obaid, G. K. Jhanani and S. Shanmugam, Valorization of pineapple peel waste for fungal pigment production using Talaromyces albobiverticillius: insights into antibacterial, antioxidant and textile dyeing properties, *Environ. Res.,* 2023, 229, 115973.

Zhao, X., J. Chai, F. Wang and Y. Jia, Optimization of Submerged Culture Parameters of the Aphid Pathogenic Fungus Fusarium equiseti Based on Sporulation and Mycelial Biomass, *Microorganisms,* 2023, 11(1), 190.

Part 3. Plant Based Meat

Chapter 1

Promising and Sustainable Meat Analogues

Abstract

The rapid increase in the global population and limited natural resources have led to a scarcity of high biological value protein. While meat is an excellent source of protein, converting plant-based protein into animal protein is not economically viable. This has driven the trend of producing healthy, meat-free foods to cater to vegetarians and promote personal well-being. Consequently, low-cost plant proteins such as textured soy protein, mushrooms, wheat gluten, and pulses are increasingly used as substitutes for animal protein. These meat-like products, which mimic the texture, flavor, color, and nutritional value of meat, can be directly substituted for animal-based products in various diets.

Keywords: meat analogues, sustainable protein, plant-based alternatives, texturization technology

Introduction

Meat has a distinct position in the food basket and has been widely consumed by the humans since pre-historic era due to readily source of energy, high quality proteins, palatability as well as images of strength and power (Fiddes, 1991; Latvala et al., 2012). The per capita consumption of meat has more than doubled over 1961–2007 and recorded rapid growth in developing countries in comparison to developed countries.

It is expected that factors such as population growth, urbanization, industrialization, increased education, and rising incomes will lead to a 72% increase in meat consumption by 2030 (Steinfeld et al., 2006; Faisla, 2008). Steinfeld et al., (2006) and Bruinsma (2009) projected that the production of animal products will double, from 229 billion kilograms for a global population of 6.0 billion in 2000 to 465 billion kilograms for a population of 9.1 billion by the year 2050. Meat production has increased 5–13% in last decade and reached close to its maximum production (Post, 2012). Recently it has been noticed that ethical and environmental concerns of meat production and consumption has been in focus (McEachern and Schroder, 2002; European Commission, 2005; Mayfield et al., 2007).

Food security is a growing concern due to the increasing global population, which has placed immense pressure on natural resources such as agricultural land and water used for growing food crops. A significant portion of the world's population faces starvation, highlighting the urgent need for increased production of staple crops (Datar & Betti, 2010). The demand for food is expected to rise by 70% by 2050, driven by the addition of 2.3 billion people to the global population (Bruinsma, 2009). Currently, food crop production accounts for 20% of global electricity consumption and 70% of freshwater usage (Green et al., 1999).

For production of meat, food animals are reared and slaughtered. The rearing of food animals for meat production is basically a method of converting vegetable protein into meat proteins through food animals. The rearing of food animals for meat needs natural resources inputs in the form of water, feed, fodder etc. The per-unit production of vegetable crops requires comparatively fewer natural inputs, such as water, land, and energy, than the production of meat. Thus, meat production is considered an inefficient utilization of plant protein into animal protein, consuming significant environmental resources like land, water, and energy (Steinfeld et al., 2006; Raney et al., 2009; Bruinsma, 2009). According to the Stockholm International Water Institute, producing 1 kg of grain-fed beef requires 5–40 times more water than producing 1 kg of cereal grains (SIWI-IWMI, 2004). One estimate suggests that producing 1 kg of beef consumes two lakh liters of water (Brown, 2009; Ye and Van Rast, 2009). Pimentel and Pimentel (2003) estimated that meat production consumes 100 times more water compared to the production of food crops. In terms of human diet, it was estimated that producing meaty diets for one person per year requires two million liters of water, which is double the amount needed for a vegetarian diet per person per year (Smil, 2000).

Meat production is also responsible for increase in the greenhouse gas (GHG) emission. The contribution of livestock to the emission of three major greenhouse gases carbon dioxide, methane and nitrous oxide, is 9%, 39% and 65%, respectively (FAO, 2006). Out of total grain and soya produced, 40% of total grains and 75% of total soya is utilized for animal feeding. Out of total area of planet, 30% is devoted for animal production which counts about 70% of the total arable land.

Furthermore, the rapid expansion of livestock farming has placed increasing pressure on available natural resources and is considered a significant driver of deforestation (Smil, 2000; Pimentel and Pimentel, 2003; Steinfeld et al., 2006; Aiking, 2011). Given the current environmental concerns and the impact on animal welfare, achieving food security targets through traditional livestock farming may be challenging. However, the nutritional needs could be more easily met through alternative sources, such as plant-based proteins (Bittman, 2008).

Post (2012) identified three major concerns related to meat production: environmental, animal welfare, and public health issues. Environmental concerns include pollution, deforestation, and the depletion of natural resources. Animal welfare issues involve cruelty and unethical treatment of animals during rearing, transportation, and slaughter. Public health concerns highlight the impact of overconsumption of meat, which is responsible for a quarter of all ischemic heart disease cases, contributing to over 1.8 million deaths annually (Key et al., 1999). Larsson and Wolk (2006) suggest that consuming 120 grams of red meat per day or 30 grams of processed meat per day significantly increases the relative risk of colorectal cancer. Additionally, foodborne pathogens found in meats, such as *Salmonella*, *Campylobacter*, and *E. coli*, are responsible for millions of illness episodes each year (CDC, 2012). Consequently, the inefficient use of valuable natural resources in meat production, especially when it competes with staple grain production, raises critical ethical and environmental concerns (McMichael et al., 2007; Hoek et al., 2011). The huge environmental and socio-economic issues, such as deforestation, food security, environmental pollution, etc. have created awareness among the consumers for sustainable methods of food productions (Aiking and de Boer, 2006). It has sparked significant interest in the possibility of creating protein-rich food from plant sources in a palatable form, shifting the mindset of policymakers and scientists toward the development of suitable alternatives to simulated meat products, with controlled texture, flavor, color, and nutritional value. These alternative products are commonly referred to as meat alternatives, meat substitutes, meat analogues, or plant-based meat. These products are in infant stage of development and at present accounts nearly 1–2% of total meat market (De Bakker and Dagevos, 2010). However, due to intrinsic cheapness of proteins from vegetable origin and nutritional factors, the use of such products is bound to increase as substitutes for expensive proteins of animal origin.

Meat Analogue

A meat analogue is a food product designed to replicate the aesthetic qualities and/or chemical characteristics of certain types of meat. These products are made from non-animal proteins and closely resemble meat in both appearance and smell. According to U.S. legislation, a food is considered a "substitute" if it is nutritionally equivalent to meat, and it is classified as nutritionally inferior if it does not meet these criteria. Meat analogues are appealing to the processing industry due to their cost advantages, stable pricing (as they are less affected by seasonal supply fluctuations), longer shelf life, and easier storage. While the term "meat analogue" is primarily associated with products made from spun protein filaments, it also encompasses a range of other products, such as textured vegetable proteins (TVPs). Meat analogues are the one of the best suitable methods in which vegetable proteins and unconventional proteins can be introduced and the underlying impetus remains the utilization of a wider range of proteins for human food (Tombs, 1974).

The evolution of protein products of non- animal origin could be of three main types, viz. meat extender, meat analogues and completely new protein products not associated with meat. Spicer (1971) reported that the use of low-cost meat analogues in both developed and developing countries evolved due to the large number of relatively poor people and the competition for food within family budgets. Since true vegetarians constitute only 1–2% of the United States population, other factors must be influencing the growing interest in meat analogues in the U.S. The reasons frequently cited were health and nutritional benefits, environmental considerations, religious beliefs, and animal rights (Food and Nutritional Science (FNS) Programme, 2000).

Key Ingredients

The vegetarian diet is generally rich in fibers, fruits and vegetables. These ingredients are known to possess health benefits. Meat analogues can be obtained from an ever-increasing number of protein sources. Vegetable proteins are currently the main source of material for meat analogues, such as glutens of wheat and globulins of groundnut, cottonseed, peanut, sesame, yeast and soybean. The main ingredients used in the preparation of meat analogues include soy protein, pulses, nuts, cereal proteins, vegetables, and mycoproteins. However, the expanding consumer choices have led to the inclusion of newer ingredients in meat analogues. Currently, the majority of production relies on soybeans, which remain one of the most affordable and widely used sources of protein raw materials. Meat analogues can be formulated to have proteins of high biological value, offering a way to enrich a monotonous diet that relies on plant-based proteins. Soy and other oilseed extracts, as well as proteins produced through fermentation using various substrates and microorganisms, are primarily used in the production of meat analogues.

Though, vegetable proteins are the main source of material for meat analogues but algae, yeast, mushroom, and bacteria would probably supplement them in near future because they multiply rapidly without land and adjust very well to consumer taste (Hegenbart, 2002). Meat analogues can also be produced by hydrating textured vegetable protein (TVP), preparing a protein base consisting of vital wheat gluten, milk protein, and water, and then combining the hydrated vegetable protein with the protein base (Howse et al., 2005). Another popular category of mock-meat products that have gained traction in recent years are those made from konjac fiber, a non-digestive dietary fiber commonly found in human diets. These products are increasingly chosen by consumers for both taste and health reasons. Howsam (1976) reported that the industrial production of meat analogues from protein-rich precursor materials - such as plant-based ingredients like cereal grain gluten (from wheat, rice, or maize, either vital or with residual starch), defatted oil seeds, and various flours, meals, and derivatives (e.g., defatted soy flour, soy protein concentrates, wheat flour) - as well as animal-based ingredients like meat by-products obtained through mechanical separation, fish meal, dried egg white, and others, either alone or in combination, should be routinely considered in manufacturing processes.

The key ingredients to be used during preparation of meat analogues are as follows:

- Soya protein
- Mushroom
- Wheat gluten
- Egg albumen
- Carbohydrates and gum
- Flavoring and other miscellaneous compounds.

Soy Protein

Soya is known to have high-quality nutritional and functional benefits. It is widely used for partial or complete replacement of meat due to its comparative nutrient contents and lesser chances of cardiovascular diseases. Furthermore, the development of soy as a meat alternative has historically been characterized by its role in both extending and substituting meat products. Beloglavec (1974) noted that meat-like products could be derived from an expanding range of protein sources, with soybeans being the most common and cost-effective option. On the Protein Digestibility Corrected Amino Acid Score (PDCAAS) scale, soy protein has been rated equivalent to animal protein, achieving a perfect score of 1.0—the highest possible rating (Hoffman & Falvo, 2004). This makes soy protein an appealing alternative for individuals seeking non-animal protein sources or those with lactose intolerance. Soy protein isolates possess a biological value comparable to that of meat, further reinforcing their nutritional adequacy. In addition to protein, soybeans are also rich in calcium and linoleic acid. They are a notable source of alpha-linolenic acid, an essential omega-3 fatty acid, contributing additional health benefits. A consistent consumption of soya foods also confers certain health benefits. Thus, soya foods that provide a good source of nutrition are especially vital for vegetarians, and provide a good substitute for meat products (Wardlaw and Kessel, 2002).

Harland (2002) analyzed market trends for soy-based food products - including meat replacers and dairy alternatives - in the UK and found that recent growth in this sector is largely driven by health-conscious consumers. Soy is particularly valuable in the manufacture of emulsion-type products due to its ability to emulsify, stabilize, texture, and hydrate. Textured vegetable protein (TVP), such as soy-based meat analogues, serves as a highly digestible and protein-rich alternative. In adults, the protein digestibility and nitrogen balance were reported at 66.1% and 1.16 g N/day, respectively, for TVP made from defatted soy flour, and 63.4% and 1.31 g N/day for TVP derived from soy protein concentrate. In comparison, beef demonstrated a higher protein digestibility of 73.2% but a lower nitrogen balance of 0.42 g N/day (Cheftel, 1986).

Soya protein isolates are widely used for preparation of soya-based meat alternatives along with wheat gluten and other vegetable proteins (Morr and Ha, 1991). The nutritional quality of products prepared from soy protein isolates is inferior to that of products made from minimally processed soy flours (Alfred et al., 2004). Soy-based meat alternatives are primarily prepared using the extrusion process, which results in a fibrous texture due to disulfide bonding, closely resembling the texture of meat (Cumming et al, 1973). Meat analogues can be produced at either low moisture conditions, fewer than 35% by use of a single-screw extruder, or at high moisture conditions, above 50% by use of a twin-screw extruder (Lin et al, 2000; Lin et al, 2002).

Huffman and Powell (1970) reported soya proteins as widely used vegetable protein in the formulation of food products. It is used in various physical forms, i.e., soya flour, soyabits, isolated soya protein, spray dried soya milk and texturized soya proteins. Ishizuka and Akoi (1986) patented simulated ground meat analogue from tofu, by adjusting the pH of a soya milk to 5.6–6.2, coagulating the soya milk to from tofu, from pressing and dehydrating the tofu to obtain a solid content range of 35–50%, shaping the pressed tofu into a form of similar to ground meat and heating the product. Williamson et al., (2005) compared tofu preloads and mycoprotein with a chicken preload and found a reduction in food intake shortly after consuming the preload at lunch. Food intake following the consumption of mycoprotein and tofu did not differ, and participants did not compensate for the reduced food intake at lunch by eating more at dinner. These findings suggested that both tofu and mycoprotein had satiating properties that persisted for several hours after a meal. These findings had significant implications for the development of foods that are low in kilojoules, but are also filling.

Mushroom

The natural protein of mushroom is fibrous in nature and its incorporation provides chewability of the products (Naylor et al., 2001). The cell walls of hyphae are rich in dietary fibers whereas polyunsaturated fatty acids are abundant in cell mem- branes. The cytoplasm is rich in high quality proteins. Mycoprotein has been shown to reduce harmful low-density lipoproteins (LDLs) while increasing beneficial high-density lipoproteins (HDLs) (Rodgers, 2001). Additionally, mushrooms can contribute significantly to meeting daily nutritional requirements by providing essential proteins, minerals, and vitamins.

The first commercial meat analogues such as burger patties and sausages with mycelia were made from the edible filamentous fungi *Fusarium graminearu*. The meaty flavor of these products attributed to the presence of sulfur containing amino acids and glutamic acid in fungi (Trincy, 1994). The taste of mushroom derived products is better than other plant derived products.

The meat alternatives are prepared from mycoprotein by mixing the fermented mushroom with egg and other seasonings/flavorings. These are widely used in European countries as alternative to beef and chicken. However, the wide spread public perception of the *Fusarium* as pathogenic and not a true mold has forced the researchers and food technologist to search a better alternative.

The preference of fungal protein over bacterial and vegetable proteins can be summarized as follows:

a. Mushroom flavor of fungal protein is more acceptable as compared to beany flavor of textured soy protein.
b. Fungal protein gave convincing meat, poultry and fish analogues by alignment of the protein filaments with fibrous or flaky texture (Anon, 1981).
c. The most of mycoproteins are tasteless, colorless and unpalatable with texture similar to meat fiber but with added flavoring and coloring it could be made into a passable imitation of fish, chicken, veal or ham.
d. Health benefits and antitumor effect of mushroom Mycoprotein was the first fungal food to be produced for human consumption marketed on a limited scale in UK.

Newmark (1980) described the process of production of processed mycelia of A35 strain of *Fusarium graminearum*. Uzunov and Colova (1972) described the organoleptic and keeping quality of mushroom as raw material for sausage. They also noted the large quantities of proteins which could replace part of the raw meat and easily mixed with meat to form a homogeneous finely chopped material. The developed product is having the light grey color and specific flavor imparted by the mushrooms. Chevrolet (1988) patented a composition for making meat-like products using edible mushrooms, a gelling agent (polysaccharides undergoing gelation that is not reversible by heating), a texturing agent (proteins), and flavorings. Kumar et al., (2011a,b) reported that replacing 22.5% of texturized soy protein with mushrooms enhanced the sensory attributes of meat analogue nuggets, improving both flavor and overall acceptability.

Wheat Gluten

Wheat gluten is a cohesive, visco-elastic complex of enigmatic complex proteins prepared as byproduct during isolation of starch from wheat flour. It is widely used in meat analogues preparation due to its binding, dough forming and leavening ability (Rizvi et al., 1980; Dayab et al., 2006). The dough forming properties of wheat flour is due to wheat gluten (Day et al., 2005). Gluten is finding its way to inclusion in meat analogues due to its functional properties as solubility, viscosity, swelling, nutritional quality), possibility of scale modifications and modest price than soya, caseinate and mushroom proteins (Sarkki, 1979; Dayab et al., 2006). It is rich in glutamine, a muscle-building amino acid, but deficient in lysine and threonine. Gluten (vital, de-vital, or modified) has seen increased use as a food ingredient, offering a range of functional properties at a more affordable price compared to competitors such as milk and soy proteins. The past five decades had seen the rise of gluten as a commodity. The large-scale industrial separation of wheat starch from gluten and the controlled drying of the gluten retained its functional properties.

Olavarria (1981) patented protein binder compositions for texturized proteins consisting of 10-20% gluten, 10-20% whey proteins, 1-5% albumin (dried egg white), and sufficient liquid (preferably water, or milk). Nguyen (1988) patented a meat analogue resembling chicken breast made from wheat gluten, wheat, or soy flour. A dry powder blend of 80% wheat gluten, 11% wheat flour, and 9% vegetable oil was mixed, and water (50% by weight) was sprayed on the mixture. The dough was then prepared and heated using conventional or microwave methods. The resulting products lacked the unpleasant flavor typically found in soy-based meat analogues. Kumar et al., (2009) and Kumar et al., (2012) found that the incorporation of wheat gluten in analogue meat nuggets improved nearly all sensory traits. Sarkki (1979) reported that the addition of wheat gluten enhanced both the functional and baking properties of analogue meat products. Adding 10–20% wheat gluten to meat substitutes has also improved the flavor and color scores of the product (Olavarria, 1981).

Egg Albumen

Egg white (albumen/glair/glaire) is the common name for clear liquid contained within an egg. The incorporation of egg albumen into meat analogues enhances both binding properties and textural qualities, contributing to a more desirable bite during consumption. It also increases the overall protein content of analogue meat nuggets. Ovalbumin, a globular monomeric phosphoglycoprotein containing free sulfhydryl groups, possesses the ability to form heat-induced gels - one of the most significant functional attributes of egg albumen (Eleya et al., 2004; Jerez et al., 2007). Upon heating, globular protein molecules partially unfold into a molten globule state, which increases viscosity compared to their compact, fully folded counterparts of similar molecular weight. This structural transformation enhances the binding properties of globular proteins within the meat analogue matrix (Damodaran and Paraf, 1997).

The incorporation of egg albumen had a marked effect on the phsicochemical properties of analogue meat products but fat levels were least influenced by the variation in incorporation of egg albumen in the products (Kumar et. al., 2010). They also observed a variable but notable impact of egg albumen incorporation on the sensory characteristics of analogue meat nuggets. Jao et al., (1980) found that the effects of egg white content and cooking time were exponentially related to product hardness, while salt content showed a linear relationship with the logarithm of hardness. Response surface analysis indicated that the use of egg white could be significantly reduced by adjusting salt concentration and cooking time. This study proposed a model for optimizing the production process of meat loaf analogues. Rizvi (1977) reported that increases in temperature and pH led to a darker background color and a more gel-like appearance. The presence of brown particles and their quantity increased with higher gluten levels, while particle size decreased with greater egg albumen content. Hardness, gumminess, and chewiness were found to be exponentially correlated with egg albumen content, pH, and their interaction. Elasticity improved with increasing gluten levels, whereas cohesiveness was primarily influenced by the amount of egg albumen.

Carbohydrates and Gum

Proteins have technological limitations such as resistance to heat treatments, compatibility with the other constituents as flavor components, etc., restricting their use in meat analogues (Lucca and Tepper, 1994; Linden and Lorient, 1999) whereas results obtained from carbohydrates are more satisfactory (Akoh, 1998). Carbohydrate gums are commonly used by the food industry as texture modifying agents in many different types of products. Starch, the food reserve polysaccharide of plants is a commonly used food hydrocolloid. Starches and maltodextrins are polymers of glucose which on hydration form gel and mimic fat like texture. Carbohydrates such as starches, gums, hydrocolloids, maltodextrin, dextrin, polydextrose, pectin, inulin, hemicelluloses, and cellulose gels are commonly used as fat substitutes in low-fat food products (Shand et al., 1990; Keeton, 1994). Starches are particularly favored not only for their functional properties but also for being cost-effective alternatives (Yang et al., 1995; Jiménez-Colmenero, 1996). Glicksman (1976) noted that hydrocolloids - or gums - used at concentrations of 1% or less can impart a wide range of desirable characteristics to food products. Pearson and Gillett (1997) described the softening of products upon the addition of sucrose, which counteracts the harsh hardening effect of salt by preventing moisture loss. This process also enhances the product's shelf life by inhibiting bacterial growth.

Flavorings and Miscellaneous Ingredients

Flavor is an important sensory attribute which determine the acceptability of the products. A meat analogue should have meaty flavor. Lot of research work is going on to develop artificial meaty flavors. Hasida (1974) reported that for flavor potentiation of meat analogues vegetable protein products (acceptability of meat analogues, undesirable soy protein odor), chemical condiments (flavor effect, glutamic acid and 5-ribonucleotides in food stuff, application for food processing) were used. Hsieh et al., (1980) developed a synthetic meat flavor by surface response methodology, consisted of an autoclaved mixture of simple sugars, amino acids, 5-nucleotides, glycoprotein, monosodium glutamate and salt with fat an optional component. Sulphur containing amino acids and simple sugars played an important role in flavor development, whereas the other components either masked the harsh sulphury taste or enhanced the meaty flavor. Anon (1981) made a mushroom concentrate, produced from fresh mushrooms, gave an intense mushroom flavor to formulated foods at levels of 1–3% and at less than 1%, the concentrate acted as a flavor enhancer, and successfully replaced replace monosodium glutamate or hydrolyzed vegetable protein. Spencer et al., (1982) patented a flavoring composition of 2-(alpha-mercaptoalkyl)-3-thiazoline, which imparts a meaty, sautéed onion flavor and aroma at concentrations of 0.5–50 ppm in food. Kibler et al., (1988) reported the production of high concentrations of 1-octen-3-ol, a major volatile flavoring component found in many mushroom species, by heating homogenized mushrooms in an aqueous medium containing a water-soluble salt of linoleic acid and oxygen. Chen and Soucie (1985), in their patented work, developed edible whey protein/xanthan gum fibers by precipitating whey protein (at least 20%, preferably >30%) from a solution by adjusting the pH to the appropriate isoelectric point and separating the fibers. The solution may also include other solubilized proteins, such as egg white, soy protein, casein, and sodium caseinate. The fibers can be stabilized by heating at or above 70°C and flavored to create meat analogues and seafood products.

Caseinate are present in milk in the form of a1, a2, b, and k- casein. Commercially, caseinate are produced in numerous compounds, viz., sodium, calcium, potassium, and magnesium caseinate for food application (Giese, 1994). Sodium and potassium caseinate are more soluble and possess better functional properties than calcium caseinate (Kinsella, 1984). Jordan (1991) reported that the functional properties of caseinate are attributed to its unique distribution of electric charges along the polymeric molecule, the presence of hydrogen bonding, and its abundance of both hydrophilic and lipophilic binding sites. Singh et al., (1997) further demonstrated that caseinates enhance functional qualities when combined with vegetable proteins in comminuted meat products. Specifically, the emulsion stability of batters containing both caseinate and refined wheat flour was significantly higher than that of batters lacking these components. Additionally, Chen and Soucie (1985) patented a method for forming a serum milk protein complex by adding xanthan gum - at a gum-to-protein ratio ranging from 1:2 to 1:10 - to whey, whey protein concentrates, or whey ultrafiltrate. The mixture was then acidified to the isoelectric point of the complex to form a fibrous structure, which could be isolated and utilized as a meat or fish analogue. Flavorings can be incorporated into the final products by adding them to the initial whey/xanthan gum mixture. Alternatively, proteins such as soybean isolates can be included in the whey/gum mixture. Yadav et al., (1986) described a method for producing vegetarian sausages from milk, which involves coagulating 1%-fat milk at 70°C with citric acid, pressing the coagulant, blending it with other ingredients, and filling it into a casing. The sausages are then cooked in pressurized steam, cooled, and stored at 6–8°C. These sausages contain approximately 5% fat, 20% protein, and 45% total solids, and they exhibit good microbiological quality.

Carrageenan is sulfated high molecular weight polysaccharides extracted from red seaweeds (Rhodophyta). These are approved for use in food products as binder and extender in United States (9CFR 318.7; 9CFR 319.15) (USDA,

1973). Carrageenan is of three types iota, kappa, and lambda. Iota and kappa carrageenan act as gelling agents. The lambda type is non gelling and function as a thickner. Pannin et al., (1974) used carrageenan as a substitute for gelatin at concentrations less than 0.3%, whereas higher concentrations resulted in undesirable off-flavors in the product. Carrageenan was also successfully used to improve the rheological properties and water-binding capacity of low-fat emulsions. Foegeding and Ramsey (1987) reported that kappa carrageenan, at concentrations of 0.5% or 1%, was most effective among the varieties in increasing hardness. Huffman et al., (1992) documented that iota carrageenan gels strongly with calcium ions to form a clear, elastic, syneresis-free gel that resets after shear. Additionally, iota carrageenan has a strong ability to retain moisture.

Major Constraints and Future Challenges

At present, the major constraints in the development of meat analogues are the organoleptic quality especially texture and taste of meat analogues available in the market is lower than the meat (McIlveen and Abraham, 1999; Sadler, 2004). The primary challenge for researchers lies in developing novel protein foods or meat analogues that closely replicate the sensory attributes of conventional meat (Kumar et al., 2012). However, the success of meat alternatives has been questioned by Hoeck et al., (2011), who argue that strong taxonomic associations with meat - as a product inherently derived from animals - pose a barrier to consumer acceptance. They suggest that consumer interest may be better captured by positioning meat alternatives as non-meat items that closely resemble meat in appearance, function similarly in meal preparation, and exhibit minimal sensory differences from traditional meat. Additionally, the price competitiveness of meat analogues remains a significant obstacle to their widespread adoption.

Modern meat replacers are generally prepared by combining globular plant proteins with egg proteins, later help the retention of globular proteins at the time of frying. These products exhibit improved conversion efficiency compared to conventional meat (Aiking, 2011). The World Bank has also endorsed the development of dairy and meat analogues as a strategy for promoting sustainable development (Goodland & Anhang, 2009). However, meat analogues - particularly certain soy-based variants - tend to contain higher levels of dietary purines than traditional meat. Consequently, individuals with hyperuricemia or those at risk of elevated serum urate levels are advised to limit their consumption of these products (Vojir and Petuely, 1982). This problem is taken care of by introduction of new base materials, isolated and concentrated proteins having low dietary purines content (Havlik et al., 2010).

Elzerman et al., (2011), described the main hindrance of the success of meat analogue is the recognition of meat analogues as the product that is being eaten instead of meat by consumers and thus appearance and other sensory attributes as taste, flavor, tenderness of meat substitutes should be similar to meat steak, etc. The success of meat analogues depends on their incorporation into meals and recognition by consumers. Meat analogues produced from vegetable sources, through the incorporation of texturing agents and shearing processes to create a fibrous texture, are often reported to have an elastic, rubbery, and hard texture, resulting in poor mouthfeel. These products may also possess off-flavors originating from plant sources and could cause allergic reactions (Kim et al., 2011).

Conclusion

Plant proteins unlike meat protein, besides providing a good source for protein are expected to reduce the intake of saturated fat and cholesterol. It also provides numerous other nutrients such as phytochemicals and fibers that are also considered desirable in the diet. Plant proteins also contain several bioactive components, such as protease inhibitors, phytosterols, saponins, and isoflavones. These compounds have been shown to exert beneficial physiological effects, including lipid-lowering properties, enhanced oxidation of LDL cholesterol, and a reduction in blood pressure. Additionally, vegetarian foods are generally more easily digested than non-vegetarian foods. Over the past few decades, there has been a growing effort to incorporate non-meat proteins into meat products, serving as extenders, fillers, or binders. Common examples include refined wheat flour, defatted soy flour, and skim milk powder. The market potential for meat analogues is particularly strong due to their affordability, health benefits for non-vegetarians, suitability for lactose-intolerant individuals, and acceptability among those adhering to religious or ethical dietary practices. Given these advantages, meat analogues are well-positioned for success, as they cater to consumers seeking nutritious and organoleptically appealing products that are entirely free of meat.

References

Aiking, H. and de Boer, J. (2006). Background, aims and scope. In: *Sustain- able Protein Production and Consumption,* pp 1–21. Aiking, H., de Boer, J. and Vereijken, J., Eds., Pigs orpeas?, Springer, Dordrecht, Netherlands.

Aking, K. (2011). Future protein supply. *Trends Food Sci. Technol.* 22: 112–120.

Akoh, C. C. (1998). Fat replacer. *Food Tech.* 52:47–53.

Allred, C. D., Allred, K. F., Ju, Y. H., Goeppinger, T. S., Doerge, D. R. and Helferich, W. G. (2004). Soy processing influences growth of estrogen- dependent breast cancer tumors in mice. *Carcinogenesis.* 25:1649–1657.

Anon. (1979). Soy, binder, flavor combination makes analog preferred by some over real diced meat. *Food Prod. Dev.* 13:42.

Anon. (1981). Mushroom concentrate acts as flavor enhancer in foods. *Food Process.* 42:22–24.

Beloglavec, D. M. (1974). Meat like products and their possible impact on the demand for meat. Monthly Bull. *Agric. Econ. Stat.* 23:1–6. www.fao.org.

Bittman, M. (2008). Rethinking the meat guzzler. In: *The New York Times-The World.* dated 27th January. pp. 1.

Borders, C. (2007). Making meat analogues from soy. *Functional ingredients.* (http:// newhope360.com/botanicals/making-meat-analogues-soy).

Hoekstra, A. Y. and Chapagain, A. K. (2007). Water footprints of nations: Water use by people as a function of their consumption pattern. *Water Resour Manag.* 21:35–48.

Hoffman, J. R. and Falvo, M. J. (2004). Protein - which is best? *J. Sports Sci. Med.* 3:118–130.

Howsam, S. (1976). Better texture for vegetable protein foods. *Food Eng. Int. J.* 2:65–78.

Howse, G. T., Sidhu, K. and Grex, D. D. (2005). Methods of preparing meat analogues, meat analogues, and foodstuffs comprising meat analogues. *Food Eng. Ingred.* 30:30–38.

Hsieh, Y. P. C., Pearson, A. M. and Magee, W. T. (1980). Development of a synthetic meat flavor mixture by using surface response methodology. *J Food Sci.* 45:1125–1135.

Huffman, D. L., Mikel, W. B., Egbert, W. R., Chen, C. and Smith, K. L. (1992). Development of lean pork sausage products. *Cereal Food World.* 37:439–442.

Lin, S., Huff, H. E. and Hsieh, F. (2002). Extrusion process parameters, sensorycharacteristics, and structural properties of a high moisture soy protein meat analog. *J. Food Sci.* 67:1066–1072.

Linden, G. and Lorient, D. (1999). Lipid chemistry-fat substitute. In: New ingredients in food processing. *Biochemistry and Agricultural Cambridge,* pp. 289–314. Wood Lead Publishing Limited.

Lucca, P. A. and Tepper, B. J. (1994). Fat replacers and the functionality of fat in foods. *Trends Food Sci. Tech.* 5:12–19.

Malav, O. P., Talukdar, S., Gokulkrishna, P. and Chand, S. (2013). Meat analogue- a review. *Crit. Rev. Food Sci. Nutr.* doi: 10.1080/10408398.2012.689381.

Mathews, K. H. J. and McConnell, M. (2011). *U. S. beef and cattle industry: Background statistics and information.* http://www.ers.usda.gov/news/BSECoverage.htm Accessed: March 27, 2012.

Mayfield, L. E., Bennett, R. M., Tranter, R. B. and Woolridge, M. J. (2007). Consumption of welfare-friendly food products in Great Britain, Italy and Sweden, an how it may be influenced by consumer attitudes to and behaviour towards, animal welfare attributes. *Int. J. Socio. Food Agric.* 15:59–73.

McEachern, M. G. and Schr€oder, M. J. A. (2002). The role of livestock pro- duction ethics inconsumer values towards meat. *J. Agric. Environ. Ethics.* 15:221–237.

McIlveen, H. and Abraham, A. C. G. (1999). Meat avoidance and the role of replacers. *Nutr. Food Sci.* 1:29–36.

McMichael, A. J., Powles, J. W., Butler, C. D. and Uauy, R. (2007). Food, livestock production, energy, climatic change, and health. *Lancet.* 370:1253–1263.

Mintel. (2001). *Meat free foods.* London, www.vegsoc.org/sslpage.aspx? pidD754.UK: November, 2001.

Morr, C. V. and Ha, E. Y. W. (1991). Off-flavors of whey protein concentrates: A literature review. *Int. Dairy J.* 1:1–11.

Nguyen, G. D. T. (1988). Meat Analogue. *European Patent Application* No: 0, 262, 272.

Rizvi, S. S., Josephson, R. V., Blaisdell, J. L. and Harper, W. J. (1970). Sepa- ration of soy-spun fiber, egg albumen and wheat gluten blend sodium dodecyl sulfate gel electrophoresis. *J. Food Sci.* 45:958–961.

Rodger, G. (2001). Production and properties of mycoprotein as a meat alternative. *Food Technol-Chicago.* 55:36–41.

Rosegrant, M., Leach, N. and Gerpacio, R. (1999). Alternative futures for world cereal and meat consumption. *Proc. Nutr. Soc.* 58:219–234.

Sadler, M. J. (2004). Meat alternatives - market developments and health benefits. *Trends Food Sci. Tech.* 15:250–260.

Chapter 2

Plant-Based Meat Alternatives

Abstract

A shift towards plant-based diets has fueled the demand for sustainable plant-based meat alternatives. This chapter discusses scientific and business aspects of their development, providing a framework to guide innovation. This chapter explores the market potential, consumer insights, scientific foundations, and patent trends related to these products. Navigating scientific literature and patent databases is essential for the informed selection of ingredients, formulations, and production processes. Key to developing appealing alternatives is the optimization of flavor, texture, binding, color, and nutritional value. The chapter also identifies research opportunities aimed at improving formulations to create consumer-acceptable products that contribute to enhanced sustainability outcomes.

Keywords: plant-based meat alternatives, formulation and processing, consumer acceptance, sustainability in food production

Introduction

The increase in global population and changing demographics are driving the huge increase in protein demand and global meat consumption is growing with rise of the middle class with more disposable incomes (Henchion et al., Citation2017). The EAT–Lancet Commission recommended that people in many countries substantially reduce their animal-source meat and dairy intake and adopt a planetary health diet comprised primarily of plant-rich fruits, vegetables, whole grains and plant-source proteins (i.e., bean, legumes and nuts) (Willett et al., Citation 2019). A shift from diets primarily based on animal proteins to plant protein-based diets has been promoted due to their favorable outcomes for public health and the environment (Aiking and de Boer, 2020), as well as concerns about animal welfare (Lusk Citation 2011). Dietary changes are necessary to reduce environmental impacts, particularly given that the production of animal-based foods accounts for approximately

83% of the world's farmland and contributes 56-58% of food-related greenhouse gas emissions, while supplying only 37% of global protein and 18% of total caloric intake (Poore & Nemecek, 2018). From a public health perspective, there is also a compelling case for shifting toward more sustainable, plant-based diets and reducing overall meat consumption (Lonnie & Johnstone, 2020). These environmental and health considerations have driven significant growth in the market for animal-free alternatives to meat, dairy, and eggs. In 2020, the U.S. retail market for plant-based meat was valued at $1.4 billion, while the global market reached $4.2 billion (Gaan Citation 2021).

There is an array of alternative protein products that industry is developing and marketing, including alternatives to beef, pork, lamb, chicken, collagen, seafood, eggs and dairy. Plant-based ingredients are used to develop various plant-based alternative meat products (i.e., meat analogues). Traditional plant-based meat alternatives included products such as tofu, tempeh, and seitan (He et al., 2020; Singh et al., 2021). The first generation of these alternatives primarily consisted of texturized vegetable proteins (TVP) combined with other ingredients (Riaz, 2011). Soy and wheat proteins were commonly used as the primary bases for these products. Over time, a wider range of protein sources has been explored, including those derived from oilseeds (e.g., canola, sunflower), cereals (e.g., rice), legumes and pulses (e.g., lupin, beans, peas, lentils) (Arrutia et al., 2020; Day & Swanson, 2013; Riaz, 2011), as well as non-animal proteins from fungi, algae, duckweed, insects, and in-vitro cultured meat (He et al., 2020; van der Spiegel, Noordam, & van der Fels-Klerx, 2013). The newer generation of plant-based meat alternatives - such as those developed by Impossible Foods™ and Beyond Meat® - continue to use texturized vegetable proteins. However, these products are formulated to more closely mimic the appearance, texture, and taste of conventional beef burgers compared to earlier versions (He et al., Citation 2020).

There have been many recent reviews on plant-based meat alternatives. An integrated conceptual framework compared nutritional consequences, potential sustainability gains of a range of plant-based and other meat alternatives and the necessary socio-technological pathways for transition to meat alternatives (van der Weele et al., Citation 2019). The authors found that there was too little attention given to plant-based meat alternatives with greater sustainability potential such as pulses (van der Weele et al., Citation 2019).

Plant proteins generally have a lower protein efficiency than animal-derived proteins and also plant-based diets are lower in micronutrients (e.g., Fe, Vit B12) (Neufingerl and Eilander Citation2021). A review examining the effects of plant-based diets on human health and climate change suggests that a plant-rich flexitarian diet is generally unlikely to negatively impact overall nutrient status. However, the nutritional adequacy of such a diet largely depends on the variety and quality of other foods consumed, as well as the individual's life stage. Certain population groups - such as children, the elderly, and nursing mothers - may be at a higher risk of nutrient deficiencies when following predominantly plant-based diets (van Vliet, Kronberg, and Provenza Citation2020). Without adequate diversity in the diet or sufficient protein and micronutrient sources to support growth and development, plant-based diets can place pregnant women at risk of nutrient deficiencies (Meulenbroeks et al., 2021). However, a well-planned vegan diet can meet the nutritional needs of pregnant and lactating women, as well as children, provided it includes all plant-based food groups and ensures the adequacy of critical nutrients (protein, omega-3 fatty acids, iron, zinc, iodine, and calcium) from a variety of sources. Additionally, proper food preparation techniques must be considered to optimize nutrient availability (Baroni et al., Citation 2019).

Reviews on molecular and physico-chemical properties of plant derived proteins, lipid and carbohydrate ingredients and the science underpinning the design and processing of sustainable plant-based alternatives to meat and dairy foods (Kyriakopoulou, Dekkers, and van der Goot Citation 2019; Kyriakopoulou et al., Citation 2021; McClements & Grossmann Citation 2021a; McClements & Grossmann Citation 2021b) provide insights into food-science related aspects of animal-free food product development. Others have examined the trends in plant-based foods and plant protein from a business perspective and suggest policy and addressing consumer beliefs, perceptions and understanding need to be addressed to realize the market opportunity (Aschemann-Witzel et al., Citation 2021).

There are many factors which influence the acceptance and success of plant-based alternative meat products in the market which need to be considered in the design of plant-based meat alternatives. This chapter focuses on the food science and technology perspectives that are critical for enhancing consumer acceptance in the development of plant-based meat alternative products. It also includes a review of the patent landscape in this field, offering insights into technological innovations that support advancements in plant-based meat alternatives. These topics are examined within the context of a conceptual framework for an innovation pathway specific to plant-based meat development. Finally, the chapter presents recommendations for future research aimed at guiding the development of next-generation plant-based meat alternatives with improved quality and consumer appeal.

Formulation and Processing of Plant-Based Meat Alternatives

The source and processing history of an ingredient can significantly affect its intrinsic functionality and its performance within the final food matrix, primarily due to process-induced interactions between ingredients. To ensure that novel alternative ingredients perform effectively in plant-based meat formulations, experimental trials with varying product compositions are necessary. This is particularly important as ingredient functionality is influenced not only by the choice of components but also by the sequence and methods used during processing. Recent reviews have explored the sources and functional properties of plant proteins, fats and oils, as well as other key ingredients such as colorants, binders, and flavorings. Additionally, the critical roles of water and structuring processes in the production of plant-based meat alternatives have been highlighted (Kyriakopoulou et al., Citation 2021). While most work to date has been focused on the processing and texturizing of the protein component, there are opportunities for improving nutritional and textural properties. Various types of delivery systems (e.g., liposomes, emulsions, nanoparticles and microgels) may be used for protecting and incorporating functional ingredients for optimal performance in plant-based meat alternatives (Domínguez et al., Citation 2021).

Texturization of Protein Enriched Fractions and Formulations

The texturized protein ingredient is the major building block which imparts the meat-like texture to a plant-based meat alternative product. The texturization process re-structures the protein-based ingredient and builds in the technical functionality and performance required in meat alternative products. The traditional texturizing process is extrusion. Geometrical factors (i.e., screw design, die type and dimension), feed composition and process variables (e.g., screw speed, in-barrel moisture, barrel temperature) affect the physical and chemical transformations that occur during extrusion. The components in the feed undergo significant changes during extrusion due to the presence of moisture, heat, pressure, and shear. Moisture, temperature, and shear conditions alter protein conformation, with the extent of these changes depending on the operating conditions during extrusion. Both low and high moisture extrusion techniques can be employed (Vatansever, Tulbek, and Riaz, 2020). For the production of meat alternative products where a fibrous structure is desired, high moisture extrusion using a twin-screw extruder is typically used. The plasticizing effects of water facilitate the re-orientation and alignment of proteins into fibers (Cheftel, Kitagawa, and Quéguiner Citation 1992; Riaz Citation 2011; Ryu Citation 2020).

The process conditions, type of protein and other components added into the formulation influence the properties of the extrudate, including their porosity and fibrousness (Arêas Citation 1992; Camire Citation1991; Day and Swanson Citation 2013). The formulation, protein concentrate/isolate or protein ingredient in combination with other components (e.g., carbohydrates, fiber) are altered by the re-structuring process. Proteins are plasticized and, in the molten state, they unfold. The proteins are denatured, undergo hydrolysis, and interact with each other or other molecules through hydrophobic, covalent, and electrostatic interactions. The starch in the formulation is gelatinized, and protein-lipid and protein-carbohydrate interactions take place. Fiber is also functionalized. This complex series of reactions results in a restructured matrix with altered textural properties, as well as changes in sensory attributes and nutrient availability (Arora et al., Citation 2020; Dekkers, Boom, and van der Goot Citation 2018; He et al., Citation 2020; Pietsch, Emin, and Schuchmann Citation 2017).

Newer techniques for texturizing proteins have been explored such as using a high temperature shear cell (Cornet et al., Citation 2022; Krintiras et al., Citation2016). A low-cost alternative to extrusion for protein texturization is mechanical elongation, a process in which protein ingredients such as zein or wheat gluten are physically stretched to form striated structures that mimic the fibrous texture of meat (Chiang et al., 2021; Mattice & Marangoni, 2020). Another emerging approach involves co-axial nozzle-assisted 3D printing, which utilizes hydrocolloid-based ionic fibers - such as carrageenan, alginate, and konjac galactomannan - combined with a soy protein base to fabricate plant-based meat alternatives with improved structural and sensory qualities (Ko et al., Citation2021).

Other Considerations

The ability to source sustainable protein to produce plant-based protein-enriched fractions and texturized protein ingredients is essential. A consistent and reliable supply of raw plant material will require access to alternative protein sources from existing or new cultivars with improved sustainability credentials or enhanced functionality. This is expected to drive the development of more localized protein ingredient supply chains. The substantial capital investment required for protein fractionation plants, along with the need for specialized engineering expertise, will foster collaboration between multiple stakeholders in the value chain. To achieve improved business and sustainability outcomes, uses for co-products of protein extraction (e.g., fiber-rich fractions) will need to be developed through collaboration with potential end users.

Recommendations for Future Food-Science Related Research

The increased investment in research, development and commercialization, have been stimulated by the growing plant-based meat market. The research in the area of plant-based meat alternatives is relatively young with opportunities to create new products and intellectual property. Navigating both patent literature and publicly available scientific research is essential for identifying knowledge gaps and guiding the development of superior formulations for plant-based meat alternatives. The growing interest in ethical consumption is one of the key drivers behind the increasing demand for meat substitutes. Among consumers, cost remains a significant factor influencing purchasing decisions. While sensory attributes such as taste, flavor, and overall eating experience are critical for market success, there is a growing recognition that delivering a nutritionally balanced product provides a competitive advantage. Broad adoption and sustained market success of plant-based meat alternatives will depend on widespread stakeholder acceptance and the resolution of several core challenges—including replicating the taste and texture of conventional meat, ensuring consistent quality and safety, enhancing nutritional value, and achieving cost-effective scalability (Choudhury et al., 2020). Despite recent advancements, replicating the complex, hierarchical structure of animal muscle tissue remains a technical challenge. However, these limitations also present opportunities for further research aimed at improving both the nutritional quality and safety of alternative meat products (Kołodziejczak et al., Citation 2021; Sha and Xiong Citation 2020; Zhang et al., Citation 2021). Some suggested areas for further food-related research for underpinning future developments in plant-based meat alternatives include research in the following areas:

Ingredients

- Exploring alternative, sustainable, and locally sourced protein supply chains - such as legumes, hemp, rapeseed, and other oilseeds - is essential for ensuring a consistent and environmentally responsible supply of protein ingredients. This includes the development and optimization of processing technologies for the efficient isolation of protein fractions. Additionally, research should focus on the valorization of by-products generated during protein fractionation, promoting a circular and low-waste approach within the protein production system.
- Exploration of alternative plant-based sources of hard fats or fat blends is a key focus in replicating the functional and sensory characteristics of animal fats. This includes the development of fats through precision fermentation, which enables the production of tailored lipid profiles that closely mimic those found in animal products. Additionally, processing technologies such as fractionation, inter-esterification, and enzymatic modification are being employed to alter the physical and nutritional properties of fats. These techniques enhance texture, melting behavior, and nutritional value, making plant-derived fats more suitable for use in meat analogues and other food applications.
- Developing alternative non-adhesive binders as substitutes for hydrocolloid additives to create cleaner labels (e.g., aquafaba) (He et al., 2021) or enzyme-crosslinked proteins with pectins (Sakai et al., Citation 2021).
- Examining different protein binders for sticking layers of protein extrudate and structured fats for improved assembly of structural ingredients by covalent crosslinking for improved texture (Herz et al., Citation 2021a).

Processing

- Assessing the influence of degree of fractionation of protein-enriched fractions from plants obtained using various fractionation techniques (e.g., dry versus wet fractionation, all aqueous processes for isolation of protein from oilseed versus conventional use of solvent extracted residues for preparation of protein concentrates and isolates).
- Investigating the effects of pre-processing treatments on protein-rich fractions - particularly thermal treatments such as heating - has gained significant attention for their role in enhancing the functionality and texturizing performance of plant-based meat alternatives. Heat-induced modifications can promote protein cross-linking, which in turn influences the structural and technological properties of these proteins (Bühler et al., 2020; Federici et al., 2021; Geerts et al., 2018). These functional changes are critical for improving the texture, consistency, and overall quality of plant-based meat products.

Encapsulated Ingredients

- Exploring the structure-flavor relationships of Maillard reaction products as flavor precursors and mode of delivery of precursors (e.g., water-in-oil or oil-in-water emulsions with various types of phospholipid emulsifiers) for regulating the rate and type of Maillard reaction products developed during cooking.
- Exploring the use of encapsulated sensitive functional ingredients—such as vitamins, omega-3 oils, and iron delivery systems—offers a promising approach to improving the stability and efficacy of plant-based meat products. Encapsulation protects these compounds from degradation during processing and storage, while also preventing undesirable interactions within the formulation, such as iron-catalyzed oxidation of fats.

Consumer Acceptance/Regulations

- Developing more nutritional plant-based meat alternative formulations for a wider range of animal-free products (plant-based meat alternative products from cows and other ruminants, poultry, seafood) which are lower in sodium, have higher protein quality, improved iron levels, and other micronutrients such as vitamin B_6 and B_{12}, including assessing in-body effects after consumption.

- Investigating consumer psychology is essential to understanding the motivational barriers that hinder the adoption of plant-based meat alternatives. Gaining insights into these psychological factors can inform strategies that promote behavioral change and facilitate the transition toward more sustainable food systems at both individual and societal levels.
- Promoting consumer understanding at various levels, including developing curriculum-friendly activity for schools to encourage the conversation about plant-based meat alternatives and their place in the wider ecosystems of sustainable food systems.
- Developing comprehensive evidence and documentation is essential for navigating and complying with food labeling and regulatory requirements across different jurisdictions. This includes addressing regulatory considerations for the incorporation of new and novel ingredients into plant-based meat alternatives, such as those derived from precision fermentation. Such efforts involve ensuring that all ingredients meet safety and quality standards, providing scientific substantiation for their intended use, and aligning with jurisdiction-specific frameworks for novel food approval, labeling transparency, allergen declarations, and nutritional claims.

Safety

- Examining the shelf-life of plant-based meats (i.e., limiting deteriorative reactions that affect technological properties) as well as food safety of extended shelf-life plant-based meat alternative products and developing strategies to address these issues.
- Conducing toxicological tests to ensure safety of new and novel ingredients.

Other Considerations

- Performing cradle-to-grave life cycle analysis for ingredients and products using agreed standard protocols.
- Promoting collaboration at both national and global levels is essential to accelerate and scale supply chain solutions that ensure a reliable and sustainable supply of plant proteins and other plant-derived ingredients. Such coordinated efforts can significantly support innovation and the development of next-generation plant-based food products.
- Addressing issues at the science-business-policy interface that warrant attention and ensure transitions to plant-based diets do not compromise food and nutrition security.

Conclusion

The market for plant-based meat alternative products is set to grow, driving the demand for more plant protein. The future demands for protein and other ingredients in the plant-based alternative formulation need to be met sustainably. Plant-based meat alternative products have the potential to reduce the environmental impacts of diets, if complemented by a reduction in meat consumption. However, the complementary and symbiotic roles for plant-source and animal-source foods for sustainable healthy diets (Comerford et al., Citation2021) have to be recognized.

The broader ramifications for research, industry and society of market transformation caused by shifts to plant-based diets need to be considered in an integrated way. A broader perspective necessitates examining the complex relationships and tensions surrounding meat and milk alternatives, including consumer preferences, value chains, and the broader food system, along with ethical and political considerations (Lonkila & Kaljonen, 2021). Additionally, the transition to plant-based diets presents significant nutritional and socio-economic challenges, particularly in low-income countries (Rasmussen et al., Citation2021). An awareness of future demand and consumer preferences, along with a deep understanding of the role of food and nutritional sciences, provides the potential to design entirely new protein supply chains. A holistic approach is needed - one that examines the entire supply chain from farm to consumer, taking into account climate change, environmental pressures, changing demographics, and power inequities. This approach is essential for plant-based meat alternative products to become more mainstream in the food supply.

References

Aiking, H., and J. de Boer. 2020. The next protein transition. *Trends in Food Science & Technology* 105:515–22. doi: 10.1016/j. tifs.2018.07.008.

Alessandrini, R., M. K. Brown, S. Pombo-Rodrigues, S. Bhageerutty, F. J. He, and G. A. MacGregor. 2021. Nutritional quality of plant-based meat products available in the UK: A cross-sectional survey. *Nutrients* 13 (12):4225. https://www.mdpi. com/2072-6643/13/12/4225. doi: 10.3390/nu13124225.

Arêas, J. A. G. 1992. Extrusion of food proteins. *Critical Reviews in Food Science and Nutrition* 32 (4):365–92. doi: 10.1080/10408399209527604.

Arora, B., A. Yoon, M. Sriram, P. Singha, and S. S. H. Rizvi. 2020. Reactive extrusion: A review of the physicochemical changes in food systems. *Innovative Food Science & Emerging Technologies* 64:102429. doi: 10.1016/j.ifset.2020.102429.

Arrutia, F., E. Binner, P. Williams, and K. W. Waldron. 2020. Oilseeds beyond oil: Press cakes and meals supplying global protein require- ments. *Trends in Food Science & Technology* 100:88–102. doi: 10.1016/j.tifs.2020.03.044.

Aschemann-Witzel, J., R. F. Gantriis, P. Fraga, and F. J. A. Perez-Cueto. 2021. Plant-based food and protein trend from a business perspec- tive: Markets, consumers, and the challenges and opportunities in the future. *Critical Reviews in Food Science and Nutrition* 61 (18):3119–28. doi: 10.1080/10408398.2020.1793730.

Augustin, M. A., and L. Sanguansri. 2015. Challenges and Solutions to Incorporation of Nutraceuticals in Foods. *Annual Review of Food Science and Technology* 6 (6):463–77. doi: 10.1146/annurev-food-022814-015507.

Bakhsh, A., S. J. Lee, E. Y. Lee, N. Sabikun, Y. H. Hwang, and S. T. Joo. 2021. A novel approach for tuning the physicochemical, tex- tural, and sensory characteristics of plant-based meat analogs with different levels of methylcellulose concentration. *Foods* 10 (3):560. doi: 10.3390/foods10030560.

Baroni, L., S. Goggi, R. Battaglino, M. Berveglieri, I. Fasan, D. Filippin, P. Griffith, G. Rizzo, C. Tomasini, M. A. Tosatti, et al., 2019. Vegan nutrition for mothers and children: Practical tools for healthcare providers. *Nutrients* 11 (1):5. doi: 10.3390/nu11010005.

Brown, P. O., V. Marjia, R. Varadan, M. Eisen, and S. Sergey. 2015. Methods and compositions for consumables. US 2015/0289541 A1. Buallay, A. 2022. Sustainability reporting in food industry: An innovative tool for enhancing financial performance. *British Food Journal* 124 (6):1939–58. doi: 10.1108/BFJ-01-2021-0053.

Bühler, J. M., B. L. Dekkers, M. E. Bruins, and A. J. van der Goot. 2020. Modifying faba bean protein concentrate using dry heat to increase water holding capacity. *Foods* 9 (8):1077. doi: 10.3390/foods9081077.

Camire, M. E. 1991. Protein functionality modification by extrusion cooking. *Journal of the American Oil Chemists' Society* 68 (3):200–5. doi: 10.1007/BF02657770.

Cheftel, J. C., M. Kitagawa, and C. Quéguiner. 1992. New protein texturization processes by extrusion cooking at high moisture levels. *Food Reviews International* 8 (2):235–75. doi: 10.1080/87559129209540940.

Chiang, J. H., S. M. Loveday, A. K. Hardacre, and M. E. Parker. 2019. Effects of soy protein to wheat gluten ratio on the physicochemical properties of extruded meat analogues. *Food Structure* 19:100102. doi: 10.1016/j.foostr.2018.11.002.

Chiang, J. H., W. Tay, D. S. M. Ong, D. Liebl, C. P. Ng, and C. J. Henry. 2021. Physicochemical, textural and structural characteristics of wheat gluten-soy protein composited meat analogues prepared with the mechanical elongation method. *Food Structure* 28:100183. doi: 10.1016/j.foostr.2021.100183.

Dekkers, B. L., R. M. Boom, and A. J. van der Goot. 2018. Structuring processes for meat analogues. *Trends in Food Science & Technology* 81:25–36. doi: 10.1016/j.tifs.2018.08.011.

Domínguez, R., M. Pateiro, P. E. S. Munekata, D. J. McClements, and J. M. Lorenzo. 2021. Encapsulation of bioactive phytochemicals in plant-based matrices and application as additives in meat and meat products. *Molecules* 26 (13):3984. doi: 10.3390/molecules26133984.

Dreher, J., C. Blach, N. Terjung, M. Gibis, and J. Weiss. 2020. Formation and characterization of plant-based emulsified and crosslinked fat crystal networks to mimic animal fat tissue. *Journal of Food Science* 85 (2):421–31. doi: 10.1111/1750-3841.14993.

Dreher, J., M. König, K. Herrmann, N. Terjung, M. Gibis, and J. Weiss. 2021. Varying the amount of solid fat in animal fat mimetics for plant-based salami analogues influences texture, appearance and sensory characteristics. *LWT - Food Science and Technology* 143:111140. doi: 10.1016/j.lwt.2021.111140.

Ewy, M. W., A. Patel, M. G. Abdelmagid, O. M. Elfadil, S. L. Bonnes, B. R. Salonen, R. T. Hurt, and M. S. Mundi. 2022. Plant-based diet: Is it as good as an animal-based diet when it comes to protein? *Current Nutrition Reports*. doi: 10.1007/s13668-022-00401-8.

Federici, E., G. W. Selling, O. H. Campanella, and O. G. Jones. 2021. Thermal treatment of dry zein to improve rheological properties in gluten-free dough. *Food Hydrocolloids* 115:106629. doi: 10.1016/j. foodhyd.2021.106629.

Fiorentini, M., A. J. Kinchla, and A. A. Nolden. 2020. Role of sensory evaluation in consumer acceptance of plant-based meat analogs and meat extenders: A scoping review. *Foods* 9 (9):1334. doi: 10.3390/foods9091334.

Gorissen, S. H. M., and O. C. Witard. 2018. Characterising the muscle anabolic potential of dairy, meat and plant-based protein sources in older adults. *The Proceedings of the Nutrition Society* 77 (1):20–31. doi: 10.1017/S002966511700194X.

Graça, J., C. A. Godinho, and M. Truninger. 2019. Reducing meat consumption and following plant-based diets: Current evidence and future directions to inform integrated transitions. *Trends in Food Science & Technology* 91:380–90. doi: 10.1016/j.tifs.2019.07.046.

Gravelle, A. J., A. G. Marangoni, and M. Davidovich-Pinhas. 2018. Chapter 14: Ethylcellulose oleogels. In *Edible oleogels,* eds. A. G. Marangoni, and N. Garti, 2nd ed., 331–62. Champaign, IL: AOCS Press.

Grossmann, L., and J. Weiss. 2021. Alternative protein sources as technofunctional food ingredients. *Annual Review of Food Science and Technology* 12 (1):93–117. doi: 10.1146/annurev-food-062520-093642.

Harnack, L., S. Mork, S. Valluri, C. Weber, K. Schmitz, J. S. Stevenson, and J. Pettit. 2021. Nutrient composition of a selection of plant-based ground beef alternative products available in the United States. *Journal of the Academy of Nutrition and Dietetics* 121 (12):2401–8. e12. doi: 10.1016/j.jand.2021.05.002.

He, J., N. M. Evans, H. Liu, and S. Shao. 2020. A review of research on plant-based meat alternatives: Driving forces, history, manufacturing, and consumer attitudes. *Comprehensive Reviews in Food Science and Food Safety* 19 (5):2639–56. doi: 10. 1111/1541-4337.12610.

Shiu, J. W. Y., A. Kale, J. Karr, and R. Fraser. 2015. *Ground meat replicas.* WO2015153666 A1.

Jahn, S., P. Furchheim, and A. M. Strässner. 2021. Plant-based meat alternatives: Motivational adoption barriers and solutions. *Sustainability* 13 (23):13271. doi: 10. 3390/su132313271.

Karmaus, A. L., and W. Jones. 2021. Future foods symposium on alternative proteins: Workshop proceedings. *Trends in Food Science & Technology* 107:124–9. doi: 10.1016/j.tifs.2020.06.018.

Kaukokonen, O., T. Förbom, and P. Kuusisto. 2020. *Cooked meat substitute and production thereof from plant material.* WO202038601 A1.

Kerler, J., C. Winkel, T. Davidek, and I. Blank. 2010. Basic chemistry and process conditions for reaction flavors with particular focus on Maillard-type reactions. In *Food flavor technology,* eds. A. J. Taylor, and R. S. T. Linforth, 2nd ed., 51–88. Oxford, UK: Wiley-Blackwell. doi: 10.1002/9781444317770.ch3.

Kivelä, R., A. Häkämies, and Z. Jiang. 2019. A method of manufacturing a textured food product and a textured food product. *EP* 3 361 880 B1.

Liu, K., and F. H. Hsieh. 2008. Protein–protein interactions during high-moisture extrusion for fibrous meat analogues and comparison of protein solubility methods using different solvent systems. *Journal of Agricultural and Food Chemistry* 56 (8):2681–7. doi: 10.1021/jf073343q.

Lonkila, A., and M. Kaljonen. 2021. Promises of meat and milk al- ternatives: An integrative literature review on emergent research themes. *Agriculture and Human Values* 38 (3):625–39. doi: 10.1007/s10460-020-10184-9.

Musa-Veloso, K., and J. Juana. 2020. Regulation and labeling of plant-based beverages and simulated meat, poultry, and egg products in Canada and the United States. *Cereal Foods World* 65 (4):44. doi: 10.1094/CFW-65-4-0044.

Nanta, P., W. Skolpap, and K. Kasemwong. 2021. Influence of hydro- colloids on the rheological and textural attributes of a gluten-free meat analog based on soy protein isolate. *Journal of Food Processing and Preservation* 45 (3):e15244. doi: 10.1111/ jfpp.15244.

Neufingerl, N., and A. Eilander. 2021. Nutrient intake and status in adults consuming plant-based diets compared to meat-eaters: A systematic review. *Nutrients* 14 (1):29. doi: 10.3390/nu14010029.

Newton, A. E., A. J. Fairbanks, M. Golding, P. Andrewes, and J. A. Gerrard. 2015. The influence of emulsion structure on the Maillard reaction of ghee. *Food Chemistry* 173:1243–9. doi: 10.1016/j.food- chem.2014.10.147.

Newton, P., and D. Blaustein-Rejto. 2021. Social and economic oppor- tunities and challenges of plant-based and cultured meat for rural producers in the US. *Frontiers in Sustainable Food Systems* 5:624270. doi: 10.3389/fsufs.2021.624270.

Niemiec, R., M. S. Jones, A. Mertens, and C. Dillard. 2021. The ef- fectiveness of COVID-related message framing on public beliefs and behaviors related to plant-based diets. *Appetite* 165:105293. doi: 10.1016/j.appet.2021.105293.

Ryu, G.-H. 2020. Chapter 7: Extrusion cooking of high-moisture meat analogues. In *Extrusion Cooking*, ed. G. M. Ganjyal, 205–24. Sawston, UK: Woodhead Publishing. doi: 10.1016/B978-0-12-815360- 4.00007-9.

Saget, S., M. Porto Costa, C. S. Santos, M. Vasconcelos, D. Styles, and M. Williams. 2021. Comparative life cycle assessment of plant and beef-based patties, including carbon opportunity costs. *Sustainable Production and Consumption* 28:936–52. doi: 10.1016/j. spc.2021.07.017.

Sakai, K., Y. Sato, M. Okada, and S. Yamaguchi. 2021. Improved func- tional properties of meat analogs by laccase catalyzed protein and pectin crosslinks. *Scientific Reports* 11 (1):16631. doi: 10.1038/s41598-021-96058-4.

Saldanha do Carmo, C., S. H. Knutsen, G. Malizia, T. Dessev, A. Geny, H. Zobel, K. S. Myhrer, P. Varela, and S. Sahlstrøm. 2021. Meat analogues from a faba bean concentrate can be generated by high moisture extrusion. *Future Foods* 3:100014. doi: 10.1016/j. fufo.2021.100014.

Schreuders, F. K. G., B. L. Dekkers, I. Bodnár, P. Erni, R. M. Boom, and A. J. van der Goot. 2019. Comparing structuring potential of pea and soy protein with gluten for meat analogue preparation. *Journal of Food Engineering* 261:32–9. doi: 10.1016/j. jfood- eng.2019.04.022.

Sha, L., and Y. L. Xiong. 2020. Plant protein-based alternatives of reconstructed meat: Science, technology, and challenges. *Trends in Food Science & Technology* 102:51–61. doi: 10.1016/j.tifs.2020.05.022.

Silverman, L. 2020. *The FDA should regulate to end the plant-based meat labeling controversy*. https://law.lclark.edu/live/blogs/136-the-fda-should-regulate-to-end-the-plant-based.

Singh, M., N. Trivedi, M. K. Enamala, C. Kuppam, P. Parikh, M. P. Nikolova, and M. Chavali. 2021. Plant-based meat analogue (PBMA) as a sustainable food: A concise review. *European Food Research and Technology* 247 (10):2499–526. doi: 10.1007/s00217-021-03810-1.

van der Weele, C., P. Feindt, J. A. van der Goot, B. van Mierlo, and M. van Boekel. 2019. Meat alternatives: An integrative comparison. *Trends in Food Science & Technology* 88:505–12. doi: 10.1016/j. tifs.2019.04.018.

van Eenennaam, A. L., and S. J. Werth. 2021. Animal board invited review: Animal agriculture and alternative meats—Learning from past science communication failures. *Animal: An International Journal of Animal Bioscience* 15 (10):100360. doi: 10.1016/j.ani- mal.2021.100360.

van Vliet, S., S. L. Kronberg, and F. D. Provenza. 2020. Plant-based meats, human health, and climate change. *Frontiers in Sustainable Food Systems* 4:128. doi: 10.3389/fsufs.2020.00128.

Zhang, L., Y. Hu, I. H. Badar, X. Xia, B. Kong, and Q. Chen. 2021. Prospects of artificial meat: Opportunities and challenges around consumer acceptance. *Trends in Food Science & Technology* 116:434–44. doi: 10.1016/j.tifs.2021.07.010.

Chapter 3

Main Factors Affecting the Functionality of Plant-Based Proteins

Abstract

The global population is projected to surpass 10 billion by 2050, increasing pressure on agricultural systems, particularly for protein production. Rising per capita protein consumption, driven by nutrition and enjoyment, has led the UN to predict a doubling of protein demand by 2050, risking a severe shortage without significant changes. Technological innovations are increasingly being pursued to enhance protein availability and promote sustainability. Over the past decade, plant-based proteins have gained significant popularity, driven by consumer demand for healthier, more sustainable, ethically produced, and cost-effective food options. However, despite their rising appeal, plant-based proteins pose several challenges, particularly in the development of innovative food formulations that match the sensory and functional qualities of animal-derived products.

Keywords: plant-based proteins, nutritional value, functional properties, consumer acceptance

Introduction

The rise in popularity of plant-based (PB) foods in recent years is attributed to the consumer's increasing demand for alternatives to animal-based (AB) products. This trend is largely driven by growing consumer awareness of the health benefits associated with plant-based (PB) foods compared to animal-based (AB) products, as well as concerns over the ethical and environmental implications of animal husbandry. Additional factors include apprehension about high cholesterol levels in AB diets, lactose intolerance, and the rising incidence of allergies related to animal proteins. As a result, the increasing popularity of PB foods has led to a heightened demand for plant-derived protein concentrates and isolates that can replicate the functional properties of animal proteins in food formulations (Figure 1). However, few consumers actually understand PB diets, often associating the term with vegetarianism and veganism (Faber et al., 2020). Despite this, a rise in consumer social consciousness has prompted a change in eating habits across all generations and now a growing percentage of eaters self-identify as flexitarians (reduced meat consumption without total elimination) (Rosenfeld 2018, Spencer 2018).

Demand for soy protein isolates actually preceded the PB demand because food formulators a decade earlier were frustrated with the cyclical price and supply demands of dairy proteins. In the early 2000s, manufacturers of dairy protein concentrates and isolates began introducing a wide range of new and improved protein products. The use of both intact and hydrolyzed forms of soy, whey, and milk protein concentrates and isolates provided food formulators with versatile ingredients that offered desirable functional properties. These proteins were increasingly incorporated into recipes aimed at creating healthy, nutritious, and clean-label food products. At the same time, foods with higher protein content benefited from a "halo effect," as consumers increasingly perceived them as healthier and more nutritionally beneficial. This led to increased consumer demand for protein-fortified products, putting pressure on manufacturers. As is common in a supply-and-demand environment, the prices of high-protein dairy products began to rise, while their supply decreased. In response, food manufacturers began reformulating products with soy protein isolates, which possessed many of the functional characteristics of dairy-based proteins (Kinsella, 1979). Similarly, this shift impacted the supply and prices of soy proteins and opened new opportunities for plant-based protein (PBP) products.

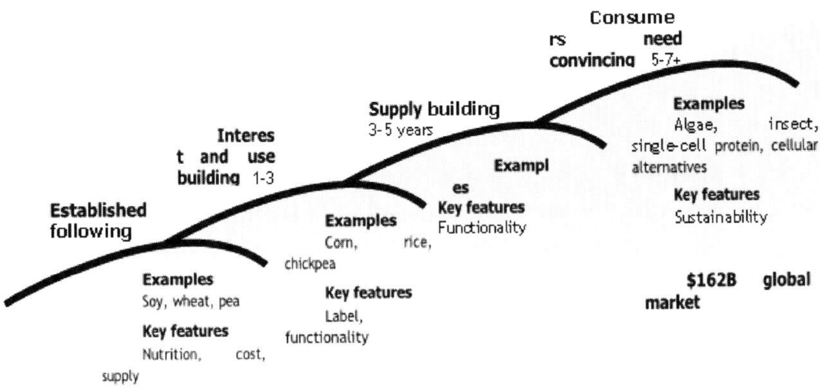

Figure 1. The evolution of protein trends in the marketplace.

This chapter discusses the positive and negative attributes of several PBPs. There has been a flurry of activity from several perspectives to launch novel PBPs for inclusion in PB products. Only a limited number of these protein sources have sufficient supply chain availability and accessible research or commercial data. Therefore, this review focuses on the most commercially relevant products: soy, wheat gluten, yellow pea, rice, and chickpea protein concentrates and isolates. For clarity, protein concentrates and isolates are defined in this context as containing more than 60% and 85% protein content, respectively.

Factors Influencing Plant-Based Proteins

Plant Genetics

The nutritional and functional properties of plant-based proteins (PBPs) can be enhanced through traditional breeding techniques as well as advanced biotechnological approaches, including gene cloning and genome editing. These strategies are crucial for meeting the increasing global demand for high-quality plant-based protein products. Additionally, novel screening technologies in plant proteomics allow for rapid assessment of protein characteristics, reducing the need for time-consuming processes such as field propagation, protein extraction, and detailed biochemical characterization (Luthria et al., 2018). Many commercial PBP manufacturers claim to be using nongenetically modified cultivars for their concentrates and isolates, with the exception of soy. Fortunately for the industry and consumers, there are numerous cultivars available for screening. Consumers are becoming more accepting of genetically modified and genetically edited products (Butkowski, 2017). This growing acceptance provides geneticists with the opportunity to remove or edit the genes responsible for negative traits (discussed below).

Several studies have screened pea (Pisum sativum L) cultivars to quantify functionality, flavor, and color after manufacturing pea protein concentrates and isolates. Protein solubility and emulsification capacity vary between cultivars (Arteaga et al., 2021). Powder color was also significantly different. Arteaga et al., (2021) reported similarities in sensory characteristics (aroma and flavor) among the cultivars and protein isolates, except for the attributes of pea-like and bitter, which varied depending on the cultivar used for protein isolation. Other studies have shown that the aroma of pea seeds varies significantly depending on the cultivar, harvest year, and processing conditions (Azarnia et al., 2011; Cui et al., 2020b). Additionally, foaming properties are influenced more by the extraction method than by the cultivar (Stone et al., 2015). Notably, higher alkaline extraction pH levels have been found to adversely affect foaming functionality (Cui et al., 2020a).

Wheat cultivars may be the most studied crops for commercial protein concentrates and isolates. The baking industry relies on protein–protein interactions to form structure in baked goods. This important functionality has been extensively studied in durum wheat (Fois et al., 2011, Tosi 2005) and semolina (Kaur 2014). The functional properties of proteins for meat analogues and meat alternatives include gelation, binding, adhesion, emulsification, water binding, and structure building (Asgar et al., 2010). Gluten has unique functional properties among other plant-based proteins (PBPs), particularly its ability to form a cohesive blend with viscoelastic properties when heat-set or plasticized (Singh & MacRitchie, 2001). Many meat alternative products have leveraged this unique property to their advantage in creating structured meat alternatives and jerky products.

The primary PBP for decades has been soy protein. Soybean genetics is extensively studied because of this crop's importance in feeding livestock and edible oil extraction. Literally thousands of cultivars exist, both non-genetically modified and genetically modified. However, the genetics of soybeans that influence protein functionality have not been characterized as extensively as those of wheat and pea proteins. Both of the latter proteins hold greater commercial significance, primarily due to their protein functionality. Numerous studies in the literature (Zhang et al., 2021) indicate that cultivars with higher production of glycinin and β-conglycinin are important for structure building in tofu, but there is limited information focusing on the improvement of functional properties such as foaming and gelation. We do know glycinin has a significant role in gel formation because of its numerous disulfide bonds, and a lesser role in emulsifying and foaming (Nishinari et al., 2014).

Food chemists have studied rice cultivars in relation to starch functionality (Patindol & Wang 2002) and nutritional properties. Protein functionality in rice has received relatively little attention, primarily due to the large molecular size of rice proteins, which results in poor aqueous solubility. Since most functional properties depend on the hydrodynamic behavior of proteins (Foegeding, 2015), this limited solubility presents a challenge. Nevertheless, a few studies have examined how different rice cultivars influence protein functionality in this globally significant crop (Rafe et al., 2016). For example, the Tarom cultivar has been reported to exhibit superior functional properties - particularly in terms of solubility, emulsifying capacity, and foaming ability - compared to the Shiroodi cultivar (Esmaeili et al., 2016). Additionally, waxy rice cultivars have been shown to form stable emulsions at neutral pH (Mun et al., 2016). The results from these three studies indicate rice cultivars should be more closely examined to identify those important for protein isolates.

Similar to rice, the functionality of chickpea starch in different cultivars has been more extensively studied than the functionality of its protein (Milán-Noris et al., 2016; Withana-Gamage et al., 2011). The use of chickpea as a plant-based protein (PBP) is increasing, and future studies will likely focus on cultivar selection in relation to its protein functionality. Commercial companies that manufacture chickpea protein concentrates and isolates have conducted extensive cultivar research but have not published their findings due to the competitive advantage this information provides. However, from the limited amount of published data, it is evident that chickpea cultivars vary in their water-holding and oil absorption capacities. Gelling and emulsifying properties also seem to differ among cultivars.

The key takeaway from this genetic discussion is the following: Genetics does have an impact on protein functionality. With the exception of soybean and wheat, there have been very few in-depth studies on the impact of genetic control on protein functionality. This creates a "gray space" for research. Pea protein isolate is being extensively studied due to its growing prominence in commercial usage, and this trend is expected to continue with the increasing number of new manufacturing plants being established for this plant-based protein (PBP). Chickpea and rice cultivars are also being screened to optimize protein synthesis and functionality. Much of this research is being carried out by industrial research groups, who typically maintain the findings as proprietary.

Supply Chain Challenges

The growing demand by consumers for more PB foods has created opportunities for entrepreneurial companies to launch new PBP concentrates and isolates from multiple raw materials. Proteins are a main constituent of many agricultural commodities, and protein manufacturers have used this to offer innovative ingredients to food formulators for improving product nutritional values, physical functionalities, and/or physiological bioactivity. Outside of the major global crops (wheat, rice, corn, and soy), farmers have not produced large quantities of other crops to meet supply chain demands as new protein entrants move into commercial markets. Take yellow pea, for example. A decade ago, most yellow peas were grown along the Canada–United States border, with a small amount also cultivated in Europe. The majority of these crops in the Western Hemisphere were shipped to Europe for processing into pea starch and fiber. As consumers began advocating for alternatives to animal-based products for ethical and sustainability reasons, meat and dairy alternatives were developed using various botanical sources, including soy, almond, rice, oat, and yellow pea. Global agronomic production has increased to meet the new demands of food formulators. However, expanding crop acreage does not happen quickly, as commercial plant breeding is a complex process. Seed stock to increase plant acreage requires extensive plant selection practices to identify top-performing candidates to produce the desired agronomic characteristics while maintaining crop yield and resistance to weeds and pests (Glenn et al., 2017). Once candidate hybrids are identified, it can take several growing seasons in both the northern and southern latitudes to obtain enough seed stock for farmers to plant the acreages required to supply raw material to protein manufacturers.

Sustainability

Increased consumption of beef and chicken as humans' main sources of protein over the past 75 years has placed pressure on global food security. This shift in nutritional preferences has driven significant growth in intensive animal farming to meet industrial demands (Grigg, 1995). Consequently, concerns have intensified over environmental impacts, including soil degradation, greenhouse gas emissions, and water pollution. The conventional meat industry faces growing criticism not only for its ecological footprint but also for the perceived ethical issues surrounding animal welfare. These concerns have played a pivotal role in the increasing popularity of plant-based (PB) meat alternatives over the past decade. However, this raises a critical question: are plant-based products (PBPs) truly more environmentally sustainable?

Scientists, journalists, celebrities, and activists have outspokenly encouraged consumers to transition from AB foods to PB foods for both healthier eating and to save the environment (Gates 2021, Willet et al., 2019). However, PB products are expensive, nutritionally inadequate, and may not be as good for the environment as often touted (Adesogan et al., 2020). There has been considerable public attention surrounding the perceived environmental and nutritional benefits of plant-based (PB) foods compared to animal-based (AB) foods. However, many scientific studies assessing the environmental impacts of PB foods are based on modeling approaches that often overlook critical variables related to dietary substitutions and nutritional adequacy (Ridoutt et al., 2017). These studies frequently rely on unverified assumptions regarding the complex interactions among food sources, processing methods, human nutrition, and environmental outcomes. To make definitive claims about the environmental superiority of PB foods, it is essential to establish a standardized and transparent knowledge framework. Such a framework would support more robust and comprehensive modeling, guiding future research on the interrelationship between diet and environmental impact (Ridoutt et al., 2017).

Plant-Based Protein Manufacture

Protein extraction from their native environments is a difficult task. PBP extraction is no exception (Kumar et al., 2021). Current methods are capital intensive, not very efficient, and often produce sidestreams and wastewater that are difficult and costly to further process. As a result, PBP manufacturers must often command prices higher than do ABP manufacturers.

Conventional PBP extraction techniques use alkalization and acidification technologies as part of their unit operations, but these methods are only efficient at removing approximately 50% of the protein present in the plant biomass (Karki et al., 2010, Kasai & Ikehara 2005). Plant-based proteins are challenging to extract from biomass due to the entrapment of protein molecules within the structural matrix of cellulose and hemicellulose fibers. To improve protein recovery, some manufacturers employ conventional solubilization techniques during the initial extraction phase, such as "salting out" proteins from an alkalized solution. This process typically involves the use of significant amounts of sodium chloride, which must be removed in subsequent stages. The resulting wastewater requires additional treatment before it can be safely discharged into municipal systems or the environment (Kumar et al., 2021). Additional inefficiencies in the extraction process arise from sidestreams rich in starches and fiber fractions. These by-products must be further concentrated and processed into marketable products to offset the costs associated with raw material acquisition. From a sustainability standpoint, it is crucial to maximize the utilization of all fractions derived from plant-based raw materials to minimize waste and enhance economic viability.

Newer methods of PBP extraction are being scaled from laboratory and pilot plant developments that improve protein extraction efficiency, digestibility, and functionality (Kumar et al., 2021, Rahman & Lamsal 2021). An additional benefit is the reduction of environmentally unfriendly waste streams. Unfortunately, these technologies can be capital intensive, require more area in a manufacturing plant, and are limited in scale at this point. Nonetheless, these technologies will become more cost-effective in time and offer more innovative PBP products in an environmentally favorable way.

Aside from extraction technologies, PBP manufacturers face other challenges to offer these products as food ingredients. Few new green-field manufacturing facilities have been built for PBPs; most companies have retrofitted existing manufacturing facilities. Additional challenges include the segregation and storage of raw materials, suboptimal facility hygiene design that fails to prevent cross-contamination - particularly among allergens - and limited food safety knowledge specific to plant-based products (PBPs). Compounding these issues is a general lack of expertise in process design aimed at maximizing production efficiency. These challenges are further exacerbated by a shortage of subject matter experts in food science and process innovation, particularly in areas directly related to the development and manufacturing of PBPs.

Nutritional Value of Plant-Based Proteins

Protein malnutrition (a.k.a. severe acute malnutrition) exists in many parts of the world and is the leading cause of death globally in children under two years of age (UNICEF/WHO/World Bank 2021). Proteins are one of the three essential macronutrients - alongside lipids and carbohydrates - that are critical for human health and longevity. Most plant-based proteins (PBPs) also contain small amounts of additional nutrients, including fiber, starches, minerals, and various sugars. The key nutritional qualities of any protein source include its total protein content, the profile of essential amino acids, digestibility, and overall nutrient composition. Upon digestion, dietary proteins supply amino acids that serve as precursors for de novo muscle protein synthesis. Therefore, from a nutritional standpoint, it is important to consider the efficiency of gastrointestinal digestion in supporting optimal biological function (Jiménez-Munoz et al., 2021).

The nine essential amino acids are especially important, as human cells cannot synthesize these at sufficient rates to meet metabolic demand, and therefore humans must obtain them through their diet. PBPs have relatively low amounts of essential amino acids and leucine contents compared to ABPs (Gorissen et al., 2018). Additionally, some plant-based proteins (PBPs) are low in lysine, cysteine, and/or methionine. In fact, a single plant protein in a formulated food product is not sufficient to provide an adequate daily source of amino acids for the consumer (Day, 2013). Fortunately, most individuals consume a variety of protein sources in their daily diet in sufficient quantities to meet the metabolic demands for muscle synthesis (Balandrán-Quintana et al., 2019).

Over the past fifty years, scientists have made significant efforts to genetically enhance key protein crops such as wheat, corn, and soy. These efforts have led to improvements in the levels of essential amino acids, including sulfur-containing ones (Kumar et al., 2020). However, increasing lysine content remains a persistent challenge. Conventional genetic and breeding methods have proven inadequate for modifying the tightly regulated lysine metabolic pathway in most plants without causing adverse effects on plant growth and crop yield (Galili & Amir, 2013). Nonetheless, two genetically modified cultivars have been successfully developed with elevated lysine levels. One gene encodes the enzyme responsible for the first step in lysine biosynthesis, while the other increases free lysine content (Kumar et al., 2020). Genome editing has been successfully applied to soybeans to enhance the levels of branched-chain amino acids in synthesized proteins (Li et al., 2015). Future genetic modification and editing of plants will improve the quality, nutritional value, and functionality of plant-based proteins (PBPs), but the bioefficacy and biosafety of these modifications will need to be confirmed, as nutritionists have raised concerns about whether PBPs are digested in a manner similar to animal-based proteins.

The nutritional equivalency of PBPs to ABPs is questioned because PBPs have less of an anabolic effect because of their lower essential amino acid content, digestibility, and levels of lysine and sulfur amino acids. Humans tend to direct plant-derived amino acids toward oxidation rather than muscle protein synthesis (Pencharz, 2016). In healthy younger individuals, this is generally not problematic, as their metabolic systems can compensate to maintain muscle homeostasis. However, in older adults, a reduced sensitivity to postprandial amino acid absorption—often referred to as anabolic resistance—can impair muscle protein synthesis, leading to inadequate replacement of muscle lost over time (Berrazaga et al., 2019, Traylor 2018).

ABPs are more digestible than PBPs (FAO 2011). This is partially explained by the greater amounts of β-sheet pleating in PBP secondary structure (ABPs either have no secondary structure or are mostly in a helical conformation). β-sheeting molecular conformation interferes with host digestive enzymes to access and digest native proteins (Carbonaro 2012). Digestion is further complicated by the fact that PBPs are intercalated with plant fibers, creating steric hindrance and essentially burying enzyme attack points (Nguyen, 2015). The digestibility of commercial PBP concentrates and isolates is slightly improved, but they still exhibit lower ileal digestibility (Davies & Jakeman, 2020). As plant-based (PB) foods become more popular among consumers, scientists will need to study the bioaccessibility and bioavailability of PBP amino acids.

Significant research gaps remain in our understanding of the digestibility and overall nutritional value of plant-based proteins (PBPs). Moreover, the manufacturing processes used in PBP production can both enhance and diminish their nutritional and functional properties. Emerging evidence suggests that mixed-protein systems - comprising both plant-based proteins (PBPs) and animal-based proteins (ABPs) - offer superior outcomes in terms of functionality and nutritional quality. As a result, this area is expected to attract substantial research interest in the coming years (Gorissen et al., 2018, Jiménez-Munoz et al., 2021).

Physical and Physiological Functionality

The use of PBPs to reduce or replace ABPs in food formulations has steadily increased during the past decade. As this trend evolved, protein manufacturers commercialized more PBP concentrates and isolates beyond the common soy and wheat gluten complexes. Proteins are commonly utilized in food formulations due to their amphiphilic nature, which enables them to interact with a wide range of molecular compounds present in food systems. These interactions allow proteins to perform various functional roles, such as emulsification, foaming, gelling, oil and water binding, structure building, and the formation of protective films. These functionalities arise from molecular interactions, including hydrophobic interactions, electrostatic forces, hydrogen bonding, and, in some cases, covalent bonding. In general, proteins interact with each other, other proteins, carbohydrates (including complex carbohydrates such as starches, fibers, and hydrocolloids), and minerals in food systems. Food processing, through various unit operations, can affect the extent of these interactions. Additionally, storage conditions may further influence these interactions as water evaporates from the finished product, or as the inherent mobility of protein molecules leads to undesirable interactions, such as the firming often observed in many nutritional bars.

Many PBPs function similarly to ABPs. Unfortunately, most do not have as complete a functionality repertoire as animal proteins. Food formulators have worked around this by combining different PBP types (mixed-protein systems) to accentuate some functional properties and align them more closely to ABP systems. Mixed-protein systems offer opportunities for the development of innovative food applications, often exhibiting synergistic functional properties (Alves & Tavares, 2019). Much of the research in this area has focused on combining dairy proteins with plant-based proteins, particularly soy and pea proteins. For instance, studies have shown that blending whey protein isolate with soy protein isolate and applying heat leads to the formation of both soluble and insoluble aggregates, which significantly affect the texture of model food systems (Roesch & Corredig, 2005). The size of these aggregates - and consequently the texture of the final product - can be modulated by adjusting the ratio of whey protein isolate to soy protein isolate. This provides food formulators with a method to increase protein content in a food system without altering the final product's texture. Similar results were observed with pea protein isolate and the whey protein β-lactoglobulin (Chihi et al., 2016). The authors noted a reduction in aggregate size when these proteins were heated together in a model food system, with the food texture differing when either protein was heated alone (Chihi et al., 2016). Other functional properties influenced by the combination of plant-based proteins (PBPs) and animal-based proteins (ABPs) include enhanced resistance of gas bubbles from coalescing in foams, increased emulsion and gel firmness, reduced gel syneresis, elevated protein concentration in gels without affecting gel strength, and improved thermal stability of proteins (Alves & Tavares 2019).

Consumer Acceptance

Consumer acceptance of PBPs is ambivalent, but their acceptability of PB foods is rapidly growing. The only proteins consumers have expressed negativity toward are wheat gluten and soy. Gluten intolerance only occurs in 0.5–1.0% of the population, but in recent years a global fad has implied gluten as a culprit protein in numerous human illnesses (Gujral et al., 2012). This has caused many consumers to avoid gluten in all foods, despite most not being clinically diagnosed with a gluten intolerance (Croall 2019). Soy protein flavor has been criticized since protein concentrates and isolates were first introduced nearly 50 years ago. While soy (and other plant-based) protein manufacturers have made significant improvements in flavor through modifications across the entire spectrum from seed genetics to food preparation, many consumers still express flavor sensitivity. Additionally, concerns have been raised about the estrogenic activity of soy isoflavones, but studies have shown that this activity is very low and tissue-specific (Rizzo & Baroni, 2018). Other positive health benefits, such as the reduction of inflammation, oxidation, and the prevention of cardiovascular disease, outweigh this concern.

The most important sensory attributes of PBP products are color, flavor, and texture. These factors have a large influence on consumer appeal and acceptance. Bitter flavors are often cited by consumers as a negative attribute of PB foods. The challenge of negative flavor perception in plant-based (PB) products is being actively addressed through the development of flavor blockers by specialized flavor companies. These compounds are designed to neutralize or mask specific molecules responsible for undesirable bitter tastes (Gaudette et al., 2016). Although the use of bitter blockers in PB foods has become increasingly common, many consumers still perceive these products as less flavorful compared to their animal-based (AB) counterparts. Looking ahead, advances in metabolomics are expected to significantly enhance the differentiation and understanding of the complex aroma and flavor profiles of PB products (Pavagadhi & Swarup, 2020). This emerging approach will enable flavor chemists to more precisely identify target compounds and formulate tailored blocking agents, ultimately improving the sensory appeal of plant-based foods.

Food scientists are also becoming more familiarized with formulating food dishes that incorporate PBP alternatives to both meat and dairy. Consumers have generally embraced the introduction of plant-based (PBP) meat alternatives, particularly in dishes that traditionally use ground meat or sausage. Importantly, meat - whether animal-based (ABP) or plant-based - is still perceived by many as a central component of a balanced meal (Mackenzie & Shanahan, 2018). The market potential for PBP meat alternatives is expected to expand further as innovations allow for the development of diverse "cuts" that emulate products from various animal species. Moreover, if scientific research can substantiate the nutritional equivalence of PBP meats to their ABP counterparts, it is likely that a broader range of consumers will adopt these alternatives as a regular part of their diets.

Climate Change

Consumers often cite that their reason for consuming more PB foods is attributed to the impact on our climate of raising livestock. One metric tracked by atmospheric scientists for indication of climate change is atmospheric carbon dioxide concentration. Global CO2 has risen from 180 ppm (glacial periods) to 384 ppm in 2009 with a consequential mean temperature increase of 0.76°C (IPCC 2007). The impact of climate change has encouraged plant scientists to identify how it is impacting crop characteristics, including protein concentration and composition.

A meta-analysis (228 studies) summarized the impact of elevated CO2 concentration on protein concentration and composition in major food crops (Taub et al., 2008). Protein concentration decreased 10–15% in wheat, rice, and barley; 14% in potatoes; and only 1.4% in soybean. More acres will need to be planted to feed our increasing global population if this trend is not reversed. Further investigation into the protein composition of crops grown in elevated carbon dioxide atmospheres and processed into concentrates or isolates has revealed significant reductions in certain amino acids. Specifically, wheat showed altered proportions of glutenin and gliadin protein classes, which are key contributors to dough and bread functionality (Högy & Fangmeier, 2008). Similar results were observed in protein subclasses from rice. (Terao et al., 2005). Interestingly, barley protein from plants grown in higher CO2 atmospheres had higher proportions of essential amino acids in their primary sequences (Manderscheid et al., 1995). All these studies have been conducted in potted plants under controlled atmospheres, and future investigations should be conducted under actual field conditions.

By inference, changes in protein concentration and composition could potentially be impacted by climate change. Consumers have shown a strong willingness to adopt plant-based protein meat alternatives, particularly for dishes that traditionally use ground meat or sausage. Encouragingly, meat - whether animal-based protein (ABP) or PBP - is still widely regarded as a central component of meals by most consumers (Mackenzie & Shanahan, 2018). As innovation in plant-based food technology advances, greater market opportunities are expected to emerge through the development of diverse "cuts" that replicate products from various animal species. Moreover, if scientific research can substantiate the nutritional equivalency of plant-based meats to their animal-based counterparts, a broader range of consumers is likely to incorporate these alternatives into their diets. There is some evidence from the microbial world that proteins adapt to increased environmental temperatures through a set of imprecise molecular signatures. Key factors involved in this adaptation include solvent accessibility, disulfide bonds, hydrophobicity, hydrogen bonds, and ionic and pi-electron interactions, which contribute to the condensed packing of protein molecules (Barik, 2020). Whether plant-based proteins (PBPs) will evolve to improve their thermoresistance for the plant's own survival is still unknown. Any changes in a plant's molecular signature are likely to be subtle, and it is probable that PBPs will retain or even enhance their physical functionality.

Safety and Allergenicity

Food allergies and intolerances have been increasing globally, and with the increased consumption of PB foods, this trend will likely continue. Allergenic responses to PB foods are usually caused by proteins and pollen cross-reactivity. People with food allergies generally know which of the "Big 9" (wheat, soy, sesame, eggs, fish, tree nuts, peanuts, milk, and shellfish) to avoid. However, with the increased usage of PBPs in many food categories, consumers with allergies, and even individuals that may not have experienced a previous allergic reaction, need to be cautious. For example, individuals with soy or peanut allergies must be especially cautious of foods containing pulse proteins. Some individuals with wheat intolerance or allergy may not be aware they will also negatively respond to barley and rye proteins. The concern about many novel proteins is their allergenicity may not yet be known. Reputable protein manufacturers are analyzing the primary sequences of novel proteins to identify potential allergenic epitopes within their protein products. Other emerging proteomic methods, with increased sensitivity, are being developed to detect allergenic potential (Verhoeckx et al., 2016). Any protein can become allergenic with repeated exposure. The allergenicity of many proteins can be reduced through processing methods such as heat treatment, enzyme hydrolysis, or high-pressure processing (Johnson et al., 2010, Kasera et al., 2015).

Formulating with Plant-Based Protein

Consumers have used their immense buying power to encourage the food industry to rapidly develop novel PB foods because they are more ethical and ecofriendly than AB foods. However, as food formulators began incorporating plant-based proteins into both existing and novel recipes, it became evident that PBPs could not simply replace animal-based proteins on a one-to-one basis. From the outset, PBPs have faced criticism for their undesirable sensory properties - often described as beany, bitter, and astringent - which detract from the overall eating experience. While the flavor and aroma profiles of soy protein concentrates and isolates have significantly improved over the past five decades, the same technical advancements have not yet been effectively applied to other botanical protein sources. Moreover, many PBPs lack the nutritional completeness and digestibility of ABPs. As a result, food developers are increasingly turning to blends of different plant proteins to create products that are both nutritionally balanced and aligned with consumer expectations for health and functionality (Cossen et al., 2021).

Besides their nutritional properties, proteins are often used as food ingredients for their functional properties. The functional efficacy of proteins is directly correlated to their composition, concentration, and interaction with other ingredients in a food system. As new PBPs enter commercialization, protein scientists continually characterize their protein systems from a primary perspective of physical functionality in a semipurified form (e.g., as concentrates and isolates). However, testing the physical functionality in food systems must be validated to confirm that any protein, including plant-based proteins (PBPs), demonstrates functionality. PBPs are functional in various circumstances, but there are instances where these properties cannot be fully utilized. For example, gelation of PBPs often requires the presence of significant sodium chloride levels in the food system. This may not align with consumer preferences from both sensory and health perspectives.

The emerging popularity of PBPs has encouraged scientists to study the interactions of PBPs with other food polymers to improve the quality and nutritional value of food formulations (Lin et al., 2017). Although PBPs generally display good interactions with other edible polymers, the characteristics of one botanical source may not be demonstrated by another. Furthermore, most of these studies have been conducted only in laboratory biphasic systems, and when translated to multiphasic food formulations, the positive interactions observed earlier in the laboratory do not always correlate with the behaviors seen in real-world conditions. Nonetheless, PBPs interact with themselves and other proteins, as well as edible polymers, to form a variety of complexes with different structures (Lin et al., 2017). These interactions can be advantageously used for fat substitution, encapsulation, and novel emulsion systems (Pickering emulsions) (Hu et al., 2016).

Two important criteria consumers use to judge a product are appearance, when purchasing the product for the first time, and sensorial experience, when deciding whether to repurchase the product. Flavor and aroma are paramount in this consumer evaluation. The scientific literature contains numerous studies on the interactions of volatile flavor compounds with ABPs and PBPs in model systems, but the mechanisms responsible are very ambiguous (Wang & Arntfield 2017). Understanding and evaluating protein–flavor interactions can be challenging. As discussed earlier in this review, PBPs are often undesirably associated with both volatile and nonvolatile flavors and aromas, making it difficult for food formulators to mitigate these properties. Multiple formula iterations must be tested using various flavor blockers and potentiators. These iterations are both costly and time-consuming, and food formulators must familiarize themselves with all PBP options to develop the best-tasting products. There are wide variances in product quality among PBP products.

In a haste to launch new products and take commercial advantage of rising trends, many new PB foods were formulated using very complex mixtures of food ingredients and the inclusion of PBPs. These products were soon met with criticism from nutritionists, minimalists, and members of the medical community, who argued that many plant-based products were less healthy than the conventional meat products they were intended to replace (Barrett, 2020). As food formulators gained a deeper understanding of plant-based proteins (PBPs), subsequent generations of plant-based foods began to feature fewer ingredients. Newer products now emphasize their “cleaner labels,” highlighting simpler, more recognizable ingredient lists as a key selling point.

Conclusion

A rapidly increasing global population is impelling scientists to develop novel food ingredients in an environmentally and ethically favorable manner to feed future generations. Plant-based proteins are rapidly gaining recognition as viable alternatives to animal-based proteins, driven by sustainability, health, and ethical considerations. However, food scientists aiming to incorporate PBPs into food formulations often encounter challenges, including undesirable off-flavors, limited functional properties, and the absence of essential nutrients. Despite the existence of nearly 400,000 species of vascular plants - of which an estimated half are edible - only about 200 are regularly consumed, and an even smaller number are utilized for protein production. This highlights the vast, untapped potential for diversification in plant protein sources. As research advances, the desirable attributes of currently commercialized PBPs will be further optimized, while undesirable traits will be progressively reduced in future generations of plant-based protein products.

References

Adesogan, A. T., Havelaar AH, McKune SL, Eilitta M, Dahl GE. 2020. Animal source foods: sustainability prob- lem or malnutrition and sustainability solution? Perspective matters. *Glob. Food Secur.* 25:100325.

Aiking, H., de Boer J. 2020. The next protein transition. *Trends Food Sci. Technol.* 105:515–22.

Aimutis, W. R. 2004. Bioactive properties of milk proteins with particular focus on anticariogenesis. *J. Nutr.*134:989S–95.

Alves, A. C., Tavares GM. 2019. Mixing animal and plant proteins: Is this a way to improve protein techno- functionalities? *Food Hydrocoll.* 97:105171.

Amagliani, L., O'Regan J, Kelly AL, O'Mahony JA. 2017. The composition, extraction, functionality and ap- plications of rice proteins: a review. *Trends Food Sci. Technol.* 64:1–12.

Anderson, J. W., Johnstone BM, Cook-Newell ME. 1995. Meta-analysis of the effects of soy protein intake on serum lipids. *N. Engl. J. Med.* 333:276–82.

Anjum, F. M., Khan MR, Din A, Saeed M, Pasha I, Arshad MU. 2007. Wheat gluten : high molecular weight glutenin subunits - structure, genetics, and relation to dough elasticity. *J. Food Sci.* 72:R56–63.

Arteaga, V. G., Kraus S, Schott M, Muranyi I, Schweiggert-Weisz, Eisner P. 2021. Screening of twelve pea (Pisum sativum L.) cultivars and their isolates focusing on the protein characterization, functionality, and sensory profiles. *Foods* 10:758–75.

Aschemann-Witzel, J., Gantriis RF, Fraga P, Perez-Cueto JA. 2020. Plant-based food and protein trend from a business perspective: markets, consumers, and the challenges and opportunities in the future. *Crit. Rev. Food Sci. Nutr.* 61(18):3119–28.

Asgar, M. A., Fazilah A, Huda N, Bhat R, Karim AA. 2010. Nonmeat protein alternatives as meat extenders and meat analogs. *Comp. Rev. Food Sci. Food Saf.* 9:513–29.

Ashaolu, T. J. 2020. Applications of soy protein hydrolysates in the emerging functional foods: a review. *Int. J. Food Sci. Technol.* 55:421–28.

Azarnia, S., Boye JI, Warkentin T, Malcolmson L, Sabik H, Bellido AS. 2011. Volatile flavor profile changes in selected field pea cultivars as affected by crop year and processing. *Food Chem.* 124:326–35.

Balandrán-Quintana, R. R., Mendoza-Wilson AM, Ramos-Clamont GM, Huerta-Ocampo JA. 2019. Plant-based proteins. In *Proteins: Sustainable Source, Processing and Application*, ed. C Galanakis, pp. 97–130. New York: Academic.

Barik, S. 2020. Evolution of protein structure and stability in global warming. Int. J. Mol. Sci. 21:9662 Barraquio VL, van de Voort FR. 1988. Milk and soy proteins: their status in review. *Can. Inst. Food Sci. Technol.* 21:477–93.

Barrett, S. 2020. Beyond Meat, Impossible Foods face a new 'fake meat' foe with long controversial his- tory. *CNBC News,* Febr. 6. https://www.cnbc.com/2020/02/06/beyond-meat-impossible-foods-face-new-powerful-fake-meat-foe.html.

Berrazaga, I., Micard V, Gueugneau M, Walrand S. 2019. The role of the anabolic properties of plant- versus animal-based protein sources in supporting muscle mass maintenance: a critical review. Nutrients 11:1825 Boland MJ, Rae AN, Vereijken JM, Meuwissen MPM, Fischer ARH, et al., 2012. The future supply of animal-derived protein for human consumption. *Trends Food Sci. Technol.* 29:62–73.

Boukid, F. 2021. Chickpea (Cicer arietinum L.) protein as a prospective plant-based ingredient: a review. *Int. J. Food Sci. Technol.* 56(11):5435–44.

Boukind, P., Rosell CM, Castellari M. 2021. Pea protein ingredients: a mainstream ingredient to (re)formulate innovative foods and beverages. *Trends Food Sci. Technol.* 110:729–42.

Butkowski, O. K., Pakseresht A, Lagerkvist CJ, Bröring S. 2017. Debunking the myth of general consumer re- jection of green genetic engineering: empirical evidence from Germany. Int. J. Consum. Stud. 41:723–34 Carbonaro M, Maselli P, Nucara A. 2012. Relationship between digestibility and secondary structure of raw and thermally treated legume proteins: a Fourier transform infrared (FT-IR) spectroscopic study. *Amino Acids* 43:911–21.

Chihi, M. L., Mession JL, Sok N, Saurei R. 2016. Heat-induced soluble protein aggregates from mixed pea globulins and β-lactoglobulin. J. Agric. Food Chem. 64:2780–91.

Christians, N., Liu D, Unrah JB. 2010. The use of protein hydrolysates for weed control. In Protein Hydrolysates in Biotechnology, ed. V. K. Pasupuleti, *AI Demain*, pp. 127–33. New York: Springer.

Cossen, A. C., Blumenthal D, Descamps N, Souchon I, Saint-Eve A. 2021. Using a mixture design and fraction- based formulation to better understand perceptions of plant-protein-based solutions. *Food Res. Int.* 141:110151.

Croall, G. I. D., Trott N, Rej A, Aziz I, O'Brien DJO, et al., 2019. A population survey of dietary attitudes towards gluten. *Nutrients* 11:1276.

Cui, L., Bandillo N, Wang Y, Ohm JB, Chen B, Rao J. 2020a. Functionality and structure of yellow pea protein isolate as affected by cultivar and extraction pH. *Food Hydrocoll.* 108:106008.

Cui, L., Kimmel J, Zhou L, Rao J, Chen B. 2020b. Identification of extraction pH and cultivar associated with aromatic compound changes in spray dried pea protein isolate using untargeted and targeted metabolomic approaches. *J. Agric. Food Res.* 2:100032.

Davies, R. W., Jakeman PM. 2020. Separating the wheat from the chaff: nutritional value of plant proteins and their potential contribution to human health. *Nutrients* 12:2410.

Day, L. 2013. Proteins from land plants: potential sources for human nutrition and food security. *Trends Food Sci. Technol.* 32:25–42.

Esmaeili, M., Rafe A, Shahidi SA, Hasan-Saraei AG. 2016. Functional properties of rice bran protein isolate at different pH levels. *Cereal Chem.* 93:58–63.

Faber, I., Castellanos-Feijoo NA, Van de Sompel L, Davydova A, Perez-Cueto FJA. 2020. Attitudes and knowl- edge towards plant-based diets of young adults across four European countries. Exploratory survey. *Appetite* 145:104498.

FAO. 2011. Dietary protein evaluation in human nutrition: report of an FAO expert consultation. *Food Nutr. Pap.* 92, FAO, Rome. http://www.fao.org/ag/humannutrition/35978-02317b979a686a57aa4593304ffc17f06.pdf.

Foegeding, E. A., Davis JP. 2011. Food protein functionality: a comprehensive approach. *Food Hydrocoll.* 25:1853-64.

Foegeding, E. A. 2015. Food protein functionality: a new model. *J. Food Sci.* 80:C2670–77.

Fois, S., Schlichting L, Marchylo B, Dexter J, Motzo R, et al., 2011. Environmental conditions affect semolina quality in durum wheat (Triticum turgidum ssp. durum L.) cultivars with different gluten strength and gluten composition. *J. Sci. Food Agric.* 91:2664–73.

Food Drug Adm. (FDA). 1999. Soy proteins and coronary heart disease. *Fed. Regist.* 64:57688–733.

Food products as sources of protein and amino acids: the case of Poland. Nutrients 10:1977 Grigg D. 1995. The nutritional transition in Western Europe. *J. Hist. Geogr.* 21:247–61.

Galili, G., Amir R. 2013. Fortifying plants with the essential amino acids lysine and methionine to improve nutritional quality. *Plant Biotechnol. J.* 11:211–22.

Glenn, K. C., Alsop B, Bell E, Goley M, Jenkinson J, et al., 2017. Bringing new plant varieties to market: plant breeding and selection practices. *Crop Sci.* 57:2906–21.

Gorissen, S. H. M., Crombag JJR, Senden JMG, Waterval WAH, Bierau J, et al., 2018. Protein content and amino acid composition of commercially available plant-based protein isolates. *Amino Acids* 50:1685–95.

Gorska-Warsewicz, H., Laskowski W, Kulykovets O, Kudlinska-Chylak A, Czeczotko M, Rejman K. 2018.

Gujral, N., Freeman HJ, Thomson ABR. 2012. Celiac disease: prevalence, diagnosis, pathogenesis and treat- ment. *World J. Gastroenterol.* 18:6036–59.

Henchion, M., Hayes M, Mullen AM, Fenelon M, Tiwari B. 2017. Future protein supply and demand: strategies and factors influencing a sustainable equilibrium. *Foods* 6:53.

Högy, P., Fangmeier A. 2008. Effects of elevated atmospheric CO2 on grain quality of wheat. *J. Cereal Sci.* 48:580–91.

Hu, Y. Q., Yin S W, Zhu J H, Qi J R, Guo J, et al., 2016. Fabrication and characterization of novel Pickering emulsions and Pickering high internal emulsions stabilized by gliadin colloidal particles. *Food Hydrocoll.* 61:300–10.

IPCC (Intergov. Panel Clim. Change). 2007. Climate change 2007: the physical science basis. *Rep., WMO/UNEP,* Paris. https://www.ipcc.ch/site/assets/uploads/2018/05/ar4_wg1_full_report-1.pdf.

Jiménez-Munoz, L. M., Tavares GM, Corredig M. 2021. Design future foods using plant protein blends for best nutritional and technological advantage. *Trends Food Sci. Technol.* 113:139–50.

Johnson, P. E., Van der Plancken I, Balasa A, Husband FA, Grauwet T, et al., 2010. High pressure, thermal and pulsed electric-field-induced structural changes in selected food allergens. *Mol. Nutr. Food Res.* 54:1701-10.

Karki, B., Lamsal BP, Jung S, van Leeuwen JH, Pometto AL, et al., 2010. Enhancing protein and sugar release from defatted soy flakes using ultrasound technology. *J. Food Eng.* 96:270–78.

Kasai, N., Ikehara H. 2005. Stepwise extraction of protein and carbohydrates from soybean seed. *J. Agric. Food Chem.* 53:4245–52.

Kasera, R., Singh AB, Lavasa S, Prasad KN, Arora N. 2015. Enzymatic hydrolysis: a method in alleviating legume allergenicity. *Food Chem. Toxicol.* 76:54–60.

Kaur, A., Singh N, Kaur S, Ahlawat AK, Singh AM. 2014. Relationships of flour solvent retention capacity, secondary structure and rheological properties with the cookie making characteristics of wheat cultivars. *Food Chem.* 158:48–55.

Kinsella, J., Damodaran S, German B. 1985. Physicochemical and functional properties of oilseed proteins with emphasis on soy proteins. *New Protein Foods* 5:107–79.

Kinsella, J. 1979. Functional properties of soy proteins. *J. Am. Oil Chem. Soc.* 56:242–58.

Kumar, K., Gambhir G, Dass A, Tripathi AK, Singh A, et al., 2020. Genetically modified crops: current status and future prospects. *Planta* 251:91.

Kumar, M., Tomar M, Potkule J, Verma R, Punia S, Koellhoffer JP, et al., 2021. Advances in the plant protein extraction: mechanism and recommendations. *Food Hydrocoll.* 115:106595.

Lam, A. C. Y., Karaca AC, Tyler RT, Nickerson MT. 2018. Pea protein isolates: structure, extraction, and func- tionality. *Food Rev. Int.* 2:126–47.

Li, S. S., Mejia SB, Lytvyn L, Stewart SE, Viguiliouk E, et al., 2017. Effect of plant protein on blood lipids: a systematic review and meta-analysis of randomized controlled trials. *J. Am. Heart Assoc.* 6(12):e006659.

Li, Z., Liu Z-B, Xing A, Moon BP, Koellhoffer JP, et al., 2015. Cas-9-guide RNA directed genome editing in soybean. *Plant Physiol.* 169:960–70.

Lin, D., Lu W, Kelly AL, Zhang L, Zheng B, Miao S. 2017. Interactions of vegetable proteins with other polymers: structure-function relationships and applications in the food industry. *Trends Food Sci. Technol.* 68:130–44.

Lonnie, M., Laurie I, Myers M, Horgan G, Russel WR, Johnstone AM. 2020. Exploring health-promoting attributes of plant proteins as a functional ingredient for the food sector: a systematic review of human intervention studies. *Nutrients* 12:2291.

Luthria, D. L., John KMM, Marupaka R, Natarajan S. 2018. Recent update on methodologies for extraction and analysis of soybean proteins. *J. Sci. Food Agric.* 98:5572–80.

Mackenzie, M., Shanahan L. 2018. Attitudes to meatless meals: a comparison of the general public and those with links to the agricultural economy. *Nutr. Food Sci.* 48:858–72.

Madureira, A. R., Pereira CI, Gomes AMP, Pintado ME, Malcata FX. 2007. Bovine whey proteins: overview on the main biological properties. *Food Res. Int.* 40:1197–211.

Madureira, A. R., Tavares T, Gomes AMP, Pintado ME, Malcata FX. 2010. Physiological properties of bioactive peptides obtained from whey proteins. *J. Dairy Sci.* 93:437–55.

Manderscheid, R., Bender J, Jager HJ, Weigel HJ. 1995. Effect of season long CO2 enrichment on cereals. *Agric. Ecosyst. Environ.* 54:175–85.

Milán-Noris, A. K., de la Rosa-Millan J, Reyes-Moreno C, Serna-Saldivar O. 2016. Physicochemical, functional properties, and digestion of isolated starches from pigmented chickpea (Cicer arietinum L.) cultivars. *Starch* 69(5–6):1600152.

Monteiro, S. R., Lopes-de-Silva JA. 2019. Critical evaluation of the functionality of soy protein isolates obtained from different raw materials. *Eur. Food Res. Technol.* 245:199–212.

Mun, S., Shin M, Kim Y. R. 2016. Emulsifying properties of proteins isolated from various rice cultivars. *Food Bioprocess Technol.* 9:813–21.

Nguyen, T. T. P., Bhandari B, Cichero J, Prakash S. 2015. Gastrointestinal digestion of dairy and soy proteins in infant formulas: an *in vitro* study. *Food Res Int.* 76:348–58.

Nishinari, K., Fang Y, Guo S, Phillips GO. 2014. Soy proteins: a review on composition, aggregation and emulsification. *Food Hydrocoll.* 39:301–18.

Patil, P., Goswami A, Kalia K, Kate A. 2020. Plant-derived bioactive peptides: a treatment to cure diabetes. *Int. J. Pept. Res. Ther.* 26:955–68.

Patindol, J., Wang Y-J. 2002. Fine structures of starches from long-grain rice cultivars with different function- ality. *Cereal Chem.* 79:465–69.

Pavagadhi, S., Swarup S. 2020. Metabolomics for evaluating flavor -associated metabolites in plant-based prod- ucts. *Metabolites* 10:197.

Pencharz, P. B., Elango R, Wolfe RR. 2016. Recent developments in understanding protein needs: How much and what kind should we eat? *Appl. Physiol. Nutr. Metab.* 41:577–80.

Rafe, A., Sadeghian A, Hoseini-Yazdi SZ. 2016. Physiochemical, functional, and nutritional characteristics of stabilized rice bran from Tarom cultivar. *Food Sci. Nutr.* 5:407–41.

Rahman, M. M., Lamsal BR. 2021. Ultrasound-assisted extraction and modification of plant-based proteins: impact on physicochemical, functional, and nutritional properties. *Curr. Rev. Food Sci. Technol.* 20:1457–80.

Rep., UNICEF, New York. https://data.unicef.org/resources/jme-report-2021 United Nations (UN). 2015. Sustainable development goals. *United Nations.* https://www.un.org/sustainabledevelopment/sustainable-development-goals/.

Ridoutt, B. G., Hendrie GA, Noakes M. 2017. Dietary strategies to reduce environmental impact: a critical review of the evidence base. *Adv. Nutr.* 8:933–46.

Rizzo, G., Baroni L. 2018. Soy, soy foods and their role in vegetarian diets. *Nutrients* 10(1):43.

Roesch, R. R., Corredig M. 2005. Heat-induced soy-whey protein interactions: formation of soluble and insoluble protein complexes. *J. Agric. Food Chem.* 53:3476–82.

Romero, A., Beaumal V, David-Briand E, Cordobes F, Guerrero A, Anton M. 2012. Interfacial and emulsifying behaviour of rice protein concentrate. *Food Hydrocoll.* 29:1–8.

Rosenfeld, D. L. 2018. The psychology of vegetarianism: recent advances and future directions. *Appetite* 131:125–38.

Rustgi, S., Shewry P, Brouns F, Deleu LJ, Delcour JA. 2019. Wheat seed proteins: factors influencing their content, composition, and technological properties, and strategies to reduce adverse reactions. *Comp. Rev. Food Sci. Food Saf.* 18:1751–69.

Shevkani, K., Singh N, Kaur A, Rana JC. 2015. Functional properties of kidney bean and field pea protein isolates: a comparative study. *Food Hydrocoll.* 43:679–89.

Silva, S. V., Malcata FX. 2004. Caseins as a source of bioactive peptides. *Int. Dairy J.* 15:1–15.

Singh, H., MacRitchie F. 2001. Application of polymer science to properties of gluten. J. Cereal Sci. 33:231–43 Spencer M, Cienfuegos C, Guinard J-X. 2018. The Flexitarian FlipTM in university dining venues: student and adult consumer acceptance of mixed dishes in which animal protein has been partially replaced with plant protein. *Food Qual. Preference* 68:50–63.

Stone, A. K., Karalash A, Tyler RT, Warkentin TD, Nickerson MT. 2015. Functional attributes of pea protein isolates prepared using different extraction methods and cultivars. *Food Res. Int.* 76:31–38.

Tang, C. H. 2017. Emulsifying properties of soy proteins: a critical review with emphasis on the role of con- formational flexibility. *Crit. Rev. Food Sci. Technol.* 57:2636–79.

Taub, D., Miller B, Allen H. 2008. Effects of elevated CO2 on the protein concentration of food crops: a meta- analysis. *Glob. Change Biol.* 14:565–75.

Terao, T., Miura S, Yanaghara T, Hirose T, Nagata K, et al., 2005. Influence of free-air CO2 enrichment (FACE) on the eating quality of rice. *J. Sci. Food Agric.* 85:1861–68.

Tosi, P., Masci S, Giovangrossi A, D'Ovidio R, Bekes F, et al., 2005. Modification of the low molecular weight (LMW) glutenin composition of transgenic durum wheat: effects on glutenin polymer size and gluten functionality. *Mol. Breed.* 16:113–26.

Traylor, D. A., Gorissen SH, Phillips SM. 2018. Perspective: protein requirements and optimal intakes in aging: Are we ready to recommend more than the Recommended Daily Allowance? Adv. Nutr. 9:171–82 UNICEF/WHO/World Bank Group. 2021. Levels and trends in child malnutrition. *Joint Child Malnutr. Estim.*

Verhoeckx, K., Brokeman H, Knulst A, Houben G. 2016. Allergenicity assessment strategy for novel food proteins and protein sources. *Food Chem.* 184:160–66.

Wang, K., Arntfield SD. 2017. Effect of protein-flavor binding on flavor delivery and protein functional properties: a special emphasis on plant-based proteins. *Flavor Fragr. J.* 32:92–101.

Willet, W., Rockstrom J, Loken B, Springmann M, Lang T, et al., 2019. Food in the Anthropocene: the EAT- Lancet Commission on healthy diets from sustainable food sources. *Lancet* 393:447–92.

Withana-Gamage, T. S., Wanasundara JPD, Pietrasik Z, Shand PJ. 2011. Physicochemical, thermal and func- tional characterization of protein isolates form Kabuli and Desi chickpea (Cicer arietinum L.): a compar- ative study with soy (Glycine max) and pea (Pisum sativum L.). *J. Sci. Food Agric.* 91:1022–31.

Chapter 4

Digestibility and Bioavailability of Plant-Based Proteins

Abstract

Plant-based proteins are gaining attention for their potential in creating meat analogues that address environmental, health, and animal welfare concerns. These products should mimic the appearance, taste, and texture of meat while offering similar or superior nutritional profiles. However, there is limited information on the digestibility and bioavailability of PBPs used in meat analogues. This chapter explores common plant protein sources and their associated production methods, with a particular focus on the role of green processing technologies in enhancing product quality. Techniques such as sonication, microwave treatment, high-pressure processing, pulsed electric fields, enzymatic hydrolysis, and fermentation are examined for their potential to improve protein digestibility and nutrient bioavailability. Careful selection of both the protein source and processing method is critical for optimizing the nutritional profile of plant-based meat analogues.

Keywords: plant-based proteins, digestibility, bioavailability, green processing technologies

Digestibility and Bioavailability

Proteins are natural polymers consisting of amino acids linked together by peptide bonds. The type, number, and sequence of amino acids along a polypeptide chain determines the native structure of a protein molecule, which influences its digestion and absorption within the human gastrointestinal tract. In addition, the digestion and ab- sorption of proteins also depends on other factors, such as their denaturation state, aggregation state, and food matrix effects. Protein digestion begins in the stomach due to the presence of pepsin in gastric fluids, but most of the digestion occurs in the small intestine, where a series of proteases, including trypsin and chymotrypsin, are active. The rate and extent of protein digestion depend on how easily the proteases can access the peptide bonds in the polypeptide chains.

The term "digestibility" refers to how well the proteases in the gastrointestinal tract (GIT) can breakdown an ingested protein into amino acids. The term bioavailability refers to the proportion of amino acids from an ingested protein that is actually absorbed and utilized by the body (Stein, Sève, Fuller, Moughan, & De Lange, 2007). In practice, bioavailability should be considered in relation to individual amino acids, as some are more readily released and absorbed from dietary proteins than others. As the primary source of nitrogen in the human diet, dietary proteins are vital for growth, tissue maintenance, and various physiological functions (Sá, Moreno, & Carciofi, 2020). During digestion, proteins are broken down into free amino acids and small peptides, which are then absorbed through the epithelial cells of the gastrointestinal tract. If a protein is not fully digested and absorbed in the upper gastrointestinal tract, certain amino acids may remain unavailable to the body, potentially resulting in nutritional deficiencies (Xie et al., 2022). Moreover, any undigested proteins reaching the colon could have adverse effects on the gut microbiome, which can also have undesirable health effects.

Due to their importance in human nutrition, several researchers have recently focused on the digestibility and/or bioavailability of the proteins in meat analogues. Zhou and co-workers compared the *in vitro* digestibility and gastrointestinal fate of animal-based with plant-based meat (Zhou, Hu, Tan, Zhang, & McClements, 2021). The authors found that meat proteins in ground beef were digested more rapidly during the gastric phase but more slowly during the intestinal phase compared to the plant proteins in soy-based meat analogues. In a separate study, Xie and colleagues observed that the digestibility of proteins in plant-based meat analogues was lower than that of real meat during the intestinal phase. However, their findings indicated variability in the gastric phase, with different digestion patterns observed between beef and pork (Xie et al., 2022). The differences in digestion behavior observed in these studies may be due to differences in the types of proteins used, the processing and cooking methods employed, and the compositions of the food matrices surrounding the proteins (Xie et al., 2022; Zhou et al., 2021). In general, the digestibility and bioavailability of proteins is known to depend on protein type, food matrix effects, and manufacturing conditions (Reynaud et al., 2021). Previous studies suggest that plant proteins have lower digestibility (75-80%) compared to animal proteins (90-95%), which has been attributed to the presence of antinutritional factors, indigestible cell walls, and specific protein structures in plants (Sá et al., 2020). Understanding the factors that reduce the digestibility of plant proteins can be used to design processing methods that improve their nutritional profile. For instance, removing or chelating antinutrients, breaking down plant cell walls, or modifying protein structure through pretreatments and processing can all be utilized to improve the digestibility and bioavailability of plant proteins. Clearly, further research is needed to better understand the relative digestibility of various meat products and their plant-based analogues, as well as the underlying factors contributing to these differences. Addressing this knowledge gap will require a combination of *in vitro* and *in vivo* methodologies, including simulated gastrointestinal tract models, cell culture systems, animal feeding trials, and human clinical studies.

Different Sources of Plant-Proteins in Meat Analogues

The growing demand for alternative proteins, specifically meat substitutes, has led to a rapid market expansion in this area (Kyriakopoulou et al., 2019). Various plant-based protein sources have been used to create meat analogues that mimic burgers, sausages, nuggets, meatballs, and whole muscle products. These products are often created using extrusion technology, but other methods are being employed such as the shear cell and 3D printing (McClements & Grossmann, 2021a; 2021b). As mentioned earlier, these products are typically designed to mimic the appearance, texture, flavor, and mouthfeel of real meat products. However, it is also critical that they are designed to match the nutritional attributes of real meat products, such as amino acid profile, digestibility, and bioavailability, otherwise there may be undesirable health consequences.

Several types of plant proteins have already demonstrated the techno-functional and nutritional properties necessary to serve as effective substitutes for meat proteins. For example, soybean and wheat proteins have been widely utilized in the development of meat analogues that replicate many of the desirable physicochemical, sensory, and nutritional characteristics of conventional meat (Ishaq et al., 2022). However, a range of other plant proteins also hold potential for use in meat analogue formulations. Despite this, limited research has been conducted on their digestibility and bioavailability, particularly within the context of meat alternatives. This section, therefore, explores various plant protein sources suitable for the formulation of meat analogues and highlights the current state of knowledge regarding their nutritional profiles.

Cereals

Cereals such as wheat, rice, hemp, corn, barley, and oats are among the primary sources of plant-based proteins in the human diet. Wheat, in particular, is valued for its gluten, which possesses viscoelastic properties that make it especially well-suited for producing meat-like structures and textures. These properties arise from the formation of a three-dimensional protein network stabilized by intramolecular (primarily gliadins) and intermolecular (primarily glutenins) disulfide bonds, along with hydrogen bonds and hydrophobic interactions. During processes such as extrusion or shear cell treatment, hydrated gluten develops these networks, leading to the formation of fibrous structures that closely mimic the texture of meat (Ahmad et al., 2022; Kyriakopoulou et al., 2021). The structural organization of protein molecules can be modified by altering the degree of gluten cross-linking through reducing conditions, high temperatures, hydrostatic pressures, changes in pH and ionic strength, isolation of protein subunits, the use of cross-linking enzymes (especially transglutaminase), and by interacting the proteins with polyphenols. All of these factors can also influence the digestibility and bioavailability of the protein molecules. It is important to note that consuming gluten-containing products poses a potential health risk for individuals with gluten sensitivity, intolerance, or allergies (Ishaq et al., 2022; Kyriakopoulou et al., 2019).

Cereal proteins can also be induced to form fibrous structures by extruding them through electrospinning syringes. For instance, zein, the main storage protein in corn, can be dissolved in ethanol and then extruded using electrospinning and mechanical elongation techniques to form individual fibers or fibrous networks. Zein fibers produced through this method have been integrated with soy protein networks to develop plant-based meat analogues that resemble chicken in texture and structure (Kyriakopoulou et al., 2021). The incorporation of zein in fibrous form enhances the texture of these analogues, contributing to a more meat-like mouthfeel. However, gels containing fibrous zein exhibit relatively low water-holding capacity, a limitation that can be addressed by incorporating hydrocolloids such as gums (Mattice & Marangoni, 2020).

The nutritional quality of food proteins depends on their amino acid profile (especially the levels of essential amino acids), as well as the fraction of these amino acids that actually get absorbed by the body (Raja et al., 2022). Protein quality is usually characterized by the digestible indispensable amino acid score (DIAAS) or the protein digestibility corrected amino acid score

(PDCAAS) (Berrazaga, Micard, Gueugneau, & Walrand, 2019; van Vliet et al., 2015). These parameters are based on the amino acid profile of the protein, as well as the fraction of essential amino acids that are actually absorbed by the human body (Joye, 2019). Typically, a DIAAS or PDCAAS of 100 or above indicates that the protein contains all of the essential amino acids required in the human diet, whereas a value below 100 suggests that the protein lacks one or more essential amino acids. Many major sources of plant proteins have DIAAS or PDCAAS values below 100. For example, cereals are deficient in lysine, while legumes are lacking in methionine. To estimate protein quality, it is therefore important to know the amino acid composition, digestibility, and bioavailability of proteins. The bioavailability of amino acids is often established by measuring the fraction that is not absorbed in a specific region of the gastrointestinal tract. For more accurate estimates of bioavailability, it is better to measure the concentration of non-absorbed amino acids at the end of the small intestine (ileum), rather than in the faces, because amino acids passing through the colon may be metabolized by microorganisms in the large intestine (Han et al., 2019).

Studies have shown that the digestibility of proteins depends on their three-dimensional conformation because this influences the ability of proteins to reach the peptide bonds. For example, a negative correlation has been observed between protein digestibility and the presence of β-sheet and β-turn structures in proteins. While these secondary structures contribute significantly to the viscoelastic properties of wheat flour, their high hydrophobicity makes them less accessible to digestive enzymes, thereby reducing their overall digestibility (Joye, 2019).

The presence of antinutritional factors in plants may also reduce the bioavailability of amino acids by inhibiting protein digestion (Vanga, Wang, & Raghavan, 2020). Phytate and dietary fibers are common compounds found in PBPs that reduce the digestibility and bioavailability of proteins by forming stable complexes with metal ions (enzyme cofactors) and proteins (Joye, 2019). In most cereals, such as rice, wheat, and millet, phytate accumulates in the aleurone layer and can be removed through milling (Joye, 2019). Phytate and phytases can be inactivated by food processing operations that involve heating and shearing. Dietary fibers may slow the digestion and absorption of proteins in the stomach and small intestine by reducing the accessibility of proteases to the proteins (Joye, 2019). For example, they may increase the viscosity of gastrointestinal fluids, form hydrogels that trap proteins inside, or create complexes with the proteins or enzymes.

Pseudo-cereal proteins have also been explored for their potential in developing meat analogues. Quinoa is a pseudo-cereal with a relatively high protein content (12-20%) that contains a high fraction of essential amino acids (39%). Quinoa is particularly notable for its high levels of essential amino acids, including lysine (6.5%) and methionine (2.6%), making it an effective complement to legume proteins, which are typically deficient in methionine (Alrosan et al., 2022; Fernandez-Lopez et al., 2021; Shokry, 2016; van Vliet et al., 2015). Due to its functional versatility, quinoa has been incorporated into various meat products as a meat extender, gelling agent, fat substitute, or binder (Kyriakopoulou et al., 2021; Shokry, 2016). Shokry (2016) demonstrated that replacing soy flour with quinoa flour in beef burgers significantly improved their water and oil holding capacities in both raw and cooked forms, while also enhancing cooking performance and sensory qualities. Furthermore, quinoa proteins are increasingly used in the development of meat analogues duc to their favorable functional properties. For example, quinoa protein has been employed as an alternative to wheat protein in gluten-free meat analogues, showing comparable water retention, emulsification, and foaming characteristics to soy protein (Fernandez-Lopez et al., 2021).. It has been successfully applied as a binder, gelling, and thickening agent in meat analogues, like plant-based sausages (Kyriakopoulou et al., 2021; Lingiardi, Galante, de Sanctis, & Spelzini, 2022).

Legumes

Pea, lentil, lupine, chickpea, soybean, fava bean, mung bean, and other types of beans are common legumes, containing 20–40% (dry weight) of dietary proteins. These legume proteins are commonly used to formulate meat analogues due to their functional attributes, such as gelling, water holding, binding, and emulsifying. Legumin and vicilin are the most common globulin legume proteins, which occur in different ratios in different legume species. These storage proteins are insoluble in pure water but are soluble in dilute aqueous salt solutions (Gu et al., 2022). The functional properties of these proteins depend on the type and ratio of globulin proteins (2S, 7S, 11S, and 15S) and the presence of non-protein compounds, such as carbohydrates (Kyriakopoulou et al., 2021). Legume proteins have been used to formulate a wide range of meat analogue products, including minced, emulsified, and whole muscle types (Gu et al., 2022; Kyriakopoulou et al., 2021). These proteins not only increase the protein content of meat analogues but also enhance their chewiness and hardness due to the formation of crosslinked protein networks (Ahmad et al., 2022).

Soybean proteins are a good source of proteins for use in plant-based meat analogues because of their abundance, high protein content, good protein quality, and desirable functional properties. They can be used as defatted soy flours (≈50% protein), protein concentrates (≈70% protein), or protein isolates (≈90% protein). The ingredient selected for a particular application depends on the required physicochemical, sensory, and nutritional properties (Ahmad et al., 2022; Kyriakopoulou et al., 2021). Studies on the bioavailability of soy protein have shown that a significant portion is degraded and its amino acids are excreted from the body as urea, resulting in a reduced amount available for skeletal muscle synthesis (Berrazaga et al., 2019). Fortifying soy protein with branched-chain amino acids - specifically leucine, isoleucine, and valine - may enhance overall protein synthesis in the body, including in skeletal muscle. This fortification can also improve the peripheral availability of dietary amino acids and decrease splanchnic extraction, particularly in elderly individuals (Berrazaga et al., 2019).

There are several potential limitations to using legume proteins to formulate meat analogues. One drawback of using soy proteins to formulate these products is that some people have soy allergies (Shokry, 2016). Moreover, legumes may contain antinutritional factors that impair protein digestibility and increase nitrogen excretion in the feces. For example, certain polyphenols can bind to and inhibit proteolytic enzymes, thereby reducing the efficiency of protein digestion. Another limitation of using legumes as functional ingredients is the presence of compounds that may cause gastrointestinal discomfort, such as raffinose. Humans lack the enzyme α-galactosidase necessary to hydrolyze raffinose in the gut, which prevents its proper digestion and absorption (Bennetau-Pelissero, 2019). Anaerobic fermentation of this and other carbohydrates by the microflora in the human intestine can result in gastrointestinal symptoms such as diarrhea. These challenges can often be mitigated by using appropriate pre-treatment and processing methods (e.g., soaking, dehulling, heating, extruding, and enzymatic hydrolysis) to remove or inactivate antinutritional factors and enhance protein digestibility and bioavailability (Gu et al., 2022).

Oilseeds

Plant proteins are also extracted from various kinds of oilseeds, including rapeseed, sunflower, and cotton seed, usually as a by-product of the edible oil processing industry (Kyriakopoulou et al., 2021). These oilseed proteins are affordable, abundant, and often have good functional attributes. For instance, sunflower proteins have been reported to exhibit water holding, emulsification, gelling, and foaming properties comparable to those of soybean proteins (Salgado, Molina Ortiz, Petruccelli, & Mauri, 2012). The cell walls of sunflower seeds are typically disrupted through pressure treatment and n-hexane extraction, processes that enhance enzymatic activity and improve protein digestibility. However, if the applied temperatures and pressures are excessively high, they can negatively impact the bioavailability of sunflower protein and its essential amino acids (Ditta & King, 2017). The presence of polyphenols and antinutritional factors in defatted oilseeds may also pose challenges, though these can be mitigated through the use of efficient extraction techniques (Kyriakopoulou et al., 2021). While phenolic compounds in extracted sunflower proteins have been shown to reduce gel hardness, this effect can be counteracted by the application of high-intensity ultrasound (Malik & Saini, 2018; Salgado et al., 2012). Rapeseed and canola proteins can also exhibit acceptable functional properties provided they are processed appropriately, typically using high temperature and pressure extrusion processes. The rapeseed and canola proteins mainly consist of cruciferin (globulin) and napin (albumin). In the presence of polysaccharides and salts, they can form coherent gels and exhibit good emulsifying and foaming characteristics. Thus, they can be used to formulate meat analogues with desirable textural attributes (Ahmad et al., 2022; Kyriakopoulou et al., 2021).

Hemp seed protein (25-30% of the seed) has a high digestibility and a good essential amino acid profile, which makes it a promising candidate for meat analogue applications. Typically, it is used as a defatted hemp seed protein meal (HPM) containing around 44% protein or a hemp seed protein isolate (HPI) containing around 84% protein. The extraction and isolation methods used affect the functional properties of HPM and HPI. Comparing the amino acid composition of HPI and HPM revealed a higher Arg/Lys ratio in HPI. The favorable functional attributes of hemp proteins make them suitable substitutes for soy proteins in meat analogues produced by high moisture extrusion (Kyriakopoulou et al., 2021; Malomo, He, & Aluko, 2014). The protein digestibility percentage and PDCAAS values for whole, meal, and dehulled hemp seed are reported to be around 84-86% and 50%, 91-98% and 47%, and 84-92% and 64%, respectively. The main limiting amino acid in all compositions was reported to be lysine. The removal of the hull increases the protein digestibility and the PDCAAS score. These results demonstrate that hemp proteins have a PDCAAS equivalent to or greater than some cereals, nuts, and some pulses (House, Neufeld, & Leson, 2010).

Nuts

Certain types of nuts also serve as valuable protein sources for developing meat analogues. For example, peanut protein concentrates exhibit favorable functional properties such as thermal gelation, binding, and foaming. High-moisture extrusion of defatted peanut flour can produce fibrous structures suitable for use in minced or emulsion-type meat substitutes, offering a protein content of approximately 55% (Kyriakopoulou et al., 2021). Walnut protein is another promising candidate, with glutelins comprising the majority of its protein content (70%), followed by globulins (18%), albumins (7%), and prolamins (5%). It is characterized by good digestibility and a well-balanced essential amino acid profile, making it a nutritionally beneficial ingredient for plant-based meat formulations. (Paz-Yepez, Peinado, Heredia, & Andre´s, 2019). Peanuts have a higher protein content than other nuts. Globulins are the most prevalent protein in peanuts, with the two main globulins (arachin and conarachin) making up approximately 87% of the total protein content (Arya, Salve, & Chauhan, 2016). The digestibility of peanut proteins is often relatively low, which may be due to alterations in protein structure or aggregation resulting from roasting (Paz-Y´epez et al., 2019). Freitas et al., (2012) reported that lysine is the limiting amino acid in baru almond, peanut, and Brazil nut, whereas cashew nut had no limiting amino acid. These nuts have a relatively high protein digestibility, but their PDCAAS vary significantly, e.g., from 57% for baru almond to 90% for cashew nut.

Chia Seeds

Chia seeds have also been used as a source of plant proteins to pro- duce meat analogues. They have a relatively high protein content (19–23%) and the proteins have good functional attributes, including water holding, emulsifying, and gelling capacities (Fernandez Lopez et al., 2021; Rendon-Villalobos,

Ortíz-Sanchez, & Flores-Huicochea, 2018). Although this protein source is rich in glutamic acid (12%), arginine (11%), and aspartic acid (6.1%), it is limiting in lysine (3.7%). Chia seed proteins have been utilized as animal fat substitutes to enhance the nutritional and sustainability properties of meat products. The high water and oil-holding capacity of chia seeds results in low cooking loss without negatively affecting sensory properties, making it an ideal protein source for use in sausages and meat analogues (Fernandez-Lopez et al., 2021; Olivos-Lugo, Valdivia-Lopez, & Tecante, 2010). Another advantage of chia protein is that it is gluten-free, which is an advantage for those with celiac disease (Rendo´n-Villalobos et al., 2018).

Olivos-Lugo et al., (2010) examined the nutritional and functional properties of chia seed proteins. They found that a protein isolate derived from chia seeds exhibited high digestibility, which was attributed to protein denaturation during the isolation process. This process involved thermal treatment at 90 °C for 10 minutes and alkaline treatment at pH 12 for 1.5 hours, both of which promoted protein unfolding. As a result, the proteins became more accessible to proteolytic enzymes, enhancing their digestibility. The authors noted that chia seeds possess a high protein content, comparable to that of lentils, peas, and chickpeas, making them a promising alternative protein source for meat analogues. The incorporation of chia seeds into both animal- and plant-based meat products primarily serves to increase dietary fiber content and reduce fat content. More research on the amino acid profile and protein digestibility and bioavailability of chia seed as a protein substitute in meat analogues is still needed.

Effect of Main Development Processes on Digestibility and Bioavailability of Meat Analogues

Plant-based meat analogues are primarily produced using extrusion technology; however, several alternative methods are currently being explored, including shear cell processing, 3D printing, electrospinning, freeze structuring, cell culturing, and wet-spinning. Broadly, all fibrous meat analogue production techniques fall into two categories: top-down and bottom-up approaches. Top-down strategies - such as extrusion, electrospinning, wet-spinning, phase separation, freeze structuring, and cell culturing - focus on generating microstructural protein fibers through the application of external force fields, allowing for efficient alignment and structuring. In contrast, bottom-up approaches, such as 3D printing, involve assembling pre-fabricated microstructural building blocks to create larger, more complex architectures that closely resemble the structure of conventional meat (Dekkers, Boom, & van der Goot, 2018; Indurkar, Pandit, Jain, & Dandekar, 2021).

During the conventional production of plant-based meats, typically involving the extrusion technique (and sometimes shear cell processing), plant-based proteins are exposed to harsh conditions. The thermal treatment and shear stress applied during this process can disrupt intra- and intermolecular bonds, leading to the denaturation of the proteins. This denaturation may weaken hydrogen and disulfide bonds, enhance hydrophobic interactions, and form reversible protein networks. As a result, the protein structure is altered, which can impact both the bioavailability and digestibility of the product (Cornet et al., 2022). Furthermore, different plant protein sources exhibit unique responses to varying temperatures. For example, when wheat protein is heated in extruders, glutenin polymers are formed through disulfide bonding, contributing to the development of viscoelastic properties and affecting the nutritional characteristics of the product. On the other hand, the fate of disulfide bonds in soy protein leads to a reduction in viscosity and the deactivation of antinutritional components (Dekkers, Emin, Boom, & van der Goot, 2018). Overall, selecting the most suitable main process for each plant protein source and optimizing its parameters seem crucial to getting meat analogues with the best bioavailability, digestibility, and functional qualities.

Green Processing Methods

The processing of PBPs meant for meat analogue production has been significantly impacted by the innovative green technologies. Green technologies and the increased public interest in chemical free products have encouraged innovation in processing methods. These cutting-edge techniques include enzymatic, sonication, microwave, high-pressure process (HPP), electric field (PEF), and ohmic heating (OH) technologies, among others. The health and safety aspects of green processes are taken into account in addition to the environmentally friendly ones (Picot-Allain, Mahomoodally, Ak, & Zengin, 2021).

In this section, we examine the impact of green processing methods on the digestibility and bioavailability of alternative proteins. In general, the amount of protein digested by digestive enzymes and absorbed by cells, relative to the total protein ingested, is determined through protein digestibility assays (Lopez, Galante, Robson, Boeris, & Spelzini, 2018). The digestibility of proteins is influenced by factors such as their structural composition, aggregation state, release from the food matrix, and the presence of anti-nutritional substances (Habiba, 2002; Nowin´ski & Ufnal, 2018). The digestibility of plant proteins can therefore be improved by using processing methods to control these various factors. We also examined the potential of several green processing technologies that could be utilized as pre-manufacturing treatments to enhance protein digestibility and bioavailability, including enzymatic treatment, sonication, microwave, high-pressure processing (HPP), pulsed electric fields (PEF), and ozone (OH) technologies.

Enzymatic Treatment

Enzymatic treatments can be used to enhance protein digestibility and bioavailability using either fermentation and/or controlled hydrolysis methods.

Fermentation Process

Fermentation processes use a variety of microbial species, especially yeasts and bacteria, to enhance the digestibility and bioavailability of plant proteins. Enzymes are generated by the microorganisms during fermentation, which selectively hydrolyze the proteins (Boye, Wijesinha-Bettoni, & Burlingame, 2012). The hydrolysis of proteins during fermentation can greatly enhance protein digestibility and bioavailability (Bartkiene, Juodeikiene, & Vidmantiene, 2012; Fawale, Gbadamosi, Ige, & Kadiri, 2017; Sá et al., 2020). For example, proteins that are typically insoluble in water and have low digestibility can have their bioavailability increased through fermentation.

Enzyme Hydrolysis

In general, the partial hydrolysis of proteins using enzymes alters their molecular structure and aggregation state, which can increase their solubility, digestibility, and bioavailability (Day, Cakebread, & Loveday, 2021; Liceaga & Hall, 2019). The digestibility and bioavailability of proteins isolated from plant sources can also be enhanced by using en- zymes that hydrolyze dietary fibers that can inhibit protein digestion, such as cellulases and hemicellulase (Dias, Abreu, Silvestre, & Schwan, 2010; Fawale et al., 2017). There are various commercial proteases or peptidases, such as Alcalase™ and

Flavorzyme™, that have been employed to improve the digestibility and bioavailability of proteins (Kapravelou et al., 2013; Vaz Patto et al., 2015). Several studies have shown that hydrolyzing plant proteins under controlled conditions can improve their functionality (Sun-Waterhouse, Zhao, & Waterhouse, 2014; Voudouris et al., 2017). As a result, manufacturers can use these enzyme treatments to enhance both the nutritional (digestibility and bioavailability) and functional (solubility, foaming, emulsifying, and gelling) properties of plant proteins in meat analogues. However, there are challenges with using enzymatically hydrolyzed proteins in meat analogues, as the process can lead to the formation of hydrophobic residues that cause bitterness (Amagliani, O'Regan, Kelly, & O'Mahony, 2017; Liceaga & Hall, 2019). Additionally, enzymes are often relatively expensive to use.

Ultrasound and Microwave Treatments

Initially, the mechanism of action of ultrasound and microwave treatments are briefly presented, and then their potential impact on the digestibility and bioavailability of plant proteins is discussed. The utilization of high-intensity ultrasound treatments for improving protein functionality has been the subject of considerable research and development efforts, but it still needs to be implemented commercially (Lo´pez-Pedrouso, Lorenzo, Zapata, & Franco, 2019). This technique has been shown to alter the structure of proteins, resulting in higher digestibility and bioavailability (Kang, Zhang, Guo, Lei, & Yang, 2022; Khatkar, Kaur, & Khatkar, 2020; Pan et al., 2020; F. Sun et al., 2022). The mechanism of action on plant-based proteins (PBPs) primarily involves microbubble cavitation in the presence of high-intensity oscillating pressure waves (Figure 1-a). These pressure waves induce acoustic cavitation, leading to the formation and collapse of bubbles within a fluid. The extremely high local temperatures and pressures associated with the rapid expansion and collapse of these bubbles cause thermo- mechanical impacts on the surrounding fluids (Lo´pez-Pedrouso et al., 2019).

Jin, Okagu, Yagoub, and Udenigwe (2021) studied the effects of ultrasound treatment on the structure and *in vitro* digestibility of buck- wheat protein isolates. Unlike most other cereals, including wheat and corn, which are deficient in lysine, buckwheat is a rich source of lysine. The application of high-intensity ultrasonic waves (20 kHz) was shown to alter the secondary structure of the proteins, as well as disrupting any protein aggregates, which led to an increase in protein digestibility. Other researchers have shown that sonication can enhance buckwheat digestibility by disruption intramolecular disulfide bonds, which promotes partial protein unfolding (Jin et al., 2021). Sonication has also been shown to alter the secondary structure of peanut proteins, which enhanced protein digestibility (Ochoa-Rivas, Nava-Valdez, Serna Saldívar, & Chuck-Hernandez, 2017).

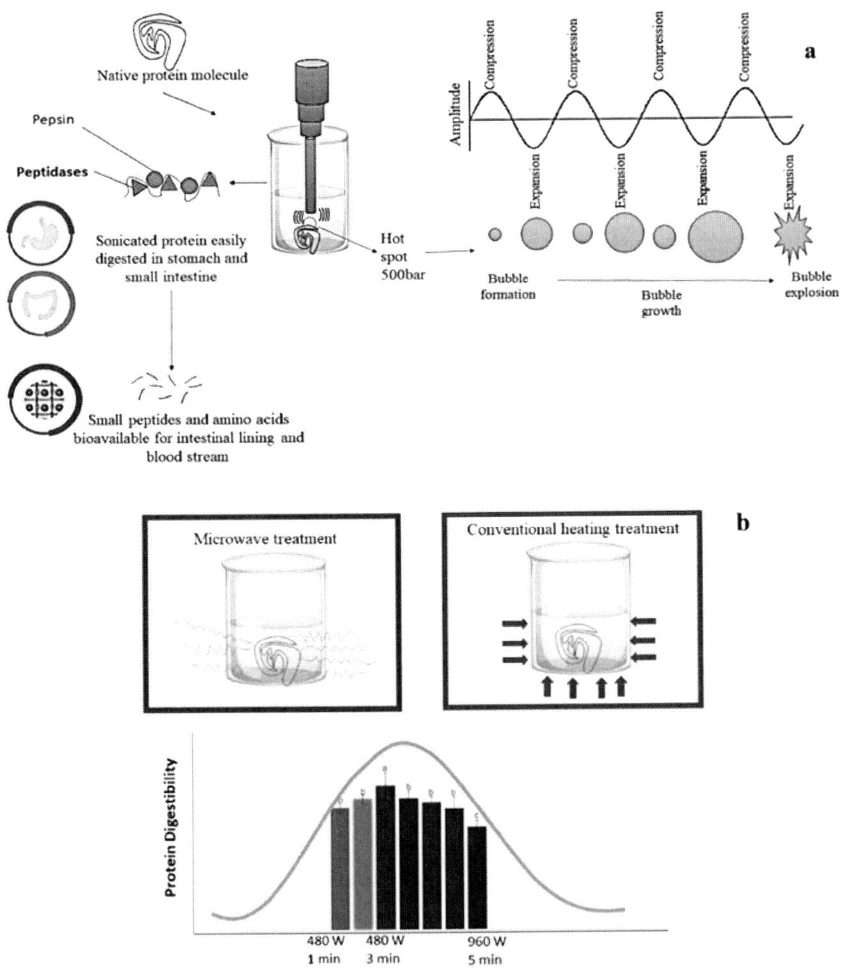

Figure 1. Effect of (a) ultrasound and (b) microwave treatment on the digestibility and bioavailability of plant-based proteins.

The ability of microwave treatments to improve the digestibility and bioavailability of plant proteins has also been examined (Drulyte & Orlien, 2019; Habiba, 2002; Sa et al., 2020). Microwave treatments induce changes in protein structure due to the heat generated by the molecular motion of polar molecules when exposed to electromagnetic waves in the frequency range of 300 MHz to 300 GHz (Divekar et al., 2017). In addition to being a more environmentally friendly technology compared to other heating methods, microwaves interact directly with proteins, allowing for more efficient energy transfer (Brittany & Hayes, 2002) (Figure 1-b). There is growing interest in using microwave processing to improve the digestibility and bioavailability of plant proteins intended for use in meat analogues. Additionally, microwave treatments have the potential to deactivate protease inhibitors and microorganisms, which could enhance the health benefits and safety of plant-based foods (Abbas & Ahmad, 2018).

Trypsin inhibitors are one of the most important antinutritional factors in soybeans. These are substances that can interact strongly with trypsin and chymotrypsin, thereby reducing protein digestibility. Vanga et al., (2020), investigated the effects of ultrasound and microwave processing on the digestibility of soymilk protein. The researchers demonstrated that a 16-minute ultrasound treatment improved soymilk protein digestibility to 84%, while a 10-minute microwave treatment at 85°C increased digestibility to 93%. The superior performance of the microwave treatment was attributed to the inactivation of trypsin inhibitors and the enhanced protein unfolding caused by the high temperatures involved, while ultrasound processing generated minimal heat. However, it is crucial to optimize processing conditions. Excessive microwave heating of soybeans has been shown to reduce protein digestibility due to increased protein denaturation and aggregation (Vanga et al., 2020) (Figure 1-b). Additionally, the water content of soybeans significantly influences the inactivation of trypsin inhibitors during processing. For example, microwave heating of dry soybeans (7% moisture) led to 70% enzyme inactivation, whereas heating wet soybeans (25% moisture) resulted in 96% enzyme inactivation (Hernandez-Infante, Sousa, Montalvo, & Tena, 1998). When processing legumes with microwaves, the radiation energy breaks down hydrogen bonds, increases dissolved ion mobility, and changes the secondary structure of proteins, enhancing their functional properties, such as water holding capacity, foaming, emulsifying, and digestibility (Negi, Boora, & Khetarpaul, 2001; Satya, Kaushik, & Naik, 2010). According to Shimelis and Rakshit (2005), the improvement provided by microwave treatments depends on how long the product is processed.

Typically, protein digestibility increases with processing time up to a certain point, after which it may decrease with extended treatment. Additionally, nonenzymatic browning can occur with prolonged treatment, negatively impacting protein quality (Sa et al., 2020). According to Pysz, Polaszczyk, Leszczyn´ska, and Piątkowska (2012), the digestibility of proteins in various broad bean varieties (Bacchus, Basta, and Windsor White) improved following microwave treatment, with the best results achieved at an intermediate processing time. It is crucial that the protein molecules partially unfold without aggregating with each other.

High-Pressure Process (HPP)

HPP is considered to be a green processing technology that can also promote modifications to the functionality and nutritional quality of proteins. Typically, high isostatic pressure (ranging from 100 to 1000 MPa) is applied to the food matrix, often transmitted through water at approximately room temperature (Lopez-Pedrouso et al., 2019). High-pressure processing (HPP) induces protein denaturation when the pressure exceeds a critical threshold (Martínez-Monteagudo & Balasubramaniam, 2016). However, it does not alter the primary structure of proteins. As a result, HPP can enhance the digestibility and bioavailability of proteins without negatively affecting their essential amino acid profile. Many researchers have investigated the effects of HPP on the functional properties of PBPs (Akharume, Aluko, & Adedeji, 2021; Avelar, Vicente, Saraiva, & Rodrigues, 2021; Queiro´s, Saraiva, & da Silva, 2018). In contrast, there are only a few studies on its impact on protein digestibility and bioavailability.

Yin, Tang, Wen, Yang, and Li (2008) examined how HPP affects the *in vitro* digestibility of bean protein isolate. They found that protein digestibility was influenced by the magnitude of pressure applied during processing. Relatively low-pressure treatments (<200 MPa) enhanced protein digestibility by promoting partial protein unfolding and deactivating antinutritional factors. In contrast, higher pressure treatments (>200 MPa) reduced protein digestibility due to excessive protein aggregation. Therefore, optimizing processing conditions is crucial for improving protein digestibility (Linsberger-Martin, Weiglhofer, Phuong, & Berghofer, 2013). Similarly, Hall and Moraru (2021) demonstrated that high-pressure processing (HPP) of fava bean or lentil proteins resulted in gastric digestibility comparable to or higher than untreated controls, although the overall protein digestibility remained unchanged.

Electric Field

Ohmic Heating (OH)

Ohmic heating (OH) is a novel food technology being efficient in processing time, energy, water, resources and cost of operation. The application of a moderate electric field (less than 1000 V/cm) leads to a uniform heat distribution within the food matrix, which induces both structural and sequential changes in proteins. The presence of hydroxyl radicals (OH) can enhance the functional and nutritional properties of plant proteins by promoting their degradation and facilitating the formation of molecular interactions or the rearrangement of disulfide bonds. Although the allergenic soybean trypsin inhibitor protein is highly thermostable and acid-stable, OH at 50 Hz (95°C for 10 minutes) can significantly reduce the immunoreactivity of this SPI allergen. However, it remains unclear whether the improvement is due to protein aggregation or electrochemical interactions (Pereira et al., 2021). In a study by Jung, Hwang, Jun, and Park (2022), the soy protein-based meat analogues were processed by OH (30 V/cm electric field). It was concluded that 3 min of cooking at 100°C yielded in similar textural quality of commercial chicken nuggets and restructured beef steaks (Jung et al., 2022).

Electric field processing modifies the functional and nutritional properties of proteins by altering denaturation mechanisms, aggregation, and interaction patterns. A study by Rodrigues, Avelar, Machado, Pereira, and Vicente examined the impact of EF on the digestibility and allergenicity of certain proteins. The pepsin resistance of β-conglycinin in soybeans, which contributes to allergenicity, is attributed to its hydrophobic regions, resulting in reduced release of bioactive peptides during protein digestion. EF processing can enhance protein digestibility by inducing structural modifications that increase the protein's susceptibility to gastrointestinal enzymes (Rodrigues et al., 2020).

However, the probability of formation of some aggregated proteins with lower digestibility or generation of cytotoxic and immunoreactive compounds should be investigated.

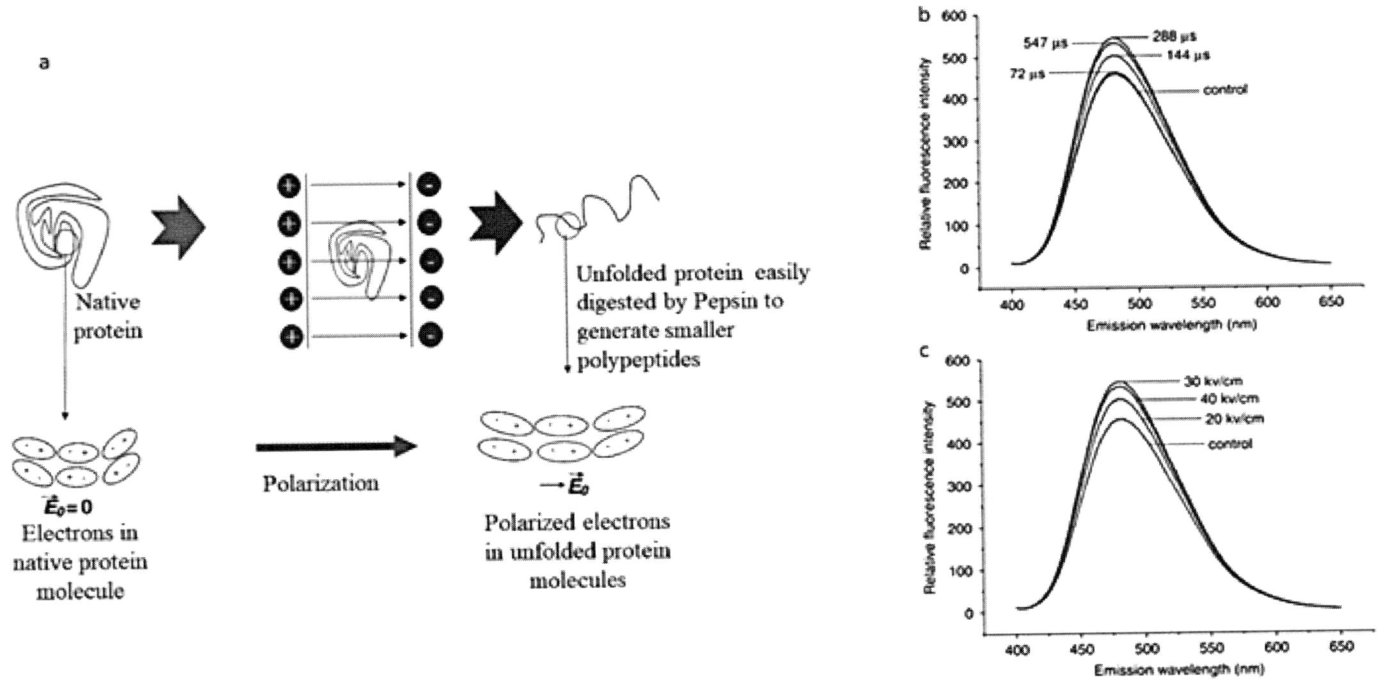

Figure 2. (a) Schematic representation of the PEF (Pulsed Electric Field) mechanism of action in plant-based proteins. Fluorescence spectra of PEF-treated soy protein isolate at (b) different treatment times and (c) varying strengths.

Pulsed Electric Field (PEF)

PEF is a non-thermal process that consists of applying electrical pulses of short time and high voltage to a material (Zhao, Yang, Tang, Zhang, & Hua, 2009). There have not been many studies investigating the impact of PEF treatments on food proteins, especially plant-based ones. Powerful PEF devices have the potential to ionize some of the chemical groups in proteins, which affects their structure and aggregation state, resulting in an impact on their digestibility and bioavailability (Zhao, Tang, Lu, Chen, & Li, 2014) (Figure 2-a). Mild PEF treatments primarily affect the secondary structure of PBPs, thus altering protein digestibility (Lopez-Pedrouso et al., 2019). According to Vanga et al., (2016), PEF processing of peanut protein led to changes in its secondary structure, which increased its *in vitro* digestibility. Protein surface hydrophobicity, a structural feature used to assess protein conformational changes, is linked to both functional properties (Kato & Nakai, 1980) and digestibility (Zhang et al., 2019). A study by Li, Chen, and Mo (2007) demonstrated that PEF treatment can influence the surface hydrophobicity of soy protein isolate, resulting in a conformational change in the protein structure (Figure 2-b,c). These changes resulted from surface hydrophobicity along with other conformational modifications such as molecular size, surface charge and polydispersity can lead to a better digestibility (X. Sun, Ohanenye, Ahmed, & Udenigwe, 2020).

Conclusions and Future Perspective

There is growing concern about the negative impact of raising livestock for meat production. Livestock farming contributes to the spread of zoonotic diseases and antibiotic resistance, while also exacerbating resource depletion, increasing greenhouse gas emissions, causing pollution, and contributing to biodiversity loss. As a result, there is increasing interest in developing plant-based alternatives to traditional meat products. These alternatives can be formulated from widely used plant protein sources, such as soybeans, wheat, corn, and peas. There is also a broad range of other protein-rich plant sources throughout the world that are not currently being utilized. Many of these plant species may contain proteins that are beneficial for formulating next-generation plant-based meat analogues. However, the nutritional and functional characteristics of plant proteins are often inferior to animal ones. Consequently, there is a need to develop effective strategies to improve the nutritional value, digestibility, bioavailability, and functional properties of plant proteins from both conventional and emerging sources. The presence of antinutritional compounds in plant proteins is a significant consideration when producing plant-based meat analogues. As such, efficient processing techniques are essential to enhance the digestibility and bioavailability of plant proteins. This review discusses various technologies used in the production of meat analogues, with a particular focus on green processing technologies aimed at improving plant protein digestibility and bioavailability.

A range of physical, chemical, and biological methods - such as fermentation, enzyme hydrolysis, pulsed electric fields, sonication, and microwave treatments - has been explored, along with their impacts on the nutritional value of plant proteins. Each of these techniques has its own set of advantages and drawbacks. For example, enzymatic hydrolysis and fermentation can break down peptide bonds, thereby improving protein digestibility. However, they can also result in the formation of undesirable bitter compounds. Similarly, ultrasound, microwave, high-pressure processing (HPP), and electric field treatments can promote partial unfolding of proteins, enhancing their digestibility and bioavailability. Nevertheless, excessive processing can lead to protein aggregation, which may negate these benefits. On the other hand, having knowledge of the relationships between each plant source and each processing method within a selected parameter configuration is crucial for producing higher-quality proteins intended for meat analogue manufacturing. However, one should take into account that improving the protein digestibility may cause a plant protein source to lose some of its nutritious qualities, therefore processing choice also needs to be given more careful consideration in this area.

Consequently, operating conditions must be optimized. Currently, there is only a limited understanding of the advantages and disadvantages of different green technologies for improving the nutritional attributes of plant proteins, as well as about their potential for commercial application. Future initiatives should definitely concentrate on improving these processing techniques to address the issue of PBPs' lower bioavailability and digestibility with the least possible negative effects on their functional and nutritive qualities.

References

Abbas, Y., & Ahmad, A. (2018). Impact of processing on nutritional and antinutritional factors of legumes: A review. *Annals Food Science and Technology,* 19, 199–215.

Ahmad, M., Qureshi, S., Akbar, M. H., Siddiqui, S. A., Gani, A., Mushtaq, M., et al., (2022). Plant-based meat alternatives: Compositional analysis, current development and challenges. *Applied Food Research,* Article 100154.

Akharume, F. U., Aluko, R. E., & Adedeji, A. A. (2021). Modification of plant proteins for improved functionality: A review. *Comprehensive Reviews in Food Science and Food Safety,* 20, 198–224.

Alrosan, M., Tan, T.-C., Easa, A. M., Gammoh, S., & Alu'datt, M. H. (2022). Recent updates on lentil and quinoa protein-based dairy protein alternatives: Nutrition, technologies, and challenges. *Food Chemistry,* Article 132386.

Amagliani, L., O'Regan, J., Kelly, A. L., & O'Mahony, J. A. (2017). The composition, extraction, functionality and applications of rice proteins: A review. *Trends in Food Science & Technology,* 64, 1–12.

Arise, A. K., Aliyu, B. N., & Ajidagba, S. D. (2020). Effect of thermal processing and fermentation on the chemical composition, protein digestibility and functional properties of bambara protein isolate. *Carpathian Journal of Food Science & Technology,* 12.

Arya, S. S., Salve, A. R., & Chauhan, S. (2016). Peanuts as functional food: A review. *Journal of Food Science & Technology,* 53, 31–41.

Avelar, Z., Vicente, A. A., Saraiva, J. A., & Rodrigues, R. M. (2021). The role of emergent processing technologies in tailoring plant protein functionality: New insights. *Trends in Food Science & Technology,* 113, 219–231.

Bartkiene, E., Juodeikiene, G., & Vidmantiene, D. (2012). Nutritional quality of fermented defatted soya and flaxseed flours and their effect on texture and sensory characteristics of wheat sourdough bread. *International Journal of Food Sciences & Nutrition,* 63, 722–729.

Bennetau-Pelissero, C. (2019). Plant proteins from legumes. In *Bioactive molecules in food* (pp. 223–265). Springer.

Berrazaga, I., Micard, V., Gueugneau, M., & Walrand, S. (2019). The role of the anabolic properties of plant-versus animal-based protein sources in supporting muscle mass maintenance: A critical review. *Nutrients,* 11, 1825.

Boye, J., Wijesinha-Bettoni, R., & Burlingame, B. (2012). Protein quality evaluation twenty years after the introduction of the protein digestibility corrected amino acid score method. *British Journal of Nutrition,* 108, S183–S211.

Brittany, L., & Hayes, D. (2002). *Microwave synthesis: Chemistry at the speed of light.* USA: Cem. Publ.

Cornet, S. H., Snel, S. J., Schreuders, F. K., van der Sman, R. G., Beyrer, M., & van der Goot, A. J. (2022). Thermo-mechanical processing of plant proteins using shear cell and high-moisture extrusion cooking. *Critical Reviews in Food Science and Nutrition,* 62, 3264–3280.

Day, L., Cakebread, J. A., & Loveday, S. M. (2021). Food proteins from animals and plants: Differences in the nutritional and functional properties. *Trends in Food Science & Technology,* 119, 428–442.

Dekkers, B. L., Boom, R. M., & van der Goot, A. J. (2018). Structuring processes for meat analogues. *Trends in Food Science & Technology,* 81, 25–36.

Dias, D. R., Abreu, C. M. P.d., Silvestre, M. P. C., & Schwan, R. F. (2010). *In vitro* protein digestibility of enzymatically pre-treated bean (Phaseolus vulgaris L.) flour using commercial protease and Bacillus sp. protease. *Food Science and Technology,* 30, 94–99.

Ditta, Y., & King, A. (2017). Recent advances in sunflower seed meal as an alternate source of protein in broilers. *World's Poultry Science Journal,* 73, 527–542.

Divekar, M. T., Karunakaran, C., Lahlali, R., Kumar, S., Chelladurai, V., Liu, X., et al., (2017). Effect of microwave treatment on the cooking and macronutrient qualities of pulses. *International Journal of Food Properties*, 20, 409–422.

Drulyte, D., & Orlien, V. (2019). The effect of processing on digestion of legume proteins. *Foods,* 8, 224.

Espinosa-Pa´ez, E., Alanis-Guzma´n, M. G., Hern´andez-Luna, C. E., Ba´ez-Gonz´alez, J. G., Amaya-Guerra, C. A., & Andr´es-Grau, A. M. (2017). Increasing antioxidant activity and protein digestibility in Phaseolus vulgaris and Avena sativa by fermentation with the Pleurotus ostreatus Fungus. *Molecules,* 22, 2275.

Fawale, O. S., Gbadamosi, S. O., Ige, M. M., & Kadiri, O. (2017). Effects of cooking and fermentation on the chemical composition, functional, and antinutritional properties of kariya (Hildergardia barteri) seeds. *Food Sciences and Nutrition,* 5, 1106–1115.

Fern´andez-Lo´pez, J., Viuda-Martos, M., & P´erez-Alvarez, J. A. (2021). Quinoa and chia products as ingredients for healthier processed meat products: Technological strategies for their application and effects on the final product. *Current Opinion in Food Science*, 40, 26–32.

Freitas, J. B., Fernandes, D. C., Czeder, L. P., Lima, J. C. R., Sousa, A. G., & Naves, M. M. V. (2012). Edible seeds and nuts grown in Brazil as sources of protein for human nutrition. *Food and Nutrition Sciences,* 3.

Fu, Y., Chen, T., Chen, S. H. Y., Liu, B., Sun, P., Sun, H., et al., (2021). The potentials and challenges of using microalgae as an ingredient to produce meat analogues. *Trends in Food Science & Technology,* 112, 188–200.

Giami, S. (2004). Effect of fermentation on the seed proteins, nitrogenous constituents, antinutrients and nutritional quality of fluted pumpkin (Telfairia occidentalis Hook). *Food Chemistry,* 88, 397–404.

Gu, J., Bk, A., Wu, H., Lu, P., Nawaz, M. A., Barrow, C. J., et al., (2022). Impact of processing and storage on protein digestibility and bioavailability of legumes. *Food Reviews International,* 1–28.

Habiba, R. (2002). Changes in anti-nutrients, protein solubility, digestibility, and HCl-extractability of ash and phosphorus in vegetable peas as affected by cooking methods. *Food Chemistry,* 77, 187–192.

Hall, A. E., & Moraru, C. I. (2021). Effect of High Pressure Processing and heat treatment on *in vitro* digestibility and trypsin inhibitor activity in lentil and faba bean protein concentrates. *Lebensmittel-Wissenschaft & Technologie*, 152, Article 112342.

Han, F., Han, F., Wang, Y., Fan, L., Song, G., Chen, X., et al., (2019). Digestible indispensable amino acid scores of nine cooked cereal grains. *British Journal of Nutrition*, 121, 30–41.

Hernandez-Infante, M., Sousa, V., Montalvo, I., & Tena, E. (1998). Impact of microwave heating on hemagglutinins, trypsin inhibitors and protein quality of selected legume seeds. *Plant Foods for Human Nutrition*, 52, 199–208.

House, J. D., Neufeld, J., & Leson, G. (2010). Evaluating the quality of protein from hemp seed (Cannabis sativa L.) products through the use of the protein digestibility-corrected amino acid score method. *Journal of Agricultural and Food Chemistry*, 58, 11801–11807.

Indurkar, A., Pandit, A., Jain, R., & Dandekar, P. (2021). Plant-based biomaterials in tissue engineering. *Bioprinting*, 21, Article e00127.

Ishaq, A., Irfan, S., Sameen, A., & Khalid, N. (2022). Plant-based meat analogs: A review with reference to formulation and gastrointestinal fate. *Current Research in Food Science*.

Jin, J., Okagu, O. D., Yagoub, A. E. A., & Udenigwe, C. C. (2021). *Effects of sonication on the in vitro digestibility and structural properties of buckwheat protein isolates*.

Joye, I. (2019). Protein digestibility of cereal products. *Foods*, 8, 199.

Jung, A. H., Hwang, J. H., Jun, S., & Park, S. H. (2022). Application of ohmic cooking to produce a soy protein-based meat analogue. *Lebensmittel-Wissenschaft & Technologie*, 160, Article 113271.

Kang, S., Zhang, J., Guo, X., Lei, Y., & Yang, M. (2022). Effects of ultrasonic treatment on the structure, functional properties of chickpea protein isolate and its digestibility *in vitro*. *Foods*, 11, 880.

Kapravelou, G., Martínez, R., Andrade, A. M., Sanchez, C., Chaves, C. L., Lopez- Jurado, M., et al., (2013). Health promoting effects of Lupin (Lupinus albus var. multolupa) protein hydrolyzate and insoluble fiber in a diet-induced animal experimental model of hypercholesterolemia. *Food Research International*, 54, 1471–1481.

Kato, A., & Nakai, S. (1980). Hydrophobicity determined by a fluorescence probe method and its correlation with surface properties of proteins. B*iochimica et Biophysica Acta (BBA) - Protein Structure*, 624, 13–20.

Ketnawa, S., & Ogawa, Y. (2021). *In vitro* protein digestibility and biochemical characteristics of soaked, boiled and fermented soybeans. *Scientific Reports*, 11, 1–11.

Khatkar, A. B., Kaur, A., & Khatkar, S. K. (2020). Restructuring of soy protein employing ultrasound: Effect on hydration, gelation, thermal, in-vitro protein digestibility and structural attributes. *Lebensmittel-Wissenschaft & Technologie*, 132, Article 109781.

Kyriakopoulou, K., Dekkers, B., & van der Goot, A. J. (2019). Plant-based meat analogues. In *Sustainable meat production and processing* (pp. 103–126). Elsevier.

Kyriakopoulou, K., Keppler, J. K., & van der Goot, A. J. (2021). Functionality of ingredients and additives in plant-based meat analogues. *Foods*, 10, 600.

Lafarga, T., Acien-Fernandez, F. G., & Garcia-Vaquero, M. (2020). Bioactive peptides and carbohydrates from seaweed for food applications: Natural occurrence, isolation, purification, and identification. *Algal Research*, 48, Article 101909.

Li, Y., Chen, Z., & Mo, H. (2007). Effects of pulsed electric fields on physicochemical properties of soybean protein isolates. *LWT–Food Science and Technology*, 40, 1167–1175.

Liceaga, A. M., & Hall, F. (2019). *Nutritional, functional and bioactive protein hydrolysates.*

Lingiardi, N., Galante, M., de Sanctis, M., & Spelzini, D. (2022). Are quinoa proteins a promising alternative to be applied in plant-based emulsion gel formulation? *Food Chemistry,* Article 133485.

Linsberger-Martin, G., Weiglhofer, K., Phuong, T. P. T., & Berghofer, E. (2013). High hydrostatic pressure influences antinutritional factors and *in vitro* protein digestibility of split peas and whole white beans. *LWT–Food Science and Technology*, 51, 331–336.

Lopez, D. N., Galante, M., Robson, M., Boeris, V., & Spelzini, D. (2018). Amaranth, quinoa and chia protein isolates: Physicochemical and structural properties. *International Journal of Biological Macromolecules,* 109, 152–159.

Lopez-Pedrouso, M., Lorenzo, J. M., Zapata, C., & Franco, D. (2019). Proteins and amino acids. In *Innovative thermal and non-thermal processing, bioaccessibility and bioavailability of nutrients and bioactive compounds* (pp. 139–169). Elsevier.

Malik, M. A., & Saini, C. S. (2018). Rheological and structural properties of protein isolates extracted from dephenolized sunflower meal: Effect of high intensity ultrasound. *Food Hydrocolloids*, 81, 229–241.

Malomo, S. A., He, R., & Aluko, R. E. (2014). Structural and functional properties of hemp seed protein products. *Journal of Food Science,* 79, C1512–C1521.

Martínez-Monteagudo, S. I., & Balasubramaniam, V. (2016). Fundamentals and applications of high-pressure processing technology. In *High pressure processing of food* (pp. 3–17). Springer.

Mattice, K. D., & Marangoni, A. G. (2020). Comparing methods to produce fibrous material from zein. *Food Research International,* 128, Article 108804.

McClements, D. J., & Grossmann, L. (2021a). A brief review of the science behind the design of healthy and sustainable plant-based foods. *Npj Science of Food*, 5.

McClements, D. J., & Grossmann, L. (2021b). The science of plant-based foods: Constructing next-generation meat, fish, milk, and egg analogs. *Comprehensive Reviews in Food Science and Food Safety*, 20, 4049–4100.

Negi, A., Boora, P., & Khetarpaul, N. (2001). Effect of microwave cooking on the starch and protein digestibility of some newly released moth bean (Phaseolus aconitifolius Jacq.) cultivars. *Journal of Food Composition and Analysis,* 14, 541–546.

Nowin´ski, A., & Ufnal, M. (2018). Trimethylamine N-oxide: A harmful, protective or diagnostic marker in lifestyle diseases? *Nutrition*, 46, 7–12.

Ochoa-Rivas, A., Nava-Valdez, Y., Serna-Saldívar, S. O., & Chuck-Hern´andez, C. (2017). Microwave and ultrasound to enhance protein extraction from peanut flour under alkaline conditions: Effects in yield and functional properties of protein isolates. *Food and Bioprocess Technology*, 10, 543–555.

Ogodo, A. C., Ugbogu, O. C., Onyeagba, R. A., & Okereke, H. C. (2017). Effect of lactic acid bacteria consortium fermentation on the proximate composition and in-vitro starch/protein digestibility of maize (Zea mays) flour. *American Journal of Microbiology and Biotechnology*, 4, 35–43.

Ojokoh, A. O., & Yimin, W. (2011). Effect of fermentation on chemical composition and nutritional quality of extruded and fermented soya products. *International Journal of Food Engineering*, 7.

Olivos-Lugo, B. L., Valdivia-Lo´pez, M. A., & Tecante, A. (2010). Thermal and physicochemical properties and nutritional value of the protein fraction of Mexican chia seed (Salvia hispanica L.). *Food Science and Technology International*, 16, 89–96.

Pan, M., Xu, F., Wu, Y., Yao, M., Xiao, X., Zhang, N., et al., (2020). Application of ultrasound-assisted physical mixing treatment improves *in vitro* protein digestibility of rapeseed napin. *Ultrasonics Sonochemistry*, 67, Article 105136.

Paz-Ye´pez, C., Peinado, I., Heredia, A., & Andr´es, A. (2019). Influence of particle size and intestinal conditions on *in vitro* lipid and protein digestibility of walnuts and peanuts. *Food Research International*, 119, 951–959.

Pereira, R. N., Rodrigues, R. M., Machado, L., Ferreira, S., Costa, J., Villa, C., et al., (2021). Influence of ohmic heating on the structural and immunoreactive properties of soybean proteins. *Lebensmittel-Wissenschaft & Technologie*, 148, Article 111710.

Picot-Allain, C., Mahomoodally, M. F., Ak, G., & Zengin, G. (2021). Conventional versus green extraction techniques - a comparative perspective. *Current Opinion in Food Science*, 40, 144–156.

Pinckaers, P. J., Trommelen, J., Snijders, T., & van Loon, L. J. (2021). The anabolic response to plant-based protein ingestion. *Sports Medicine*, 51, 59–74.

Pysz, M., Polaszczyk, S., Leszczyn´ska, T., & Piątkowska, E. (2012). Eff ect of microwave fi eld on trypsin inhibitors activity and protein quality of broad bean seeds (Vicia faba var. major). *Acta Scientiarum Polonorum Technologia Alimentaria*, 11, 193–198.

Queiro´s, R. P., Saraiva, J. A., & da Silva, J. A. L. (2018). Tailoring structure and technological properties of plant proteins using high hydrostatic pressure. *Critical Reviews in Food Science and Nutrition*, 58, 1538–1556.

Raja, K., Kadirvel, V., & Subramaniyan, T. (2022). Seaweeds, an aquatic plant-based protein for sustainable nutrition-a review. *Future Foods, Article* 100142.

Rawiwan, P., Peng, Y., Paramayuda, I. G. P. B., & Quek, S. Y. (2022). Red seaweed: A promising alternative protein source for global food sustainability. *Trends in Food Science & Technology.*

Rendo´n-Villalobos, J. R., Ortíz-Sa´nchez, A., & Flores-Huicochea, E. (2018). Nutritionally enhanced foods incorporating Chía seed. In *Therapeutic foods* (pp. 257–281). Elsevier.

Reynaud, Y., Buffie`re, C., Cohade, B., Vauris, M., Liebermann, K., Hafnaoui, N., et al., (2021). True ileal amino acid digestibility and digestible indispensable amino acid scores (DIAASs) of plant-based protein foods. *Food Chemistry*, 338, Article 128020.

Rodrigues, R. M., Avelar, Z., Machado, L., Pereira, R. N., & Vicente, A. A. (2020). Electric field effects on proteins - Novel perspectives on food and potential health implications. *Food Research International*, 137, Article 109709.

Sa, A. G. A., Moreno, Y. M. F., & Carciofi, B. A. M. (2020). Food processing for the improvement of plant proteins digestibility. *Critical Reviews in Food Science and Nutrition*, 60, 3367–3386.

Salgado, P. R., Molina Ortiz, S. E., Petruccelli, S., & Mauri, A. N. (2012). Functional food ingredients based on sunflower protein concentrates naturally enriched with antioxidant phenolic compounds. *Journal of the American Oil Chemists' Society*, 89, 825–836.

Satya, S., Kaushik, G., & Naik, S. (2010). Processing of food legumes: A boon to human nutrition. *Mediterranean Journal of Nutrition and Metabolism*, 3, 183–195.

Shimelis, E., & Rakshit, S. (2005). Effect of microwave heating on solubility and digestibility of proteins and reduction of antinutrients of selected common bean (Phaseolus volgaris L.) varieties grown in Ethiopia. *Italian Journal of Food Science*, 17.

Shokry, A. M. (2016). The usage of quinoa flour as a potential ingredient in production of meat burger with functional properties. *Middle East J. Appl. Sci*, 6, 1128–1137.

Stein, H. H., Seve, B., Fuller, M., Moughan, P., & De Lange, C. (2007). Invited review: Amino acid bioavailability and digestibility in pig feed ingredients: Terminology and application. *Journal of Animal Science,* 85, 172–180.

Sun, F., Li, B., Guo, Y., Wang, Y., Cheng, T., Yang, Q., et al., (2022). Effects of ultrasonic pretreatment of soybean protein isolate on the binding efficiency, structural changes, and bioavailability of a protein-luteolin nanodelivery system. *Ultrasonics Sonochemistry*, 88, Article 106075.

Sun, X., Ohanenye, I. C., Ahmed, T., & Udenigwe, C. C. (2020). Microwave treatment increased protein digestibility of pigeon pea (Cajanus cajan) flour: Elucidation of underlying mechanisms. *Food Chemistry,* 329, Article 127196.

Sun-Waterhouse, D., Zhao, M., & Waterhouse, G. I. (2014). Protein modification during ingredient preparation and food processing: Approaches to improve food processability and nutrition. *Food and Bioprocess Technology,* 7, 1853–1893.

Ultrasonics Sonochemistry, 70, Article 105348.

van Vliet, S., Burd, N. A., & van Loon, L. J. (2015). The skeletal muscle anabolic response to plant-versus animal-based protein consumption. *Journal of Nutrition,* 145,

Vanga, S. K., Singh, A., Kalkan, F., Gariepy, Y., Orsat, V., & Raghavan, V. (2016). *Effect of thermal and high electric fields on secondary structure of Peanut protein.*

Vanga, S. K., Wang, J., & Raghavan, V. (2020). Effect of ultrasound and microwave processing on the structure, in-vitro digestibility and trypsin inhibitor activity of soymilk proteins. *Lebensmittel-Wissenschaft & Technologie,* 131, Article 109708.

Vaz Patto, M. C., Amarowicz, R., Aryee, A. N., Boye, J. I., Chung, H.-J., Martin- Cabrejas, M. A., et al., (2015). Achievements and challenges in improving the nutritional quality of food legumes. *Critical Reviews in Plant Sciences,* 34, 105–143.

Voudouris, P., Tenorio, A. T., Lesschen, J. P., Kyriakopoulou, K., Sanders, J. P., van der Goot, A. J., et al., (2017). Sustainable protein technology: An evaluation on the STW Protein programme and an outlook for the future. *Wageningen Food & Biobased Research.*

Willett, W., Rockstrom, J., Loken, B., Springmann, M., Lang, T., Vermeulen, S., et al., (2019). Food in the anthropocene: The EAT-lancet commission on healthy diets from sustainable food systems. *Lancet,* 393, 447–492.

Xie, Y., Cai, L., Zhao, D., Liu, H., Xu, X., Zhou, G., et al., (2022). Real meat and plant-based meat analogues have different *in vitro* protein digestibility properties. *Food Chemistry,* 387, Article 132917.

Yakubu, C. M., Sharma, R., Sharma, S., & Singh, B. (2022). Influence of alkaline fermentation time on *in vitro* nutrient digestibility, bio-& techno-functionality, secondary protein structure and macromolecular morphology of locust bean (Parkia biglobosa) flour. *Lebensmittel-Wissenschaft & Technologie*, 161, Article 113295.

Yin, S. W., Tang, C. H., Wen, Q. B., Yang, X. Q., & Li, L. (2008). Functional properties and *in vitro* trypsin digestibility of red kidney bean (Phaseolus vulgaris L.) protein isolate: Effect of high-pressure treatment. *Food Chemistry,* 110, 938–945.

Zhang, J., Liu, L., Liu, H., Yoon, A., Rizvi, S. S., & Wang, Q. (2019). Changes in conformation and quality of vegetable protein during texturization process by extrusion. *Critical Reviews in Food Science and Nutrition*, 59, 3267–3280.

Zhao, W., Tang, Y., Lu, L., Chen, X., & Li, C. (2014). Pulsed electric fields processing of protein-based foods. *Food and Bioprocess Technology*, 7, 114–125.

Zhao, W., Yang, R., Tang, Y., Zhang, W., & Hua, X. (2009). Investigation of the protein - protein aggregation of egg white proteins under pulsed electric fields. *Journal of Agricultural and Food Chemistry,* 57, 3571–3577.

Zhou, H., Hu, Y., Tan, Y., Zhang, Z., & McClements, D. J. (2021). Digestibility and gastrointestinal fate of meat versus plant-based meat analogs: An *in vitro* comparison. *Food Chemistry,* 364, Article 130439.

Chapter 5

Synthetic Biology Prospects for Plant-Based Meat Production

Abstract

Plant-based meat can help reduce the strain on resources, the environment, and health caused by traditional meat production. However, it still lags behind real meat in aspects like appearance, texture, flavor, and nutrition, requiring novel methods for improvement. This chapter begins by addressing the key challenges associated with the development of plant-based meat products. It then explores current processing technologies aimed at improving texture simulation, enhancing nutrition, and refining flavor. Additionally, the chapter examines the role of synthetic biology in the creation of essential ingredients, such as enzymes and functional additives, that contribute to the overall quality and sensory appeal of plant-based meats.

Keywords: synthetic biology, plant-based meat, cell engineering, meat analogues

Introduction

Consumption of adequate amounts of high-quality protein plays a particularly important role in human health (Wu et al., 2014). With the continuous growth of population and general improvement of consumption level, the demand for animal meat protein has been increasing (Colgrave et al., 2021). However, the large amount of carbon emissions produced by the animal farming process makes people choose to reduce the intake of animal food to relieve the pressure on the environment (Kumar, Abubakar et al., 2022). Moreover, the potential health hazards and risk of infectious diseases associated with long-term overconsumption of animal-derived foods have led people to favor plant-based foods (Bryant, 2022). These new challenges and concepts in the development of human society have promoted the rapid development of plant-based foods. Plant-based meat can alleviate the pressure on resources, the environment, health, ethics, and other aspects associated with traditional meat production, while also addressing issues related to meat supply and carbon emissions (Kumar, Mehta et al., 2022).

Plant-based meat is typically formulated the plant protein as the main raw material, combining with several different components such as water, fat, polysaccharide, flavors, colors, and other agents (Kyriakopoulou, Keppler, & van der Goot, 2021). In recent years, plant-based meat has attracted more and more attention from consumers, and its sales performance in the market has also gradually increased. Beyond Meat, a producer of plant-based meat substitutes in the United States, became the first plant-based meat company to go public in 2019. The company was valued at $3.8 billion on its IPO day, marking the best-performing public offering by a major U.S. company in nearly two decades (Choudhury, Singh, Seah, Yeo, & Tan, 2020). It is estimated that the global plant-based meat market generates about USD 21 billion in 2025 (Zhou, Hu, Tan, Zhang, & McClements, 2021).

However, at present, there is still a big gap between plant-based meat and real meat in terms of texture, taste, flavor and nutrition. This chapter was firstly aimed to introduce current situation and challenge of plant-based meat. Then, the latest advances regarding a comprehensive assessment of processing technologies for improving the characteristic attributes of plant-based meat were overviewed. Finally, the potential opportunities of improving the overall quality performance of plant-based meat via synthetic biology was analyzed.

Current Challenges of Plant-Based Meat

Animal meat products are typically favored by consumers due to the characteristic chewy texture provided by their fiber structure (Joshi & Kumar, 2015). Today, commercially available plant-based meat alternatives can generally replicate the fibrous texture and taste of various animal meat products. However, challenges remain in achieving the same overall texture and quality as traditional meat, with some discrepancies still evident in the final product. In terms of mimicking animal meat, some of the reconstituted ground meat products have been commercially produced, but the whole meat (e.g., steak) has a complex hierarchy of muscle tissue, adipose tissue, and connective tissue surrounding muscle fibers. The complexity of muscle fibers presents significant challenges in fully understanding their physical, chemical, and functional properties. One key research direction for the future is finding ways to transform plant globular proteins into meat-like fibers that satisfy the texture requirements of meat products. To address these challenges, it is essential to replicate whole meat by combining various plant-based proteins—such as soy, wheat, and pea—with fats, insoluble fibers, and other ingredients to create a texture and flavor that closely resemble meat. Emerging techniques, such as 3D printing, hold promise in producing plant-based products that closely mimic the appearance, texture, and taste of steak.

Unlike animal proteins (e.g., meat) as complete proteins, plant proteins are often considered incomplete proteins because they lack one or more essential amino acids (Mariotti & Gardner, 2019). For instance, soybean is an excellent source of lysine, which can complement lysine-deficient cereals to produce nutritionally balanced protein products. It is crucial to consider the amino acid profiles of various protein sources and incorporate a diverse range of these sources into plant-based meat processing to ensure the adequate intake of all essential amino acids. Additionally, a previous study reported that, compared to beef, the proteins in beef analogues are digested more rapidly in the stomach but less quickly in the small intestine (Zhou et al., 2021). Additionally, considering the nutritional differences between plant protein and animal protein, it is necessary to explore appropriate approach for enhancing the nutrition of plant-based meat and reducing its nutritional loss in the food processing. Firstly, improving the content of active substances to optimize the ingredients such as vitamins and minerals to enhance its nutritional value. Secondly, improving the recovery rate of nutritional compounds during processing, such as through filtration or centrifugation, can help minimize nutrient loss. Thirdly, excessive processing can negatively affect the properties of plant protein-based products. It is important to note that techniques like high-pressure processing and microwave-assisted heating can help preserve the nutritional value of plant-based meat products during production. Lastly, inhibiting or reducing anti-nutritional factors can enhance the development of plant-based food processing.

The challenge for mimicking meat flavor in plant-based meat is to replicate the complex flavor profile of meat, which includes the Maillard reaction and the presence of certain amino acids and fats (Sun et al., 2022a, Sun et al., 2022b). During the processing of plant-based foods, flavor precursors such as reducing sugars and amino acids can be added to promote the formation of specific flavor-active compounds (Li & Li, 2020). Additionally, microbial fermentation can be used to create a flavor profile that closely mimics that of meat. Building on this approach, the encapsulation methods and treatment conditions are optimized based on the volatility of flavor substances. This combination of techniques helps to better preserve the flavor compounds, ultimately achieving a more complex and multi-dimensional flavor profile for plant-based foods.

The color of meat is a result of complex chemical reactions that occur during cooking and processing, and is influenced by a range of factors, including the type of meat, the cooking method, and the presence of certain compounds such as myoglobin and hemoglobin. To replicate the color of traditional meats in plant-based alternatives, manufacturers must develop methods that mimic the chemical reactions responsible for the characteristic hues, as well as simulate the presence of myoglobin and hemoglobin. This presents a significant challenge, as plant-based ingredients naturally lack these compounds, and finding suitable substitutes can be difficult. Common approaches for achieving the desired color in plant-based meats often involve the use of natural colorants, such as beet juice, beet powder, and carrot pigment extracts, to mimic the appearance of myoglobin (Hong, Shen, & Li, 2022). However, these methods can be costly and may not always result in a meat-like color that is convincing or realistic.

Processing Technologies of Plant-Based Meat

Structuring Process of Plant-Based Meat

Extrusion technology is the most common structuring process for the conversion of plant proteins into fibrous extrudes, which is for later processing into alternative products for meat (Sun, Fu, Chang, Li, & Fang, 2022). Depending on the moisture content of the raw material, the extrusion technology can be divided into low-moisture extrusion (20-40% moisture) and high-moisture extrusion (40-80% moisture) (Zhang, Chen, Kaplan, & Wang, 2022). Low-moisture extrusion products offer the advantages of strong flexibility and ease of storage. Textured soy protein produced through low-moisture extrusion can serve as a scaffold that enables cultured cells to grow into beef-like products. Several start-up companies in the EU are leveraging this technology, positioning themselves to make a significant impact on the meat industry in the coming years. In contrast, high-moisture extrusion products exhibit a fibrous structure and texture that closely resemble meat, without the need for rehydration. Currently, high-moisture extrusion is gradually being adopted in the production of plant-based meats within the food industry, as it offers a more meat-like texture compared to low-moisture extrusion (Singh et al., 2021). Previous studies have used the high moisture extrusion technology to produce plant-based meat with legume and/or cereal proteins as raw materials, which can mimic steamed chicken (Chiang et al., 2021), and can also be made similar to beef by changing the ratio of raw materials and moisture content. Shear cell as a new technology has been reported to produce anisotropic fibrous meat analogues with a cell-like geometry (e.g., cone-in-cone or couette cell) (Krintiras, Göbel, Van der Goot, & Stefanidis, 2015). A meat-like fiber structure can be formed through a simple combination of shear force and high temperature in the gap between the cones, followed by cooling and resting at room temperature. Additionally, electrospinning technology enables the fabrication of micro- and nanoscale fibers by creating an electric force between the spinneret and the ground surface, which has attracted researchers for meat analogue applications. 3D printing technology can also be applied to "meat" manufacturing, using a paste-like material consisting of protein powder, additional ingredients, and water as edible ink, to form a fiber structure through a layer-by-layer application (Ramachandraiah, 2021). However, for electrospinning and 3D printing technology, it is difficult to achieve from laboratory to large scale production for expanding applications.

Texture Simulation of Plant-Based Meat

Enzymatic cross-linking strategy during extrusion can contribute to form fibrous structure. Transglutaminase (TG) mainly catalyzes the acyl transfer reaction between the γ-hydroxylamine group of glutamine residue in protein and primary amine compound (acyl receptor), which leads to covalent cross-linking of protein. Previous studies have shown that addition of appropriate TG improves texture properties of plant-based extrudates by controlling intermolecular/molecular forces and forms denser and more homogeneous 3D network gels during cross-linking, while the addition of excess TG is not conducive to this improvement (Qin et al., 2022; Zhang et al., 2021). Moreover, laccase can be utilized to enhance the gelation properties of protein-polysaccharide complexes, facilitating the creation of recombinant systems. The cross-linking effect of laccase improves the thermal stability of shear-induced gels and can also be employed to form double-net gels. For instance, soy protein isolate can form a double network gel with corn fiber glue by combining laccase-catalyzed cross-linking and heat treatment. This gel not only enhances the texture of the food but also holds promise for controlling the release of nutrients, offering significant potential to revolutionize the future of plant-based meat products (Yan, Jia, Yan, & Yin, 2021).

Food carbohydrates have been used to enhance the viscosity and pseudoplasticity of plant-based meat. For example, the heat induced gelation of methylcellulose is considered as binding agent for plant-based meat. Addition of methylcellulose into plant-based patties can contribute to form uniformity gel and bind water during cooking, leading to improved overall performance and consumers' acceptability of products (Bakhsh et al., 2021). In addition, fibers from citrus and black quinoa significantly reduced the cooking loss of cooked bologna sausages, which improved textural perception during oral processing (Fernández-López et al., 2020; Powell, Sebranek, Prusa, & Tarté, 2019).

Fat as a high-valued food ingredient contributes to many sensory qualities of meat products, such as physical appearance, flavor, juiciness and chewing sensation. The incorporation of fat plays a crucial role in imparting both technological properties and sensory characteristics to food products. As such, the development of fat mimetics in plant-based meat is essential for enhancing its overall performance. Fat mimetics, such as the structuring of vegetable liquid oils through emulsion templates, oleogelation using gelators, and structured emulsions (Patel, Nicholson, & Marangoni, 2020), can form a physical barrier around the oil. This barrier helps to reduce or modify the undesirable flavors of vegetable oils, which may not be suitable for meat products. Furthermore, a previous study has shown that replacing beef tallow with HPMC oleogels at 50% and 100% in meat patty formulations significantly reduced cooking loss and resulted in a much softer texture (Oh, Lee, Lee, & Lee, 2019).

Flavor Improvement of Plant-Based Meat

Plant-based meat manufacturers have started using umami compounds (e.g., glutamate, inosinate, and guanylate) or heme-iron from soy leghemoglobin to create meaty flavor. However, plant-based protein materials have undesirable flavor, such as bean and bitter flavor (Wang et al., 2022). Legumes are a major source of plant-based protein due to their high protein content and rich nutrient profile. However, the presence of off-flavors, such as the beany flavor and grassy taste, often makes legumes less appealing to consumers, thereby limiting the broader adoption and consumption of legume-based proteins and products. To address this issue, two primary methods can be employed to reduce or eliminate these undesirable flavors. One such method involves the modification of proteins through controlled biological treatments. For instance, fermentation is commonly used to enhance both the flavor and nutritional value of plant proteins (Arbab Sakandar et al., 2021). The metabolic activity of fermentation strains can lead to the production of new aromatic compounds, while simultaneously reducing or eliminating some of the undesirable flavor compounds (Figure 2). For instance, Arteaga et al., found that fermentation of pea protein isolates using Lactobacillus casei reduced the intensity of bitterness after 48 h, which might be attributed to its strong activity peptidase against bitter peptides. On the other hand, fermentation using *Lactobacillus perolen* resulted in a buttery aroma in pea protein isolates, likely due to the high concentrations of diacetyl produced *by L. perolen* (Arteaga, Leffler, Muranyi, Eisner, & Schweiggert-Weisz, 2021). However, there is limited literature on fermentation processes that lead to a significant reduction in the beany flavor, which is the predominant odor of plant proteins. Therefore, the related content needs to be studied in more depth to inhibit the generation of the off-flavors or decrease its content substantially to improve consumers' acceptability of plant proteins.

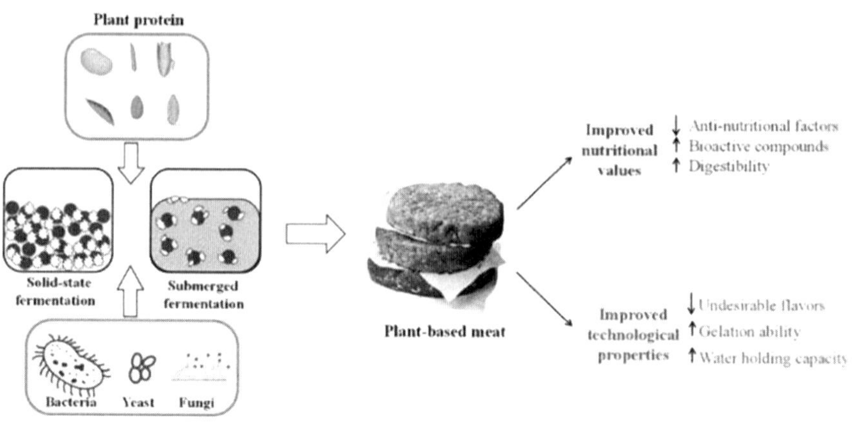

Figure 1. The role of fermentation in enhancing the nutrition and flavor of plant-based meat.

Another method is odor masking technology. Adding condiments to plant-based meat products can have the effect of masking off-flavors and simulating the flavor of meat (Mittermeier-Kleßinger, Hofmann, & Dawid, 2021). Common condiments used in plant-based meat products include several key ingredients. First, natural spices such as basil and fennel are often added to impart specific flavors, which can help mask off-flavors. Second, precursors of the Maillard reaction, including monosaccharides, amino acids, flavor nucleotides, and thiamine, are incorporated into plant-based meat products. These compounds promote the Maillard reaction during preparation and cooking, resulting in the formation of meat-like volatiles and flavors. Third, yeast extract, which is resistant to high temperatures, is commonly used. The amino compounds in yeast extract undergo Maillard reactions with reducing sugars in the product during high-temperature cooking, further enhancing the meat-like taste and aroma of the final product.

Nutritional Enhancement of Plant-Based Meat

Different from animal products, plant-based meat diet may lead to deficiencies in vitamins B2, B12, iodine, zinc, calcium, potassium, selenium and other nutrients, which need to be supplemented according to the nutrient defects of plant raw materials (McClements & Grossmann, 2021). Proteins from plants, such as soybeans, wheat, and peas, are the primary raw materials used in the production of meat analogues. However, the amino acid composition of most plant-based protein sources does not fully meet human nutritional requirements. For example, wheat protein is low in lysine, and pea protein is deficient in methionine and cysteine. To address this, plant proteins are often combined with other protein sources, such as algal protein or edible fungus protein, to create meat analogues that provide a more balanced amino acid profile for human nutrition. Additionally, the digestibility and bioavailability of plant proteins are often limited by the presence of anti-nutritional factors. These factors can be reduced through processes such as soaking, cooking, or fermentation, which help improve the nutritional quality of plant-based meat products. For example, solid state fermentation has been shown to have the ability to improve nutritional properties, because the microorganisms involved can produce enzymes that degrade anti-nutritional factors (Figure 1). Take phytic acid as an example, phytic acid is commonly found in legumes, but the presence of phytic acid affect the absorption of minerals as well as proteins, so efforts should be made to reduce or remove phytic acid. The use of microbial inoculants for solid-state fermentation to reduce phytic acid content is an effective method. For example, in the study of the effects of *Aspergillus sojae*, *Aspergillus ficuum* and their co-cultured solid-state fermentation on antinutritional factors in lupin flour (Olukomaiya et al., 2020), by comparing unfermented lupin flour, *Aspergillus sojae* fermented lupin flour, *Aspergillus ficuum* fermented lupin flour and co-fermented lupin flour, it was found that the phytic acid content was significantly reduced. Among them, the use of *Aspergillus ficuum* fermented flour had the best effect, and the phytic acid content was reduced by 73%.

Synthetic Biology Approaches for Developing Plant-Based Meat

The development of synthetic biology has enabled scientists to engineer cells, a process now being applied in food manufacturing. In the context of plant-based meat, these engineered cells can transform renewable raw materials into key ingredients such as color, taste, and flavor, helping to replicate the appearance and characteristics of real meat. Hemoglobin or myoglobin as the molecule that carries iron in animals, is responsible for the color and flavor of meat (Shleikin & Medvedev, 2014). Hence, hemoglobin can be used as functional food indigent in meat analogues. Cell factories able to synthesize heme or hemoglobin have been designed (Zhao, Choi, & Lee, 2018), which provide an alternative production method to extraction from plant tissue or animal blood. Impossible Foods, Inc. has developed a heme derived from soy plants to be used as an ingredient in their ground beef analogue, imparting a distinctive meaty color and flavor. Additionally, patatin, a protein found in potatoes, holds potential as a gelling agent in plant-based meat products due to its excellent gelation properties. Patatin storage protein with excellent gelation properties can be overexpressed in *Pichia pastoris* GS115. To improve secretory expression of patatin storage protein in *P. pastoris* GS115, several strategies were used. Furthermore, *P. pastoris*-derived patatins had better gelation properties with a much lower gel-forming temperature and similar gel strength compared with potato-extracted patatins (Dai, Wu, Zeng, Rao, & Zhou, 2022). Enzymes have been shown to improve the functional properties of proteins. However, considering extrusion under high temperature and pressure is the most common processing technology for the conversion of plant proteins into fibrous extrudes, which is for later processing into alternative products for meat. Extreme high temperatures can cause an enzyme to denature and cease functioning. Therefore, enzymes involved in plant protein-based meat production (such as glutamine transaminase and laccase) must be modified to enhance their thermal stability. This can be achieved through the identification and transformation of stability- and activity-related sites in enzyme molecules. Techniques such as amino acid homologous sequence alignment, crystal structure analysis, and molecular dynamics simulations are used to develop heat-resistant engineered enzymes with high activity. Moreover, gene editing methods applicable to microorganisms and a microbial fermentation method for high-efficient preparation of high-activity and heat-resistant's enzyme need to be established.

From our point of view, current food technology for plant-based meat processing, combining with synthetic biology technology to produce key ingredients (e.g., hemoglobin, enzyme, vitamins), which can produce plant-based meat with similar texture, flavor, nutrition and other characteristics of animal meat.

Conclusion

In terms of quantity, quality and sustainable supply, traditional protein production methods are unable to meet the future needs of human life. Plant-based protein as low-cost and sustainable production has been attracted the attention, while research about plant-based meat with high-quality is in urgent need of innovation and development. The application of synthetic biology has enabled the large-scale production of key ingredients, such as vitamins, minerals, and enzymes, in plant-based foods. This technology allows for the optimization of these ingredients to enhance the overall quality of plant-based products, tailoring them to both the raw material characteristics and consumer demands. Looking ahead, one promising direction for future research in synthetic biology for plant-based meat production is the development of specific functional proteins. These proteins, with properties such as water-holding and fiber-forming capabilities, could be designed to improve the texture and nutritional quality of plant-based protein meats. Through rational design, these functional proteins can enhance the processing of plant-based proteins, contributing to more desirable product characteristics. Additionally, establishing efficient production systems for these key ingredients in food-grade hosts, such as *Bacillus subtilis*, will be crucial for advancing plant-based meat production. However, the security of technology and products in the manufacturing process should be ensured. Potential biosafety risks need to be fully predicted, so as to form independent, objective technical systems for production safety or nutritional evaluation. Taken together, combing current existing technology and synthetic biology technology may provide scientific support for production of plant-based meat with high-quality.

References

Arbab Sakandar, H., Chen, Y., Peng, C., Chen, X., Imran, M., & Zhang, H. (2021). Impact of fermentation on antinutritional factors and protein degradation of legume seeds: A review. *Food Reviews International*, 1–23. https://doi.org/10.1080/87559129.2021.1931300.

Bakhsh, A., Lee, S. J., Lee, E. Y., Sabikun, N., Hwang, Y. H., & Joo, S. T. (2021). *A novel approach for tuning the physicochemical, textural, and sensory characteristics of plant-based meat analogs with different levels of methylcellulose concentration.*

Bryant, C. J. (2022). Plant-based animal product alternatives are healthier and more environmentally sustainable than animal products. *Future Foods.*, Article 100174. https://doi.org/10.1016/j.fufo.2022.100174.

Chiang, J. H., Tay, W., Ong, D. S. M., Liebl, D., Ng, C. P., & Henry, C. J. (2021). Physicochemical, textural and structural characteristics of wheat gluten-soy protein composited meat analogues prepared with the mechanical elongation method. *Food Structure,* 28, Article 100183. https://doi.org/10.1016/j.foostr.2021.100183.

Choudhury, D., Singh, S., Seah, J. S. H., Yeo, D. C. L., & Tan, L. P. (2020). Commercialization of plant-based meat alternatives. *Trends in Plant Science,* 25(11), 1055–1058. https://doi.org/10.1016/j.tplants.2020.08.006.

Colgrave, M. L., Dominik, S., Tobin, A. B., Stockmann, R., Simon, C., Howitt, C. A., et al., (2021). Perspectives on future protein production. *Journal of Agricultural and Food Chemistry*, 69(50), 15076–15083. https://doi.org/10.1021/acs.jafc.1c05989.

Dai, Z., Wu, X., Zeng, W., Rao, Y., & Zhou, J. (2022). Characterization of highly gelatinous patatin storage protein from Pichia pastoris. *Food Research International,* 162, Article 111925. https://doi.org/10.1016/j.foodres.2022.111925.

Fernandez-Lopez, J., Lucas-Gonzalez, R., Roldan-Verdú, A., Viuda-Martos, M., Sayas-Barbera, E., Ballester-Sanchez, J., et al., (2020). Effects of black quinoa wet-milling coproducts on the quality properties of bologna-type sausages during cold storage. *Foods,* 9(3), 274. https://doi.org/10.3390/foods9030274.

Hong, S., Shen, Y., & Li, Y. (2022). Physicochemical and functional properties of texturized vegetable proteins and cooked patty textures: Comprehensive characterization and correlation analysis. *Foods,* 11(17), 2619. https://doi.org/10.3390/foods11172619.

Joshi, V. K., & Kumar, S. (2015). Meat Analogues: Plant based alternatives to meat products-A review. *International Journal of Food and Fermentation Technology,* 5(2), 107–119. https://doi.org/10.5958/2277-9396.2016.00001.5.

Krintiras, G. A., Go¨bel, J., Van der Goot, A. J., & Stefanidis, G. D. (2015). Production of structured soy-based meat analogues using simple shear and heat in a Couette Cell. *Journal of Food Engineering*, 160, 34–41. https://doi.org/10.1016/j.jfoodeng.2015.02.015

Kumar, P., Abubakar, A. A., Verma, A. K., Umaraw, P., Adewale Ahmed, M., Mehta, N., et al., (2022a). New insights in improving sustainability in meat production: Opportunities and challenges. *Critical Reviews in Food Science and Nutrition,* 1–29. https://doi.org/10.1080/10408398.2022.2096562.

Kumar, P., Mehta, N., Abubakar, A. A., Verma, A. K., Kaka, U., Sharma, N., et al., (2022b). Potential alternatives of animal proteins for sustainability in the food sector. *Food Reviews International,* 1–26. https://doi.org/10.1080/87559129.2022.2094403.

Kyriakopoulou, K., Keppler, J. K., & van der Goot, A. J. (2021). Functionality of ingredients and additives in plant-based meat analogues. *Foods,* 10(3), 600. https://doi.org/10.3390/foods10030600.

Li, X., & Li, J. (2020). The flavor of plant-based meat analogues. *Cereal Foods World,* 65 (4), Article 0040. https://doi.org/10.1094/CFW-65-4-0040.

Mariotti, F., & Gardner, C. D. (2019). Dietary protein and amino acids in vegetarian diets-A review. *Nutrients,* 11(11), 2661. https://doi.org/10.3390/nu11112661.

McClements, D. J., & Grossmann, L. (2021). The science of plant-based foods: Constructing next-generation meat, fish, milk, and egg analogs. *Comprehensive Reviews in Food Science and Food Safety,* 20(4), 4049–4100. https://doi.org/10.1111/1541-4337.12771.

Mittermeier-Kleßinger, V. K., Hofmann, T., & Dawid, C. (2021). Mitigating off-flavors of plant-based proteins. *Journal of Agricultural and Food Chemistry,* 69(32), 9202–9207. https://doi.org/10.1021/acs.jafc.1c03398.

Oh, I., Lee, J., Lee, H. G., & Lee, S. (2019). Feasibility of hydroxypropyl methylcellulose oleogel as an animal fat replacer for meat patties. *Food Research International,* 122, 566–572. https://doi.org/10.1016/j.foodres.2019.01.012.

Olukomaiya, O. O., Adiamo, O. Q., Fernando, W. C., Mereddy, R., Li, X., & Sultanbawa, Y. (2020). Effect of solid-state fermentation on proximate composition, anti-nutritional factor, microbiological and functional properties of lupin flour. *Food Chemistry*, 315, Article 126238. https://doi.org/10.1016/j.foodchem.2020.126238.

Patel, A. R., Nicholson, R. A., & Marangoni, A. G. (2020). Applications of fat mimetics for the replacement of saturated and hydrogenated fat in food products. *Current Opinion in Food Science,* 33, 61–68. https://doi.org/10.1016/j.cofs.2019.12.008.

Powell, M. J., Sebranek, J. G., Prusa, K. J., & Tart´e, R. (2019). Evaluation of citrus fiber as a natural replacer of sodium phosphate in alternatively-cured all-pork Bologna sausage. *Meat Science,* 157, Article 107883. https://doi.org/10.1016/j.meatsci.2019.107883.

Qin, J., Zhao, Y., Zhou, J., Zhang, G., Li, J., & Liu, X. (2022). Rheological properties of transglutaminase-treated concentrated pea protein under conditions relevant to high-moisture extrusion processing. *Frontiers in Nutrition,* 9, Article 970010. https://doi.org/10.3389/fnut.2022.970010.

Ramachandraiah, K. (2021). Potential development of sustainable 3D-printed meat analogues: A review. *Sustainability*, 13(2), 938. https://doi.org/10.3390/su13020938.

Sensory profile, functional properties and molecular weight distribution of fermented pea protein isolate. *Current Research in Food Science*, 4, 1–10. https://doi.org/10.1016/j.crfs.2020.12.001.

Shleikin, A. G., & Medvedev, Y. V. (2014). Role of peroxidation and heme catalysis in coloration of raw meat. *Acta Scientiarum Polonorum Technologia Alimentaria*, 13(2), 123–127. https://doi.org/10.17306/J.AFS.2014.2.1.

Singh, M., Trivedi, N., Enamala, M. K., Kuppam, C., Parikh, P., Nikolova, M. P., et al., (2021). Plant-based meat analogue (PBMA) as a sustainable food: A concise review. *European Food Research and Technology*, 247, 2499–2526. https://doi.org/10.1007/s00217-021-03810-1.

Sun, A., Wu, W., Soladoye, O. P., Aluko, R. E., Bak, K. H., Fu, Y., et al., (2022b). Maillard reaction of food-derived peptides as a potential route to generate meat flavor compounds: A review. *Food Research International*, 151, Article 110823. https://doi.org/10.1016/j.foodres.2021.110823.

Sun, C., Fu, J., Chang, Y., Li, S., & Fang, Y. (2022a). Structure design for improving the characteristic attributes of extruded plant-based meat analogues. *Food Biophysics*, 1–13. https://doi.org/10.1007/s11483-021-09692-w.

Wang, Y., Tuccillo, F., Lampi, A. M., Knaapila, A., Pulkkinen, M., Kariluoto, S., et al., (2022). Flavor challenges in extruded plant-based meat alternatives: A review. *Comprehensive Reviews in Food Science and Food Safety*, 21(3), 2898–2929. https://doi.org/10.1111/1541-4337.12964.

Wu, G., Fanzo, J., Miller, D. D., Pingali, P., Post, M., Steiner, J. L., et al., (2014). Production and supply of high-quality food protein for human consumption: Sustainability, challenges, and innovations. *Annals of the New York Academy of Sciences*, 1321(1), 1–19. https://doi.org/10.1111/nyas.12500.

Yan, J., Jia, X., Yan, W., & Yin, L. (2021). Double-network hydrogels of corn fiber gum and soy protein isolate: Effect of biopolymer constituents and pH values on textural properties and microstructures. *Foods*, 10(2), 356. https://doi.org/10.3390/foods10020356.

Zhang, J., Chen, Q., Kaplan, D. L., & Wang, Q. (2022). High-moisture extruded protein fiber formation toward plant-based meat substitutes applications: Science, technology, and prospect. *Trends in Food Science & Technology*, 128, 202–216. https://doi.org/10.1016/j.tifs.2022.08.008.

Zhang, J., Chen, Q., Liu, L., Zhang, Y., He, N., & Wang, Q. (2021). High-moisture extrusion process of transglutaminase-modified peanut protein: Effect of transglutaminase on the mechanics of the process forming a fibrous structure. *Food Hydrocolloids*, 112, Article 106346. https://doi.org/10.1016/j. foodhyd.2020.106346.

Chapter 6

The Prospects of 3D Printing in Meat Materials

Abstract

Three-dimensional (3D) printing is advancing the food industry, particularly in meat production, where it could address issues like raw material waste and contamination. However, challenges remain, including the rheological properties of the printing ink, such as shear thinning, viscosity, and yield stress, which affect print quality. Meat materials, with their complex fibrous structures, cannot be printed directly, making printability a key limitation. Additionally, the functionality of printers, along with the singular nutritional and sensory properties of formulations, restrict the applications of 3D printing. However, advances in cultured meat, by-product utilization, and new technologies present opportunities for expanding 3D printing in the meat industry.

Keywords: 3D printing, meat materials, food manufacturing, innovative technologies

Introduction

Meat is a nutrient-rich food containing many trace elements such as iron, zinc, and B vitamins and is the primary source of complete protein in most people's diets. Throughout history, humans have consistently consumed meat, and anthropologists suggest that adequate meat intake played a key role in the evolution of larger brains. As both the economy and population grew, the demand for meat continued to rise (Roser, 2017). However, this development is increasingly viewed as environmentally unsustainable, as significant challenges arise from livestock farming and meat production processes. Firstly, livestock breeding requires vast amounts of land and forage, leading to the degradation of pastures and a dramatic reduction in forest cover. Additionally, data from the United Nations indicate that livestock farming is responsible for approximately 15% of global greenhouse gas emissions, contributing to a broader system of carbon emissions that negatively impacts the global climate (K. Handral, Hua Tay, Wan Chan & Choudhury, 2022). In addition, during the production of raw materials, meat is cut and trimmed to varying degrees to suit consumer preferences, and despite possessing the same nutritional value, at least 30% of the meat is usually sold as a low-value by-product or even discarded as waste, resulting in significant waste and food waste contamination (Dick, Bhandari & Prakash, 2019a; Welin, 2013; Zhao et al., 2021). Therefore, the meat industry urgently needs new technologies to link these problems.

3D printing, also known as additive manufacturing, is a technology based on computer numerical control, involving three-dimensional modeling, model slicing, information processing, and layer-by-layer printing to form a 3D solid (Guo et al., 2019). Hull developed the first commercial 3D printer in 1986 (Hull, 1986), and since then, 3D printing has gained widespread recognition, gradually becoming widely used in industrial design, model making, automotive, and aviation industries (Kelly et al., 2019; Lu et al., 2018). The original 3D printing technology was developed based on photo-hardened polymers to manufacture complex metal and ceramic plastic parts rather than food materials. In 2007, an extrusion-based printer designed by researchers at Cornell University introduced 3D printing to the food industry (Malone and Lipson, 2007).

3D printing technology, as an emerging innovation in the food industry, offers a promising solution to the environmental and sustainability challenges faced by the traditional meat industry. When applied to the production of cell-cultured meat, 3D printing can create products that closely replicate the nutritional profile and taste of conventional meat, while eliminating the need for traditional farming. This approach not only reduces land use but also minimizes waste emissions, making it a more sustainable alternative. Moreover, the raw materials for 3D printing are often powders or pastes, which allow for the creation of custom-shaped food products in a relatively short time without the need for complicated pre-processing or cutting. This process significantly reduces raw material waste and helps mitigate food waste pollution. In addition, 3D printing is efficient, economical, innovative, and personalized to meet the unique needs of children, pregnant women, the elderly, soldiers, and other special populations nutritional need (Bulut and Candoğan, 2022; Prakash et al., 2019). Accordingly, it has received wide attention.

However, the meat materials used for 3D printing can be regarded as a polyphase colloid system with water as a continuous phase and protein, carbohydrate, and inorganic salt as the dispersed phase. This system has a complex rheological property and is not naturally printable, resulting in great challenges to 3D printing. Therefore, this chapter first summarized the challenges faced by 3D printing in the meat industry, including printability, simultaneous precision cooking, 3D printer development, and product nutritional sensory limitations, and discussed the methods and research directions that can be used to deal with these challenges. This chapter provides a detailed exploration of the application opportunities of 3D printing in the meat industry. It reviews the current research on cell-cultured meat and discusses the feasibility of developing meat by-products using 3D printing. Additionally, it summarizes the latest technologies for evaluating the printability of meat. The chapter highlights the challenges and opportunities associated with the application of food 3D printing technology to meat materials, offering insights into the design of printable materials and suggesting strategies to accelerate the development of meat printing materials.

3D Printing of Traditional Meat Materials

The leading 3D printing methods include selective laser sintering, binder jetting, inkjet printing, and extrusion printing (Demei et al., 2022). Extrusion 3D printing is the most appropriate technology for food materials and has been widely used compared with other technologies (Varvara et al., 2021). Extrusion-based 3D printing is carried out through computer-controlled extrusion of ink from a nozzle, building up layer by layer on the printer base. During the printing process, the nozzle's motion path is guided by a preset 3D model, enabling the extruded material to form a specific shape after curing (Sun et al., 2015; Sun et al., 2018). According to the printing characteristics, the food can be divided into raw printable food materials (e.g., chocolate (Kim et al., 2017), cheese, and mashed potatoes (He et al., 2020a, He et al., 2020b)) and non-raw printable food materials (e.g., meat, vegetables, and fruits). The unique fibrous structure of meat presents a significant challenge in 3D printing, necessitating pre-treatment or the use of additives to ensure smooth extrusion from the nozzle while maintaining proper self-support and formability. While limited studies have been conducted on the development of 3D-printed meat products, most of the existing research has primarily focused on surimi materials, which are known for their excellent adhesion, gelation properties, and softness (Chen et al., 2021a, Chen et al., 2021b; Oyinloye and Yoon, 2022; Wang et al., 2018a, Wang et al., 2018b; Xie et al., 2022). Several studies have focused on meats with a solid fiber structure and low viscosity, such as beef and chicken breast. Therefore, further development of printable raw materials and the expansion of printable meat types are essential for advancing the application of 3D printing in meat products.

Challenges of 3D Printing in Meat Material

Poor Printability

Improving the rheological properties of meat materials is a key challenge for meat printing. Meat is a complex polyphase colloidal system that exhibits complex rheology and is usually plastic or pseudoplastic (shear thinning). Complex structural units result from the interactions between particles in the dispersed phase and with other colloidal molecules that are added to control the properties of the system, such as the addition of gelatin to chicken, to improve its rheological properties and gelation ability (Bulut et al., 2022). For extrusion-based 3D printing, the rheological properties of the material - such as shear thinning behavior, viscosity, and yield stress - are crucial for determining whether the material can be printed successfully and ensuring the final product's quality (Jiang et al., 2019). Controlling rheology is essential for both the processing of the material and the regulation of texture and sensory perception. The 3D printing process can be divided into four stages: start-up flow, continuous extrusion, recovery, and deposition. In general, the 3D printing ink must exhibit good extrusion, recovery, and self-supporting capabilities. The yield stress represents the minimum force required to initiate fluid flow and is closely linked to the ink's extrusion performance during the 3D printing process. The minimum pressure required to start the flow of ink with yield stress can be described as follows:

$$P_{\mathrm{min}}=(4\mathrm{L}/\mathrm{D})\ \tau_{\mathrm{yield}} \quad (1)$$

where P_{min} indicates the minimum pressure required, L indicates the nozzle length, D indicates the nozzle diameter, and τ_{yield} indicates the yield stress of inks.

During the extrusion phase, the shear force at the nozzle suddenly becomes high, and the material under gravity experiences low shear action again after extrusion. The ink should therefore have shear thinning properties with high shear recoverability, exhibit low viscosity during extrusion for smooth extrusion, and have sufficiently high viscosity after deposition to adhere to the previously deposited ink layer. The shear-thinning behavior of the sample is evaluated by fitting a power-law model to the obtained viscosity measurements (Equation (2)) as expressed below:

$$\eta=\kappa\cdot\gamma^{n-1} \quad (2)$$

where η is the material viscosity (Pa·s), γ is the shear rate (s^{-1}), k is the consistency index (Pa·s^n), and n is the flow behavior index(Paxton et al., 2017). For fluid shear thinning with $n < 1$, a smaller n indicates higher shear thinning behavior.

The recovery stage is the transition phase prior to a deposition where the ink undergoes a change from high to low shear. Ideally, the ink should have the ability to recover viscosity and resist deformation immediately after extrusion. Researchers often simulate the extrusion to the recovery process by alternating strain scans to evaluate the shear recovery behavior of inks (Zeng et al., 2021). During deposition, the ink layers stack up but do not merge and collapse between the layers, thus requiring good self-supporting capabilities. The main rheological parameters involved in the deposition stage are the yield stress, complex modulus (G*), storage modulus (G'), and elastic modulus (G") at the deposition temperature. The yield stress at the deposition stage reflects the resistance of the ink to external stresses, such as the gravity and forces exerted by its top material layer, at ambient temperature. G* indicates solid-like properties, which reflect the resistance to compressive deformation and mechanical strength. G' and G" reflect the viscoelasticity of the ink. The results of the present study indicate that materials with a highly structured or more solid-like behavior have a good self-supporting ability.

Extruded 3D-printed food materials must maintain low viscosity and yield stress during the extrusion process to ensure smooth flow. After extrusion, these materials need to exhibit sufficient viscosity and yield stress to retain their shape and support the connection of subsequent layers. Furthermore, the ink should provide strong mechanical support and gel strength to preserve the stability of the spatial structure (Chen et al., 2022), preventing the product from collapsing over time.

Meat, as a complex colloidal system, lacks inherent printing capabilities. To achieve high-quality 3D printing, it is necessary to enhance the rheological and gel properties of the meat paste during the development of the ink. Currently, the printability of meat materials is primarily improved through pre-treatment processes that promote interactions between colloids in the minced meat system or by adding other colloids to the meat. Importantly, when adjusting the ink formulation, the ink's ability to withstand post-treatment processes should also be considered. Ideally, the final product should transfer mass evenly, maintain its shape, and experience minimal moisture loss during maturation.

Pre-Processing to Improve Printability

Pre-Ultrasound Treatment

As an emerging technology, ultrasound has been widely used in the food industry. Mechanical effects such as cavitation, agitation, and turbulence that are produced by ultrasound destroy muscle fibers contribute to the release of viscous substances (e.g., meat protein, fat, and glycogen) and increase the degree of fiber swelling, thus improving the water retention, softness, and viscosity of meat (Amiri et al., 2018). The protein and fat in meat may dissolve and disperse into small molecules under the action of ultrasound and interact with each other through electrostatic action and van der Waals force. This interaction increases the stability of the whole colloid system.

Myofibrillar protein, the primary protein in meat, plays a crucial role in determining the functional properties of meat products (Chang et al., 2012; Li & Sun, 2002; Sun et al., 2021; Wang et al., 2021). Alarcon-Rojo et al., (2015) discovered that ultrasound treatment enhances the rheological and gel properties of myofibrillar protein. Similarly, Zhang et al., found that combining ultrasound treatment with endogenous transglutaminase (TGase) significantly improves the gel strength of surimi. Therefore, ultrasound treatment could be used as a simple and efficient pre-treatment method for 3D printing inks to improve the printability of minced meat. Besides, ultrasound participates in enhancing mass transfer, improving heat transfer during cooking, and reducing cooking losses and cooking time(Leal-Ramos et al., 2011; Pohlman et al., 1997).

Microwave Treatment

Microwave heating (MH) absorbs microwave energy and simultaneously converts it into heat by electromagnetic induction of materials (Sumnu, 2001). Microwave heating is a promising method for both pre-treatment and post-treatment of 3D printing materials due to its fast heating speed, high energy efficiency, non-contact nature, and hygienic properties (He et al., 2020). Microwaves have been used to enhance the printability of materials such as soy protein (Bhattacharya and Jena, 2007) and potato starch (Przetaczek-

Rożnowska et al., 2019). The thermal effect of microwave denatures the proteins in the minced meat system and reversibly associates them to form soluble aggregates, thus changing some of the proteins from sol to gel and increasing the self-supporting capacity of the meat filling (Bohr and Bohr, 2000; Cai et al., 2018; Sun et al., 2020). By controlling the time and power ratio of sol and gel through microwave control, the fluidity and mechanical strength of meat fillings can be adjusted. However, microwave pre-treatment in the 3D printing of meat products has been less explored, with most studies focusing primarily on surimi materials. For example, Zhao et al., (2021) employed a synergistic approach combining microwave-focused heating and transglutaminase (TGase) to achieve the self-gelatinization of surimi during printing. Under these conditions, solid products with high resistance to deformation were successfully obtained. The findings of these studies suggest that microwave heating can effectively promote gel formation in 3D printing applications.

High-Pressure Treatment

High-pressure treatment has been a promising area of research in meat processing because of its potential to prolong meat shelf life and improve meat quality (Chattong, Apichartsrangkoon & Bell, 2007). Additionally, pressure treatment induces changes in the molecular composition of meat, affecting the structural and functional properties of proteins. It promotes protein adsorption at the fat droplet/water interface, enhancing emulsion formation, increases the thermal stability of proteins, facilitates gelation, and binds free water, resulting in denser and more uniform gels (Wang et al., 2020; Yang et al., 2021; Zhou et al., 2018). High pressure also improves the texture and printability of meat by enhancing elasticity, water retention, and adhesion. While a limited number of studies have explored the application of high pressure in meat product printing, the potential of high-pressure technology in 3D printing of meat products requires further investigation and development.

Adjusting Formulations to Improve Printability

The rheological properties of raw materials (e.g., shear-thinning properties, viscosity, and yield stress) remarkably affect the 3D printing process and the quality of the printed product. Protein, fat, and carbohydrates are essential components of a balanced diet, and they have different rheological properties and gelation characteristics (Liu et al., 2019; Liu et al., 2018; Liu et al., 2020). Therefore, the rheological characteristics of raw materials need to be understood for an effective 3D printing. By altering the proportion of meat to non-meat ingredients in ink formulations, the rheology and printability of the ink can be improved, and nutrient-rich goods can be developed.

Lipids

Lipids, also referred to as lipid-like molecules, are a class of bio-organic compounds that are poorly soluble in water but highly soluble in non-polar solvents. Most lipids are esters and their derivatives, formed from the combination of fatty acids and alcohols. The chain length and degree of saturation of the fatty acids significantly influence the physical and functional properties of the lipids, thereby affecting the outcomes of 3D printing processes. Notably, raw meat comes with animal fat, and the fat content and composition vary greatly depending on the type and part of the meat. During meat ink processing, meat pieces are crushed and sheared to form minced meat. At this stage, the fat is sheared into microspheres and evenly adsorbed onto the protein surface. When the ink is heated, the proteins are denatured and cross-linked to form a 3D gel network, while the fat fills the network pores or co-polymerizes with the protein to form an emulsified gel. As a result, lipids in meat products improve the flow behavior of the meat material, reduce viscosity, and enhance ink flow and extrudability. During gel formation, lipids act as effective fillers, forming emulsion gels with proteins that have a uniform structure, good formability, and self-supporting properties. Wu et al., found that fat globules interact with proteins through disulfide bonds, thus affecting the formation of gel networks and improving gel hardness and water retention (Wu, Xiong, Chen, Tang & Zhou, 2009a). The incorporation of functional lipids into 3D printing presents an innovative approach for designing nutritious and healthy personalized foods. However, excessive fat content can alter the fluid behavior of meat paste, leading to increased cooking loss and deformation. Arianna and colleagues found that the cooking loss and shrinkage of beef ink are directly proportional to the lard content. Additionally, the addition of fat results in a more viscous fluid behavior in the ink, which, while beneficial for extrusion, can negatively impact the continuity of extrusion and the material's self-supporting properties (Dick, Bhandari, & Prakash, 2019b). Therefore, selecting appropriate lipids and other materials is crucial to enhancing the printability of meat ink. For instance, Xie et al., developed a high-thixotropy cod protein ink by blending flaxseed oil with inulin and soybean dietary fiber (Xie et al., 2022a, Xie et al., 2022b).

Hydrophilic Colloid

Hydrocolloids, also known as edible gums, are a heterogeneous group of long-chain polymers, which mostly include polysaccharides and a few proteins. These compounds are water-soluble and can be fully hydrated under specific conditions to form viscous dispersions and/or gels. The inclusion of hydrocolloids in meat products enhances the water retention of gels, improves their mechanical strength, and stabilizes hybrid dispersion systems (Liu & Xu, 2019; Perez-Mateos et al., 2001; Ramirez et al., 2002; Wang et al., 2018a, 2018b). In 3D printing, the addition of exogenous colloids interacts with the macromolecules in minced meat, thereby increasing the stability of the meat colloids and improving their rheological properties. Hydrocolloids primarily act as thickening and gelling agents. The thickening effect mainly influences the continuity and homogeneity of the ink during extrusion. Hydrocolloids have a certain viscosity due to their molecular structure, which contains a higher number of hydrophilic groups, such as hydroxyl and carboxyl. The presence of these groups makes the protein or polysaccharide partly dissolved in the solution, forming a viscous colloidal solution. Based on Stokes' law, the greater the viscosity, the smaller the settling rate of colloidal particles, the smaller the degree of mutual agglomeration between molecules, and the more stable the system. High-viscosity inks exhibit good deposition and self-supporting properties in 3D printing; however, they can also clog the nozzle tip and hinder extrusion. On the other hand, low-viscosity inks may disrupt flow and cause agglomeration of solid particles, resulting in poor mechanical properties that make them difficult to mold (Müller et al., 2020). To address these challenges, hydrocolloids should be added to adjust the extrudability of meat materials.

Gelling mainly improves the self-supporting properties of the ink. The hydrophilic groups in hydrogels form gels through hydrogen bonding, hydrophobic association, and cation-mediated cross-linking physical association. These gels can be classified as cold-curing, heat-curing, and cation-mediated gels in terms of their gelation-forming mechanism. For example, starch as a cold-curing hydrogel has a high viscosity and shear stability, and considering the interaction between starch molecules, intermolecular interactions are enhanced at high starch concentrations and the probability of forming hydrogen bonds increases. This phenomenon contributes to the formation of a denser gel network and enhanced self-supporting properties of the ink (Goldstein et al., 2010). Starch can be used as an additive to other food materials to enhance their printability (Feng et al., 2019). Beef, chicken breast, and other meats typically have very low viscosity, making them unsuitable for printing. Therefore, they are often combined with starch and lipids to create printable inks.

However, the effect of hydrogels in meat systems is not always positive, and the effect is related to the type of hydrocolloid and the way and amount of addition. The addition of xanthan gum, Kappa-carrageenan, and alginate alone may sometimes deteri0orate the texture of meat products (Hachmeister and Herald, 1998; mittal & Barbut, 1994). The addition of small amounts (0.2% and 0.5%) of horny gum may enhance gel elasticity. However, higher concentrations of carrageenan can reduce the elasticity of sausages (Ayadi et al., 2009). Moreover, the combination of multiple hydrocolloids often results in altered rheological properties, improving the overall quality of the product. This combination also offers the added benefits of reducing fat content and lowering production costs. Cross-linking can occur between different hydrocolloids, leading to precipitation or gelation. Hydrocolloids with opposite charges may combine to form precipitates. By utilizing the synergistic effects of gels, the gel structures can be enhanced at lower addition levels (Fan et al., 2020; Yasin et al., 2016).

Transglutaminase (TGase)

TGase is an extracellular acyltransferase that catalyzes the formation of a non-disulfide covalent bond between the ε-amino group of a lysine residue and the γ-carboxamide group of the glutamine residue on the heavy chain of myosin. This reaction changes the physical and chemical properties of the meat matrix (Cando et al., 2016). Dong et al., (2020) demonstrated that microbial transglutaminase (MTGase) catalyzes the acyl transfer reaction during the production of surimi, improving its printability and forming a tight, uniform gel structure. TGase, a temperature-dependent enzyme, has a dynamic temperature range of 0–60°C and a pH activity range of 4.5–8.0. By adjusting the reaction temperature and pH using other processing techniques, the cross-linking effect and reaction process can be controlled (Kieliszek & Misiewicz, 2014). For instance, Zhao et al., (2021) reported that synergistic treatment of surimi with TGase and microwave heating led to the production of ϵ-(γ-Glu)-Lys through the microwave's synergistic effect, while TGase primarily enhanced gel formation. Trespalacios and Pla (2007) investigated the combined effect of MTGase and high pressure on the functional properties of chicken gels, observing significant improvements in hardness and chewiness. Additionally, MTGase is often combined with NaCl or phosphate in 3D meat printing to provide sufficient protein substrate, thereby enhancing protein cross-linking levels (Sadeghi-Mehr et al., 2018).

Meat-Related 3D Printers Need Further Development

Lack of Capacity to Ensure Food Safety at All Links

The increasing demand for meat alternatives has driven the development of 3D printing technologies. For instance, microbial growth can be inhibited, and spoilage of meat products can be prevented during the printing process by maintaining a printing temperature below 4°C. However, the need for cooling during printing leads to increased energy consumption, which negatively impacts the sustainability of large-scale food production. Therefore, the development of energy-efficient and environmentally friendly, automatic temperature-controlled 3D printers could serve as a foundation for the commercialization of food 3D printing. In addition, during printing, the ink comes into contact with several parts of the printer, such as the extruder, nozzle, and piston, which are located inside the printer, making them difficult to clean. Subsequently, this configuration may promote the growth and colonization of different microorganisms, such as *Staphylococcus aureus* and *Escherichia coli* (Severini et al., 2018). Muro-Fraguas et al., have developed acrylic acid and tetraethyl orthosilicate coatings applied by plasma polymerization to reduce the biofilm formation of *Pseudomonas aeruginosa*, *E. coli*, and *Listeria monocytogenes* on 3D printing contact surfaces (Muro-Fraguas et al., 2020). However, antimicrobial surfaces may form a layer of dead cell surface debris after some time, thus remarkably decreasing the lethality efficiency of the antimicrobial surface. The fabrication of self-cleaning surfaces is one of the most anticipated alternatives to slow down the rate of bacterial adhesion to surfaces. The application of superhydrophobic-based systems in 3D printers is highly desirable. Superhydrophobic materials have considerable surface roughness (Yoon et al., 2014), which forms cavities when water comes in contact with the surface. This property repels aqueous residues, spills, and droplets that carry bacterial suspensions, thus reducing the effective contact area between the suspension (Oh et al., 2019) and the coated surface. This, in turn, improves the efficiency of disinfection and cleaning of devices and surfaces (Ghasemi and Niakousari, 2020). Meat colloids are hydrophilic colloidal systems with water as the dispersed phase, and superhydrophobic coatings can effectively reduce ink adhesion, enhancing the hygienic design of 3D printers.

The balance between printer printing efficiency and cost should be focused on. Most current 3D printers for food are equipped with single or dual printheads, and the feed tubes and cartridges can only hold a limited amount of material, making them unsuitable for large-scale batch production.

Additionally, the high cost of these printers makes them unaffordable for small food factories and home kitchens. To enhance the practicality of food printers, there is a need for the development of continuous feeding systems and multi-jet printers. Most food printers primarily shape the ink or food material into its desired form. However, raw meat products require not only special pre-treatments and formulations to achieve the proper extrusion viscosity but also post-processing using traditional methods like convection or conduction. This additional processing limits the ability to combine multiple materials into a single food object effectively (Hertafeld et al., 2018). In response to these limitations, combining 3D printing technology with advanced thermal processing or texture modification technologies can enhance the development of printers with both pre- and post-processing capabilities, maximizing the benefits of 3D printing (Demei et al., 2022). Gunduz et al., combined high-frequency amplitude ultrasound on 3D printing nozzles to enable the 3D printing of high-viscosity materials without pre-processing the recipe to adjust the viscosity (Gunduz et al., 2018). However, only a few specialized studies have integrated 3D printing with new technologies, and limited data are available to support the need to fully consider food safety and production economics to address the potential problems in the application of these technologies (Dankar et al., 2018).

Lack of Simultaneous Precision Cooking Ability

Post-processing issues are among the main challenges that face the 3D meat printing industry today. The need for simultaneous maturation capabilities provides new challenges for printer development. Unlike chocolates and candies, most meat materials cannot be eaten raw and need to be processed using traditional convection or conduction methods after printing, thus limiting their ability to combine multiple materials in a single food object. Traditional cooking methods tend to cause damage to the print shape and cooking loss (De Pilli and Alessandrino, 2020; Dick et al., 2021), and the steaming process moisture, as a heat transfer medium, can over-penetrate the ink and damage its structure and shape (Dankar et al., 2020; Dong et al., 2019). The loss of surface moisture and excessive protein denaturation caused by baking can result in uneven heating, ultimately affecting the taste of the product (Severini et al., 2016). Traditional heating methods are limited in that they can only heat specific areas and lack precise control over the heating process. When printing complex multi-layer products, the heating time and conditions vary for different layers of printing materials, and traditional printing methods cannot accurately target the heat required for cooking. The combination of modern cooking technology with the printer to achieve simultaneous maturation during printing can meet the heat requirements of different food materials. Laser cooking technology allows for heat targeting and transfer, thus providing more uniform and efficient heating than conduction without physical contact with the food. Therefore, this technique is an ideal cooking method for 3D printing (Anonymous et al., 2021; Blutinger et al., 2019; Blutinger et al., 2021; Ma and Tao, 2005; Vaskoska et al., 2020). By incorporating a blue laser ($\lambda = 445$ nm), near-infrared (NIR) laser ($\lambda = 980$ nm), and mid-infrared (MIR) laser ($\lambda = 10.6$ μm) into the printer, the researcher's laser-cooked food with approximately 50% less cooking loss than oven-baked food and enabled layer-by-layer cooking. Hertafeld et al., integrated an infrared light heating mechanism into a 3D printer, enabling it to selectively cook and print multi-material food objects. This innovation allowed for the creation of complex food patterns with increased compositional variety (Hertafeld et al., 2018). However, the application of synchronous cooking technology in 3D printing presents several challenges. Improvements are needed in controlling the cooking space and laser intensity, as well as in developing a synchronous cooking method specifically suited for meat. Additionally, the safety of this new technology requires further investigation.

Improvement of Nutritional and Sensory Properties

Ideally, 3D printed products should have the same or even better sensory properties and nutritional values than traditional foods to replace traditional foods in the market gradually. However, limited studies have focused on the sensory evaluation of 3D printed products and have primarily focused on the improvement and evaluation of texture and formability of single meat raw materials (e.g., chicken, beef, and pork). At the same time, mixed material printing with high nutritional value is lacking (Pant et al., 2021). Therefore, the currently developed printed products can only meet a single requirement. For example, the development of geriatric foods primarily focuses on improving soft texture, with an emphasis on hardness and chewiness, while the sensory evaluation of color and flavor remains inadequate. In addition, printability is not the only criterion that determines product quality. Post-treatment methods significantly impact the structural stability and consumer acceptability of meat analogues. Traditional post-treatment techniques, such as frying, steaming, and baking, can alter both the structure (e.g., shrinkage and moisture loss) and functionality (e.g., protein denaturation and vitamin degradation) of the product to varying degrees. In cases where cold-curing gels are used in printing to preserve structure, the high temperatures of post-treatment processes often lead to the melting of the gel and subsequent structural collapse. Therefore, it is crucial to consider not only the ink's printability but also the stability of the post-treatment process and the sensory properties of the final product. Studying a personalized and nutritionally balanced diet is the prominent advantage of 3D printing. Many bioactive substances such as curcumin (Chen et al., 2021a, Chen et al., 2021b) and anthocyanin have been applied in the 3D printing of plant-based materials, while meat materials are rarely involved. Therefore, nutritional health and flavor need to be combined for the development of more functional meat products (Ghazal et al., 2021; Ghazal et al., 2019).

Opportunities of 3D Printing in Meat Applications

3D Printing of Cultured Meat Products

The 3D printing of complex structures such as muscles, skin, bones, and cartilage with biomaterials is called bio-printing. Bio-printing is a new technology that is based on tissue engineering, and this field is still being developed for food applications. Unlike cell-cultured meat, 3D printing of cultured meat is a more complex process. It involves achieving tissue maturity by precisely adhering stem cells and biomaterials to scaffolds, allowing them to grow and form cultured meat (Ramachandraiah, 2021). The 3D printing of meat primarily consists of initial cells, culture medium, and scaffolds. To begin, the cells must possess the ability for self-renewal, infinite proliferation, and differentiation in order to develop into the necessary cell types for meat production, such as muscle cells, fat cells, and chondrocytes. The culture medium plays a critical role in driving cell growth, proliferation, and differentiation while providing essential nutrients to support cell development, tissue regeneration, and maturation. Key components of the growth medium, such as growth factors and animal serum, are crucial for tissue maturation. These components are typically derived from animal embryos (Andreassen et al., 2020; Andreassen et al., 2020; Kolkmann et al., 2020). The 3D scaffold is essential to the structure of cultured meat, which determines the physiological similarity between cultured meat and real meat. Therefore, the scaffold used in 3D printing must have good sensory quality and food safety and ensure the sufficiency of surface area and porosity to support cell adhesion and proliferation (Datar and Betti, 2010). Ben-Arye et al., used soybean protein as a 3D scaffold to support cells and imitate extracellular matrix to create bovine muscle tissue, thus replicating the feeling and texture of meat (Ben-Arye et al., 2020). Fibrinogen, gelatin, and collagen are commonly used as 3D printing scaffolds to enhance tissue stiffness and meet the structural requirements of meat fibers (Ahmad et al., 2021; Duan et al., 2013; MacQueen et al., 2019). In the future, 3D printing technology for cultured meat could potentially address the challenges related to land use, resource consumption, and environmental impact associated with traditional farming practices. This advancement could open up new opportunities for the sustainable development of the meat industry.

Exploitation of Meat By-Product Ink

The exploitation of meat by-products and trimmings is a crucial challenge for meat printing. Meat by-products mainly include body parts other than muscle, such as offal, skin, feet, and fat, accounting for 52% and 66% of the live weight of cattle and pigs, respectively. Some of these by-products, such as heart, liver, and kidney, have high nutritional value but have an unpleasant fishy taste and are generally unacceptable to consumers, making them low-value products in the market (Zou et al., 2021). The unpleasant odor in liver is primarily caused by the release of iron bound to proteins, which produces a metallic scent, as well as the decomposition of proteins and lipid oxidation. These processes generate a variety of organic compounds, including alcohols, aldehydes, ketones, amines, low-grade fatty acids, sulfides, and other substances that contribute to the distinctive smell (Xiong et al., 2017). Undesirable flavors can be effectively eliminated using various methods. Chemical approaches, such as antioxidant treatments (Liu et al., 2021) and acid-base salt treatments (Yarnpakdee et al., 2012), can help neutralize these odors. Physical methods, including β-cyclodextrin adsorption (Yu et al., 2016), microencapsulation (Serfert et al., 2010), and fermentation, as well as enzyme treatments (Li et al., 2020), have also proven effective. However, these techniques often involve the disruption of liver tissue, which may affect the texture and structure of the product. 3D printing can customize paste ink into various shapes, indicating that the inclusion of various deodorizers in the raw material handling process has unique advantages. Thus, the use of 3D printing technology has great potential in the development of consumer-acceptable meat by-products. Notably, in addition to parts such as liver, some trimmings can be used as low-value products or even discarded because of the presence of large amounts of sinew and the connective tissue or poor meat quality, such as beef neck meat. The presence of these connective tissues directly affects the uniformity and adhesion of the ink, thus limiting their application in 3D printing. The issue of raw material waste in the meat industry can be addressed by focusing on the development of by-products. By doing so, sustainability can be enhanced, and the value of 3D printing in the industry can be fully realized.

New Technology Combined with Modeling Methods to Improve Printing Accuracy

Predicting Ink Extrusion Behavior Using Computational Fluid Dynamics (CFD)

CFD refers to computers for building mathematical models to simulate and studying the motion of fluids in a specific geometry with boundary conditions (Oyinloye and Yoon, 2021a, Oyinloye and Yoon, 2021b). The evaluation criterion is the agreement between numerical simulation results and experimental results performed under specific conditions. CFD has been used to simulate food-processing operations such as drying, cooking, sterilization, and freezing (Norton and Sun, 2006; Singh and Muthukumarappan, 2017). CFD can also be applied to predict chemical reactions, mechanical motions, phase changes, and heat and mass transfer in food processing (Jiang et al., 2021). In 3D printing processes, CFD techniques are employed to investigate the impact of material formulation and printing parameters on extrusion behavior, helping to predict the optimal deposition formulation. CFD mainly involves the selection of suitable methods to discretize the simulation of the fluid continuum, which mainly includes finite difference (FD), finite element (FE), and finite volume (FV) (Oyinloye and Yoon, 2021). In meat 3D printing, the finite element method (FEM) is widely used because of its ability to solve problems that involve non-Newtonian fluids and nonlinear flow. Oyinloye et al., successfully optimized the parameters of surimi 3D printing using FEM. They demonstrated that the nozzle diameter significantly affects the fluid properties (e.g., pressure, velocity, and shear rate) in the flow field and the residual stress and deformation of the printed sample(Oyinloye et al., 2022).

Improving Ink Printing Accuracy with Low Field Nuclear Magnetic Resonance (LF-NMR)

As the most critical chemical components in meat and meat products, the water content and water holding capacity are directly related to the edible quality of meat and meat products, such as color, tenderness, juiciness, and flavor (Guanghong, 2008). As a fast and non-destructive testing method, LF-NMR can detect much information about moisture in meat and meat products by analyzing the spin relaxation characteristics of hydrogen core (oil/gas/water) in the magnetic field. The relaxation time and corresponding peak area of food ink, as measured by LF-NMR, are closely associated with its shear-thinning and viscoelastic properties (Phuhongsung et al., 2020; Xu et al., 2020). As a result, LF-NMR is commonly utilized as a standard method for evaluating ink printability in 3D printing. Many researchers have applied LF-NMR parameters to polynomial regression models to predict 3D printability, and high correlation coefficients (R^2) were obtained for both 3D printing accuracy and stability. Therefore, LF-NMR is an excellent non-destructive tool for the fast and accurate prediction of rheological properties and the assessment of 3D printability of food inks (Liu et al., 2022). Although LF-NMR has been widely used to predict the printability of non-meat materials (Phuhongsung et al., 2020), meat materials have not been extensively studied. Therefore, the scope of its application should be expanded, and robust technical support should be provided to enhance the development of meat inks.

Predicting Ink Printability Using NIR Spectroscopy

NIR light is an electromagnetic wave with wavelengths between ultraviolet and visible spectrophotometer (UV–vis) and middle infrared (MIR). NIR correlated with sample rheological properties, modeled using appropriate methods, have proven to be effective in predicting ink printability and improving printing accuracy. First, the NIR spectroscopy technology obtains the spectra of the samples containing hydrogen groups by absorbing the frequency doubling and combining frequencies of the vibrations of hydrogen groups (–CH, –NH, and –OH) (Alexandrakis et al., 2012; Ammor et al., 2009). Traditional analytical methods are then employed to determine the properties or data of the samples. A correction model can be developed using stoichiometric techniques, such as principal component analysis, partial least squares regression, and multiple linear regression, to establish a correlation between the spectra and the data. Based on the calibration model, the composition or content of the sample can be predicted by integrating it with the near-infrared (NIR) spectrum of the sample. At present, NIR spectroscopy has been widely used for the quantitative analysis of meat products, including the prediction of meat water holding capacity, tenderness, and shear force (Samuel et al., 2011), which are strongly related to rheological properties. In comparison with low field nuclear magnetic resonance and rheological properties detection, NIR spectroscopy has the advantages of fast and easy operation. It shows a good correlation in predicting rheological and printing properties of purple sweet potato past (Phuhongsung et al., 2020). Therefore, NIR spectroscopy could be applied to predict the printability of meat inks in the future. Using models such as partial least squares, principal component regression, and artificial neural networks, the rheological properties of slurries can be effectively predicted with appropriate NIR spectral parameters. This allows for the indirect yet rapid prediction of the 3D printability of slurries.

Conclusion

As reviewed, meat materials are non-native printable materials, which need to be pretreated or formulated to improve printing characteristics. Pre-treatment technology provides the feasibility of direct printing of "additive-free" raw materials, and grafting pre-treatment technology (e.g., ultrasound, infrared) onto 3D printers can achieve the rapid printing various materials and meet consumer demand for "additive-free" printed products. However, limited studies have focused on pre-processing technologies, and the lack of valid data require more in-depth research. Many researchers have conducted extensive studies on printable meat formulations, resulting in the development of various products such as minced fish, chicken snacks, and beef burgers. However, current research is still limited by several factors, including the use of single-ingredient compositions, low levels of product innovation and functionality, and a lack of post-processing and sensory evaluation.

In the future, advancements in rapid and convenient detection technologies will offer new opportunities for the development of 3D printing inks for meat products. Evaluation techniques such as LF-NMR, NIR spectroscopy, and CFD have already proven to be crucial in the development of meat inks. However, only a limited number of studies have been published on the 3D printing of meat materials. With the continued advancement of new technologies, the development of printable meat inks will no longer pose a significant challenge.

References

Ahmad, K., Lim, J., Lee, E., Chun, H., Ali, S., Ahmad, S. S., Shaikh, S., Choi, I., 2021. Extracellular matrix and the production of cultured meat. *Foods* 10 (12).

Alarcon-Rojo, A. D., Janacua, H., Rodriguez, J. C., Paniwnyk, L., Mason, T. J., 2015. Power ultrasound in meat processing. *Meat Sci.* 107, 86–93.

Alexandrakis, D., Downey, G., Scannell, A. G. M., 2012. Rapid non-destructive detection of spoilage of intact chicken breast muscle using near-infrared and fourier transform mid-infrared spectroscopy and multivariate statistics. *Food Bioprocess Technol.* 5 (1), 338–347.

Amiri, A., Sharifian, P., Soltanizadeh, N., 2018. Application of ultrasound treatment for improving the physicochemical, functional and rheological properties of myofibrillar proteins. *Int. J. Biol. Macromol.* 111, 139–147.

Ammor, M. S., Argyri, A., Nychas, G. E., 2009. Rapid monitoring of the spoilage of minced beef stored under conventionally and active packaging conditions using Fourier transform infrared spectroscopy in tandem with chemometrics. *Meat Sci.* 81 (3), 507–514.

Andreassen, R. C., Pedersen, M. E., Kristoffersen, K. A., Beate Rønning, S., 2020a. Screening of by-products from the food industry as growth promoting agents in serum-free media for skeletal muscle cell culture. *Food Funct.* 11 (3), 2477–2488.

Anonymous, Gracia, A., Sepulveda, E., 2021. Apparatus and Method for Heating and Cooking Food Using Laser Beams and Electromagnetic Radiation. *Official Gazette of the United States Patent and Trademark Office Patents.*

Ayadi, M. A., Kechaou, A., Makni, I., Attia, H., 2009. Influence of carrageenan addition on Turkey meat sausages properties. *J. Food Eng.* 93 (3), 278–283.

Banerjee, S., Bhattacharya, S., 2012. Food gels: gelling process and new applications. *Crit. Rev. Food Sci. Nutr.* 52 (4), 334–346.

Ben-Arye, T., Shandalov, Y., Ben-Shaul, S., Landau, S., Zagury, Y., Ianovici, I., Lavon, N., Levenberg, S., 2020. Textured soy protein scaffolds enable the generation of three-dimensional bovine skeletal muscle tissue for cell-based meat. *Nature Food* 1 (4), 210–220.

Bhattacharya, S., Jena, R., 2007. Gelling behavior of defatted soybean flour dispersions due to microwave treatment: textural, oscillatory, microstructural and sensory properties. *J. Food Eng.* 78 (4), 1305–1314.

Blutinger, J. D., Meijers, Y., Lipson, H., 2019. Selective laser broiling of Atlantic salmon. *Food Res. Int.* 120, 196–208.

Blutinger, J. D., Tsai, A., Storvick, E., Seymour, G., Liu, E., Samarelli, N., Karthik, S., Meijers, Y., Lipson, H., 2021. Precision cooking for printed foods via multiwavelength lasers. *NPJ Sci. Food* 5 (1).

Bohr, H., Bohr, J., 2000. Microwave-enhanced folding and denaturation of globular proteins. *Phys. Rev.* 61 (4), 4310–4314.

Bulut, E. G., Candog˘an, K., 2022. Development and characterization of a 3D printed functional chicken meat based snack: optimization of process parameters and gelatin level. *LWT* 154, 112768.

Cai, L., Feng, J., Cao, A., Zhang, Y., Lv, Y., Li, J., 2018. Denaturation kinetics and aggregation mechanism of the sarcoplasmic and myofibril proteins from grass carp during microwave processing. *Food Bioprocess Technol.* 11 (2), 417–426.

Cando, D., Borderías, A. J., Moreno, H. M., 2016. Combined effect of aminoacids and microbial transglutaminase on gelation of low salt surimi content under high pressure processing. *Innovat. Food Sci. Emerg. Technol.* 36, 10–17.

Cao, C. N., Yuan, D. X., Kong, B. H., Chen, Q., He, J. J., Liu, Q., 2022.Effect of different carrageenan incorporation forms on the gel properties and *in vitro* digestibility of frankfurters. *Food Hydrocolloids* 129.

Chang, H., Xu, X., Zhou, G., Li, C., Huang, M., 2012. Effects of characteristics changes of collagen on meat physicochemical properties of beef semitendinosus muscle during ultrasonic processing. *Food Bioprocess Technol.* 5 (1), 285–297.

Chattong, U., Apichartsrangkoon, A., Bell, A. E., 2007. Effects of hydrocolloid addition and high pressure processing on the rheological properties and microstructure of a commercial ostrich meat product "Yor" (Thai sausage). *Meat Sci.* 76 (3), 548–554.

Chen, C., Zhang, M., Guo, C. F., Chen, H. Z., 2021a. 4D Printing of lotus Root Powder Gel: Color Change Induced by Microwave, vol. 68. *Innovative Food Science & Emerging Technologies.*

Chen, H., Zhang, M., Yang, C., 2021b. Comparative analysis of 3D printability and rheological properties of surimi gels via LF-NMR and dielectric characteristics. *J. Food Eng.* 292, 110278.

Chen, Y., Zhang, M., Sun, Y., Phuhongsung, P., 2022. Improving 3D/4D printing characteristics of natural food gels by novel additives: a review. *Food Hydrocolloids* 123, 107160.

Dankar, I., Haddarah, A., Omar, F. E. L., Sepulcre, F., Pujol`a, M., 2018. 3D printing technology: the new era for food customization and elaboration. *Trends Food Sci. Technol.* 75, 231–242.

Dankar, I., Haddarah, A., Sepulcre, F., Pujola`, M., 2020. Assessing mechanical and rheological properties of potato puree: effect of different ingredient combinations and cooking methods on the feasibility of 3D printing. *Foods* 9.

Datar, I., Betti, M., 2010. Possibilities for an *in vitro* meat production system. *Innovat. Food Sci. Emerg. Technol.* 11 (1), 13–22.

De Pilli, T., Alessandrino, O., 2020. Effects of different cooking technologies on biopolymers modifications of cereal-based foods: impact on nutritional and quality characteristics review. *Crit. Rev. Food Sci. Nutr.* 60 (4), 556–565.

Demei, K., Zhang, M., Phuhongsung, P., Mujumdar, A.S., 2022. 3D food printing: controlling characteristics and improving technological effect during food processing. *Food Res. Int.* 156, 111120.

Dick, A., Bhandari, B., Dong, X., Prakash, S., 2020a. Feasibility study of hydrocolloid incorporated 3D printed pork as dysphagia food. *Food Hydrocolloids* 107, 105940.

Dick, A., Bhandari, B., Dong, X., Prakash, S., 2020b. Feasibility study of hydrocolloid incorporated 3D printed pork as dysphagia food. *Food Hydrocolloids* 107, 105940.

Dick, A., Bhandari, B., Prakash, S., 2019a. 3D printing of meat. Meat Sci. 153, 35–44.

Dick, A., Bhandari, B., Prakash, S., 2019b. Post-processing feasibility of composite-layer 3D printed beef. *Meat Sci.* 153, 9–18.

Dick, A., Bhandari, B., Prakash, S., 2021. Printability and textural assessment of modified-texture cooked beef pastes for dysphagia patients. *Future Foods* 3, 100006.

Dong, X., Huang, Y., Pan, Y., Wang, K., Prakash, S., Zhu, B., 2019. Investigation of sweet potato starch as a structural enhancer for three-dimensional printing of Scomberomorus niphonius surimi. *J. Texture Stud.* 50 (4), 316–324.

Dong, X., Pan, Y., Zhao, W., Huang, Y., Qu, W., Pan, J., Qi, H., Prakash, S., 2020. Impact of microbial transglutaminase on 3D printing quality of Scomberomorus niphonius surimi. *LWT* 124, 109123.

Duan, B., Hockaday, L. A., Kang, K. H., Butcher, J. T., 2013. 3D Bioprinting of heterogeneous aortic valve conduits with alginate/gelatin hydrogels. *J. Biomed. Mater. Res.* 101 (5), 1255–1264.

Eilert, S. J., Blackmer, D. S., Mandigo, R. W., Calkins, C. R., 1993. Meat batters manufactured with modified beef connective tissue. *J. Food Sci.* 58 (4), 691–696.

Fan, R., Zhou, D., Cao, X., 2020. Evaluation of oat beta-glucan-marine collagen peptide mixed gel and its application as the fat replacer in the sausage products. *PLoS One* 15 (5).

Feng, C., Zhang, M., Bhandari, B., 2019. Materials properties of printable edible inks and printing parameters optimization during 3D printing: a review. *Crit. Rev. Food Sci. Nutr.* 59 (19), 3074–3081.

Ghasemi, A., Niakousari, M., 2020. Superwettability-based systems: basic concepts, recent trends and future prospects for innovation in food engineering. *Trends Food Sci. Technol.* 104, 27–36.

Ghazal, A. F., Zhang, M., Bhandari, B., Chen, H., 2021. Investigation on spontaneous 4D changes in color and flavor of healthy 3D printed food materials over time in response to external or internal pH stimulus. *Food Res. Int.* 142.

Ghazal, A. F., Zhang, M., Liu, Z., 2019. Spontaneous color change of 3D printed healthy food product over time after printing as a novel application for 4D food printing. *Food Bioprocess Technol.* 12 (10), 1627–1645.

Goldstein, A., Nantanga, K. K. M., Seetharaman, K., 2010. Review: molecular interactions in starch-water systems: effect of increasing starch concentration. *Cereal Chem.* 87 (4), 370–375.

Guanghong, Z., 2008. *Meat Science and Technology*. China Agricultural Press.

Gunduz, I. E., McClain, M. S., Cattani, P., Chiu, G. T. C., Rhoads, J. F., Son, S. F., 2018. 3D printing of extremely viscous materials using ultrasonic vibrations. *Addit. Manuf.* 22, 98–103.

Guo, C., Zhang, M., Bhandari, B., 2019. Model building and slicing in food 3D printing processes: a review. *Compr. Rev. Food Sci. Food Saf.* 18 (4), 1052–1069.

Hachmeister, K. A., Herald, T. J., 1998. Thermal and rheological properties and textural attributes of reduced-fat Turkey batters. *Poultry Sci.* 77 (4), 632–638.

Handral, H. K., Hua Tay, S., Wan Chan, W., Choudhury, D., 2022. 3D Printing of cultured meat products. *Crit. Rev. Food Sci. Nutr.* 62 (1), 272–281.

He, C., Zhang, M., Fang, Z., 2020a. 3D printing of food: pretreatment and post-treatment of materials. *Crit. Rev. Food Sci. Nutr.* 60 (14), 2379–2392.

He, C., Zhang, M., Guo, C., 2020b. 4D printing of mashed potato/purple sweet potato puree with spontaneous color change. *Innovat. Food Sci. Emerg. Technol.* 59.

Hertafeld, E., Zhang, C., Jin, Z., Jakub, A., Russell, K., Lakehal, Y., Andreyeva, K., Bangalore, S.N., Mezquita, J., Blutinger, J., Lipson, H., 2018. Multi-material three-dimensional food printing with simultaneous infrared cooking. 3D Print. *Addit. Manuf.* 6 (1), 13–19.

Hull, C. W., 1986. Apparatus for Production of Three-Dimensional Objects by Stereolithography. *Google Patents.*

Jiang, H., Zheng, L., Zou, Y., Tong, Z., Han, S., Wang, S., 2019. 3D food printing: main components selection by considering rheological properties. *Crit. Rev. Food Sci. Nutr.* 59 (14), 2335–2347.

Jiang, Q., Zhang, M., Mujumdar, A. S., 2021. Novel evaluation technology for the demand characteristics of 3D food printing materials: a review. *Crit. Rev. Food Sci. Nutr.* 1–16.

Kang, Z. L., Zou, X. L., Meng, L., Li, Y. P., 2021. Effects of NaCl and soy protein isolate on the physicochemical, water distribution, and mobility in frankfurters. *Int. J. Food Sci. Technol.* 56 (12), 6572–6579.

Kelly, B. E., Bhattacharya, I., Heidari, H., Shusteff, M., Spadaccini, C. M., Taylor, H. K., 2019. Volumetric additive manufacturing via tomographic reconstruction. *Science* 363 (6431), 1075.

Kieliszek, M., Misiewicz, A., 2014. Microbial transglutaminase and its application in the food industry. *A review. Folia Microbiol.* 59 (3), 241–250.

Kim, H. W., Bae, H., Park, H. J., 2017. Classification of the printability of selected food for 3D printing: development of an assessment method using hydrocolloids as reference material. *J. Food Eng.* 215, 23–32.

Kim, T., Yong, H. I., Jang, H. W., Kim, Y., Sung, J., Kim, H., Choi, Y., 2020. Effects of hydrocolloids on the quality characteristics of cold-cut duck meat jelly. *J. Anim. Sci. Technol.* 62 (4), 587–594.

Kolkmann, A. M., Post, M. J., Rutjens, M. A. M., van Essen, A. L. M., Moutsatsou, P., 2020. Serum-free media for the growth of primary bovine myoblasts. *Cytotechnology* 72 (1), 111–120.

Leal-Ramos, M. Y., Alarcon-Rojo, A. D., Mason, T. J., Paniwnyk, L., Alarjah, M., 2011. Ultrasound-enhanced mass transfer in Halal compared with non-Halal chicken. J. *Sci. Food Agric.* 91 (1), 130–133.

Li, B., Sun, D., 2002. Novel methods for rapid freezing and thawing of foods – a review. *J. Food Eng.* 54 (3), 175–182.

Li, J., Yeh, A., 2003. Effects of starch properties on rheological characteristics of starch/meat complexes. J. Food Eng. 57 (3), 287–294.

Li, K., Liu, J., Fu, L., Li, W., Zhao, Y., Bai, Y., Kang, Z., 2019. Effect of gellan gum on functional properties of low-fat chicken meat batters. *J. Texture Stud.* 50 (2), 131–138.

Li, Y., Zhou, W., Cao, Y., Gong, X., Li, J., Lu, X., Dai, Y., IOP, 2020. Analysis of volatile components of Tilapia enzymolysis solution after different deodorization treatments. In: *2020 5th International Conference On Materials Science, Energy Technology And Environmental Engineering*, vol. 571. 5th International Conference on Materials Science, Energy Technology and Environmental Engineering (MSETEE).

Liu, J., Xu, B., 2019. A comparative study on texture, gelatinisation, retrogradation and potential food application of binary gels made from selected starches and edible gums. *Food Chem.* 296, 100–108.

Liu, Y., Huang, Y., Wang, Z., Cai, S., Zhu, B., Dong, X., 2021. Recent advances in fishy odour in aquatic fish products, from formation to control. *Int. J. Food Sci. Technol.* 56 (10), 4959–4969.

Liu, Y., Sun, Q., Wei, S., Xia, Q., Pan, Y., Liu, S., Ji, H., Deng, C., Hao, J., 2022. LF-NMR as a tool for predicting the 3D printability of surimi-starch systems. *Food Chem.* 374, 131727.

Liu, Z., Bhandari, B., Prakash, S., Mantihal, S., Zhang, M., 2019. Linking rheology and printability of a multicomponent gel system of carrageenan -xanthan-starch in extrusion based additive manufacturing. *Food Hydrocolloids* 87, 413–424.

Liu, Z., Bhandari, B., Prakash, S., Zhang, M., 2018. Creation of internal structure of mashed potato construct by 3D printing and its textural properties. *Food Res. Int.* 111, 534–543.

Liu, Z., Dick, A., Prakash, S., Bhandari, B., Zhang, M., 2020. Texture modification of 3D printed air-fried potato snack by varying its internal structure with the potential to reduce oil content. *Food Bioprocess Technol.* 13 (3), 564–576.

Lu, B., Lan, H., Liu, H., 2018. Additive manufacturing frontier: 3D printing electronics. *Opto-Electron. Adv.* 1 (1).

Ma, L., Tao, Y., 2005. An infrared and laser range imaging system for non-invasive estimation of internal temperatures in chicken breasts during cooking. *Transac. ASAE* 48 (2), 681–690.

MacQueen, L. A., Alver, C. G., Chantre, C. O., Ahn, S., Cera, L., Gonzalez, G. M., O'Connor, B. B., Drennan, D. J., Peters, M. M., Motta, S. E., Zimmerman, J. F., Parker, K. K., 2019. Muscle tissue engineering in fibrous gelatin: implications for meat analogs. *NPJ Sci. Food* 3 (1).

Malone, E., Lipson, H., 2007. Fab@Home: the personal desktop fabricator kit. *Rapid Prototyp. J.* 13 (4), 245–255.

Mittal, G. S., Barbut, S., 1994. Effects of carrageenans and xanthan gum on the texture and acceptability of low-fat frankfurters. *J. Food Process. Preserv.* 18 (3), 201–216.

Müller, M., Fisch, P., Molnar, M., Eggert, S., Binelli, M., Maniura-Weber, K., Zenobi-Wong, M., 2020. Development and thorough characterization of the processing steps of an ink for 3D printing for bone tissue engineering. Mater. *Sci. Eng. C* 108, 110510.

Muro-Fraguas, I., Sainz-García, A., Fern´andez Go´mez, P., Lo´pez, M., Múgica-Vidal, R., Sainz-García, E., Toledano, P., Sa´enz, Y., Lo´pez, M., Gonz´alez-Raurich, M.

Norton, T., Sun, D., 2006. Computational fluid dynamics (CFD) – an effective and efficient design and analysis tool for the food industry: a review. *Trends Food Sci. Technol.* 17 (11), 600–620.

Oh, J. K., Liu, S., Jones, M., Yegin, Y., Hao, L., Tolen, T. N., Nagabandi, N., Scholar, E. A., Castillo, A., Taylor, T. M., Cisneros-Zevallos, L., Akbulut, M., 2019. Modification of aluminum surfaces with superhydrophobic nanotextures for enhanced food safety and hygiene. *Food Control* 96, 463–469.

Oyinloye, T. M., Yoon, W. B., 2021a. Application of computational fluid dynamics (CFD) simulation for the effective design of food 3D printing (A review). Processes 9.

Oyinloye, T. M., Yoon, W. B., 2021b. Stability of 3D printing using a mixture of pea protein and alginate: precision and application of additive layer manufacturing simulation approach for stress distribution. *J. Food Eng.* 288, 110127.

Oyinloye, T. M., Yoon, W. B., 2022. Investigation of flow field, die swelling, and residual stress in 3D printing of surimi paste using the finite element method. *Innovat. Food Sci. Emerg. Technol.* 78, 103008.

Pan, Y., Sun, Q., Liu, Y., Wei, S., Xia, Q., Zheng, O., Liu, S., Ji, H., Deng, C., Hao, J., 2021. The relationship between rheological and textural properties of shrimp surimi adding starch and 3D printability based on principal component analysis. *Food Sci. Nutr.* 9 (6), 2985–2999.

Pant, A., Lee, A. Y., Karyappa, R., Lee, C.P., An, J., Hashimoto, M., Tan, U., Wong, G., Chua, C. K., Zhang, Y., 2021. 3D food printing of fresh vegetables using food hydrocolloids for dysphagic patients. *Food Hydrocolloids* 114.

Patana-Anake, C., Foegeding, E. A., 1985. Rheological and stability transitions in meat batters containing soy protein concentrate and vital wheat gluten. *J. Food Sci.* 50 (1), 160–164.

Paxton, N., Smolan, W., Bo¨ck, T., Melchels, F., Groll, J., Jungst, T., 2017. Proposal to assess printability of bioinks for extrusion-based bioprinting and evaluation of rheological properties governing bioprintability. *Biofabrication* 9 (4), 44107.

Perez-Mateos, M., Hurtado, J.L., Montero, P., Fernandez-Martin, F., 2001. Interactions of kappa-carrageenan plus other hydrocolloids in fish myosystem gels. *J. Food Sci.* 66 (6), 838–843.

Phuhongsung, P., Zhang, M., Devahastin, S., 2020. Investigation on 3D printing ability of soybean protein isolate gels and correlations with their rheological and textural properties via LF-NMR spectroscopic characteristics. *LWT* 122, 109019.

Pietrasik, Z., 2003. Binding and textural properties of beef gels processed with κ-carrageenan, egg albumin and microbial transglutaminase. *Meat Sci.* 63 (3), 317–324.

Pohlman, F. W., Dikeman, M. E., Zayas, J. F., Unruh, J. A., 1997. Effects of ultrasound and convection cooking to different end point temperatures on cooking characteristics, shear force and sensory properties, composition, and microscopic morphology of beef longissimus and pectoralis muscles. *J. Anim. Sci.* 75 (2), 386–401.

Prakash, S., Bhandari, B. R., Godoi, F.C., Zhang, M., 2019. Chapter 13 - future outlook of 3D food printing. In: Godoi, F. C., Bhandari, B. R., Prakash, S., Zhang, M. (Eds.), *Fundamentals of 3D Food Printing and Applications*. Academic Press, pp. 373–381.

Prieto, M., Alvarez-Ordo´n˜ez, A., Gonza´lez-Marcos, A., Alba-Elías, F., 2020. Atmospheric pressure cold plasma anti-biofilm coatings for 3D printed food tools. *Innovat. Food Sci. Emerg. Technol.* 64, 102404.

Przetaczek-Roz˙nowska, I., Fortuna, T., Wodniak, M., Łabanowska, M., Pająk, P., Kro´likowska, K., 2019. Properties of potato starch treated with microwave radiation and enriched with mineral additives. *Int. J. Biol. Macromol.* 124, 229–234.

Ramachandraiah, K., 2021. Potential development of sustainable 3D-printed meat analogues: a review. *Sustainability* 13.

Ramírez, J. A., Barrera, M., Morales, O.G., Va´zquez, M., 2002. Effect of xanthan and locust bean gums on the gelling properties of myofibrillar protein. *Food Hydrocolloids* 16 (1), 11–16.

Ramirez, J. A., Barrera, M., Morales, O. G., Vazquez, M., 2002. Effect of xanthan and locust bean gums on the gelling properties of myofibrillar protein. *Food Hydrocolloids* 16 (1), 11–16.

Roser, H. R. A. M., 2017. Meat and Dairy Production. Sadeghi-Mehr, A., Raudsepp, P., Brüggemann, D. A., Lautenschlaeger, R., Drusch, S., 2018. Dynamic rheology, microstructure and texture properties of model porcine meat batter as affected by different cold-set binding systems. *Food Hydrocolloids* 77, 937–944.

Samuel, D., Park, B., Sohn, M., Wicker, L., 2011. Visible-near-infrared spectroscopy to predict water-holding capacity in normal and pale broiler breast meat. *Poultry Sci.* 90 (4), 914–921.

Serfert, Y., Drusch, S., Schwarz, K., 2010. Sensory odour profiling and lipid oxidation status of fish oil and microencapsulated fish oil. *Food Chem.* 123 (4), 968–975.

Severini, C., Derossi, A., Azzollini, D., 2016. Variables affecting the printability of foods: preliminary tests on cereal-based products. *Innovat. Food Sci. Emerg. Technol.* 38, 281–291.

Severini, C., Derossi, A., Ricci, I., Caporizzi, R., Fiore, A., 2018. Printing a blend of fruit and vegetables. New advances on critical variables and shelf life of 3D edible objects. *J. Food Eng.* 220, 89–100.

Shi, Y., Tu, L., Yuan, C., Wu, J., Li, X., Wang, S., Chen, H., Chen, X., 2022. Regulatory mechanisms governing collagen peptides and their 3D printing application for frozen surimi. *J. Food Sci.* 87 (6), 2692–2706.

Singh, S. K., Muthukumarappan, K., 2017. Rheological characterization and CFD simulation of soy white flakes based dough in a single screw extruder. *J. Food Process. Eng.* 40 (2), e12368.

Sumnu, G., 2001. A review on microwave baking of foods. *Int. J. Food Sci. Technol.* 36 (2), 117–127.

Sun, J., Zhou, W., Huang, D., Fuh, J. Y. H., Hong, G. S., 2015. An overview of 3D printing technologies for food fabrication. *Food Bioprocess Technol.* 8 (8), 1605–1615.

Sun, J., Zhou, W., Yan, L., Huang, D., Lin, L., 2018. Extrusion-based food printing for digitalized food design and nutrition control. *J. Food Eng.* 220, 1–11.

Sun, X., Ohanenye, I.C., Ahmed, T., Udenigwe, C. C., 2020. Microwave treatment increased protein digestibility of pigeon pea (Cajanus cajan) flour: elucidation of underlying mechanisms. *Food Chem.* 329.

Sun, Y., Ma, L., Fu, Y., Dai, H., Zhang, Y., 2021. The improvement of gel and physicochemical properties of porcine myosin under low salt concentrations by pulsed ultrasound treatment and its mechanism. *Food Res. Int.* 141, 110056.

Trespalacios, P., Pla, R., 2007. Synergistic action of transglutaminase and high pressure on chicken meat and egg gels in absence of phosphates. Food Chem. 104 (4), 1718–1727.

Varvara, R., Szabo, K., Vodnar, D. C., 2021. 3D food printing: principles of obtaining digitally-designed nourishment. *Nutrients* 13 (10), 3617.

Vaskoska, R., Ha, M., Tran, H. T. T., Khoshelham, K., White, J. D., Warner, R. D., 2020. Evaluation of 3D laser scanning for estimation of heating-induced volume shrinkage and prediction of cooking loss of pork cuboids compared to manual measurements. *Food Bioprocess Technol.* 13 (6), 938–947.

Wang, C., Virgilio, N., Wood-Adams, P. M., Heuzey, M., 2018a. Protein/polysaccharide-based hydrogels prepared by vapor-induced phase separation. *Macromol. Chem. Phys.* 219 (7).

Wang, L., Zhang, M., Bhandari, B., Yang, C., 2018b. Investigation on fish surimi gel as promising food material for 3D printing. *J. Food Eng.* 220, 101–108.

Wang, Y., Rashid, M. T., Yan, J., Ma, H., 2021. *Effect of Multi-Frequency Ultrasound Thawing on the Structure and Rheological Properties of Myofibrillar Proteins from Small Yellow Croaker,* vol. 70. Ultrasonics Sonochemistry.

Wang, Y., Zhou, Y., Wang, X., Li, P., Xu, B., Chen, C., 2020. Water holding capacity of sodium-reduced chicken breast myofibrillar protein gel as affected by combined CaCl2 and high-pressure processing. *Int. J. Food Sci. Technol.* 55 (2), 601–609.

Welin, S., 2013. Introducing the new meat. Problems and prospects. *Etikk i Praksis* 7 (1), 24–37.

Wu, M., Xiong, Y. L., Chen, J., Tang, X., Zhou, G., 2009a. Rheological and microstructural properties of porcine myofibrillar protein-lipid emulsion composite gels. *J. Food Sci.* 74 (4), E207–E217.

Xie, Y., Yu, X., Wang, Y., Yu, C., Prakash, S., Zhu, B., Dong, X., 2022a. Role of dietary fiber and flaxseed oil in altering the physicochemical properties and 3D printability of cod protein composite gel. *J. Food Eng.* 327, 111053.

Xie, Y., Yu, X., Wang, Z., Yu, C., Prakash, S., Dong, X., 2022b. The synergistic effects of myofibrillar protein enrichment and homogenization on the quality of cod protein gel. *Food Hydrocolloids* 127, 107468.

Xiong, G., Gao, X., Zheng, H., Li, X., Xu, X., Zhou, G., 2017. Comparison on the physico-chemical and nutritional qualities of normal and abnormal colored fresh chicken liver. *Anim. Sci. J.* 88 (6), 893–899.

Xu, L., Gu, L., Su, Y., Chang, C., Wang, J., Dong, S., Liu, Y., Yang, Y., Li, J., 2020. Impact of thermal treatment on the rheological, microstructural, protein structures and extrusion 3D printing characteristics of egg yolk. *Food Hydrocolloids* 100, 105399.

Yang, G., Tao, Y., Wang, P., Xu, X., Zhu, X., 2022. Optimizing 3D printing of chicken meat by response surface methodology and genetic algorithm: feasibility study of 3D printed chicken product. *LWT* 154, 112693.

Yang, H., Han, M., Wang, H., Cao, G., Tao, F., Xu, X., Zhou, G., Shen, Q., 2021. HPP improves the emulsion properties of reduced fat and salt meat batters by promoting the adsorption of proteins at fat droplets/water interface. *LWT–Food Sci. Technol.* 137.

Yarnpakdee, S., Benjakul, S., Kristinsson, H. G., Maqsood, S., 2012. Effect of pretreatment on lipid oxidation and fishy odour development in protein hydrolysates from the muscle of Indian mackerel. *Food Chem.* 135 (4), 2474–2482.

Yasin, H., Babji, A.S., Ismail, H., 2016. Optimization and rheological properties of chicken ball as affected by κ-carrageenan, fish gelatin and chicken meat. *LWT - Food Sci. Technol.* (Lebensmittel-Wissenschaft -Technol.) 66, 79–85.

Yoon, S.H., Rungraeng, N., Song, W., Jun, S., 2014. Superhydrophobic and superhydrophilic nanocomposite coatings for preventing Escherichia coli K-12 adhesion on food contact surface. *J. Food Eng.* 131, 135–141.

Yu, X., Chen, L., Sheng, L., Tong, Q., 2016. Volatile compounds analysis and off-flavors removing of porcupine liver. *Food Sci. Technol. Res.* 22 (2), 283–289.

Zeng, X., Li, T., Zhu, J., Chen, L., Zheng, B., 2021. Printability improvement of rice starch gel via catechin and procyanidin in hot extrusion 3D printing. *Food Hydrocolloids* 121, 106997.

Zhang, Y., Zeng, Q., Zhu, Z., 2011. Effect of ultrasonic treatment on the activities of endogenous transglutaminase and proteinases in Tilapia (Sarotherodon Nilotica) surimi during gel formation. *J. Food Process. Eng.* 34 (5), 1695–1713.

Zhao, L., Zhang, M., Chitrakar, B., Adhikari, B., 2021. Recent advances in functional 3D printing of foods: a review of functions of ingredients and internal structures. *Crit. Rev. Food Sci. Nutr.* 61 (21), 3489–3503.

Zhao, Z., Wang, Q., Yan, B., Gao, W., Jiao, X., Huang, J., Zhao, J., Zhang, H., Chen, W., Fan, D., 2021. Synergistic effect of microwave 3D print and transglutaminase on the self-gelation of surimi during printing. *Innovat. Food Sci. Emerg. Technol.* 67, 102546.

Zhou, Y., Wang, W., Ma, F., Li, P., Chen, C., 2018. High-pressure pretreatment to improve the water retention of sodium-reduced frozen chicken breast gels with two organic anion types of potassium salts. *Food Bioprocess Technol.* 11 (3), 526–535.

Zou, Y., Shahidi, F., Shi, H., Wang, J., Huang, Y., Xu, W., Wang, D., 2021. Values-added utilization of protein and hydrolysates from animal processing by-product livers: a review. *Trends Food Sci. Technol.* 110, 432–442.

Chapter 7

Novel Meat Analogue Foodstuffs

Abstract

Plant-based meat analogues, designed to replicate traditional meat, have gained significant traction in the protein food market, experiencing rapid growth in both the retail and food service sectors globally. This chapter examined the ingredients used in modern meat analogues, evaluated their nutritional profiles, and compared them to traditional meat. The findings indicate that these products qualify as ultra-processed foods due to their ingredients. However, their nutrient profiles closely resemble those of traditional meat products, effectively simulating the characteristics of meat. Despite being made from diverse ingredients and undergoing extensive processing, modern meat analogues offer a comparable nutrient composition to traditional meat, making them a viable alternative for consumers.

Keywords: meat analogues, plant-based proteins, nutritional composition, processed foods

Nutrition of Meat Analogues

Meat analogues are formulated in a manner where the main priority of the manufacturer is to create a product with similar organoleptic properties of the targeted meat product, while meeting macro-nutrient specifications of the targeted meat product (Joshi, 2015; Kumar, 2017; Kyriakopoulou, 2019; Malav, 2015). In terms of macro-nutrient specifications, the amount of protein is likely the first macro-nutrient of importance. From a nutritional and functional perspective, it is also important to maintain similar levels of moisture and lipids. Therefore, from a macronutrient standpoint, meat analogue products closely resemble the traditional meat products they aim to simulate. The following section discusses the composition of nutrients, providing a comparison between meat analogues and the traditional meat products they mimic.

Beef Burger Products

Beef burger simulation products are at the forefront of modern meat analogue marketplace, with several of the most popular retail and food service products being represented in this category. To begin the characterization, the nutrient values of cooked ground beef (93% lean, 7% fat) and the McDonald's beef patty are used as standards for traditional meat products. These products are considered representative of a retail beef product that is unseasoned (cooked ground beef) and a fully prepared food service product that is seasoned during cooking (McDonald's beef patty). According to the USDA Nutrient Database (U.S. Department of Agriculture, 2019), cooked ground beef (93% lean, 7% fat) contains 182.00 kcal per 100 g, 25.56 g of total protein per 100 g, 8.01 g of total fat per 100 g, 3.29 g of saturated fat per 100 g, 84.00 mg of cholesterol per 100 g, no carbohydrates or dietary fiber, 72.00 mg of sodium per 100 g, and 2.82 mg of iron per 100 g. The McDonald's beef patty has the following nutrient specifications: 266.67 kcal/100 g, 23.33 g/100 g total protein, 20.00 g/100 g total fat, 8.33 g/100 g saturated fat, 83.33 mg/100 g cholesterol, no carbohydrates and no dietary fiber, 400.00 mg/100 g of sodium, and 3.33 mg/100 g of iron. It is quite clear to see in the comparison of these two products that the McDonald's beef patty, a fully prepared and seasoned food service product, has a greater fat content indicated by both the total fat content and the saturated fat content, and that the grill seasoning used for preparation of the McDonald's beef patty increased the sodium content by approximately 5.5, or 550%, compared with unprepared ground beef.

The four meat analogue products evaluated in this chapter that simulate beef burger products differed considerably in their nutrient composition but were in general similar to that of a fully prepared food service product (McDonald's beef patty). An interesting finding is that two modern meat analogue products, the Beyond Burger and the Impossible Burger, have similar total fat content to the McDonald's beef patty, with values of 15.93 g/100 g and 12.39 g/100 g, respectively, compared to 20.00 g/100 g in the McDonald's beef patty. In terms of saturated fat, the Beyond Burger and Impossible Burger contain 5.31 g/100 g and 7.08 g/100 g, respectively, compared to 8.33 g/100 g in the beef patty. While the presence of saturated fats in plant-based products might initially seem surprising, an examination of the fats and oils used in these products - such as coconut oil and cocoa butter - helps explain why these figures are representative of the fat composition in these meat analogues. Total carbohydrates and dietary fiber were, as expected, greater for the meat analogue products, which are primarily derived from the plant-based ingredients used in their formulations. There is a general consensus that increased dietary fiber offers health benefits (Anderson, 2009). However, it is important to also consider the addition of other carbohydrates, such as starches and sugars, when discussing the nutritional advantages. Another key consideration is the level of sodium in meat analogue products, which ranged from 327.43–609.38 mg per 100 g. This is particularly concerning given that we are evaluating only the burger or patty in this case, excluding other components of the sandwich (e.g., bun and condiments). Overall, it can be confidently concluded that, aside from the potential benefits of higher dietary fiber, modern meat analogue products that simulate beef burgers do not offer significant nutritional advantages when compared to traditional beef burgers on a macro-nutrient basis.

Beef Meatball Products

Two meatball products, one meat analogue product (Gardein meatless meat balls) and one traditional product (Tyson fully cooked homestyle meatballs), were used to form a comparison of nutrient specifications for this evaluation. In terms of protein content, both the meat analogue product and the traditional meat product have the same value of 15.56 g per 100 g. However, the meat analogue product exhibits a lower energy value (166.67 kcal per 100 g compared to 300.00 kcal per 100 g), a lower fat content (7.78 g per 100 g versus 16.47 g per 100 g), a lower saturated fat content (0.56 g per 100 g versus 5.88 g per 100 g), and no cholesterol (0.00 mg per 100 g versus 47.06 mg per 100 g). The meat analogue also has a higher value for total carbohydrates (10.00 g per 100 g versus 5.88 g per 100 g), a greater dietary fiber content (3.33 g per 100 g versus 1.18 g per 100 g), and a similar sodium content (355.56 mg per 100 g versus 352.94 mg per 100 g). Notably, the meat analogue has a much higher iron content (8.33 mg per 100 g versus 2.12 mg per 100 g) compared to the traditional meat product. Overall, it can be concluded that the meat analogue product (Gardein meatless meat balls) would likely be a healthier (lower calorie, lower fat, greater dietary fiber) option when compared with the traditional meat product (Tyson fully cooked homestyle meatballs).

Pork Ham Products

Two boneless ham products, one meat analogue product (Tofurky ham roast with glaze) and one traditional product (Hormel cure 81 classic boneless ham), were used to form a comparison of nutrient specifications. In terms of protein content, both products have similar protein content, which differed by less than two percentage units (20.37 g/100 g versus 18.45 g/100 g). The meat analogue product demonstrated several notable differences in comparison to the traditional meat product. It had a higher energy value (203.7 kcal per 100 g versus 105.95 kcal per 100 g), a greater fat content (5.56 g per 100 g versus 3.57 g per 100 g), but a lower saturated fat content (0.46 g per 100 g versus 1.19 g per 100 g). Additionally, the meat analogue contained no cholesterol (0.00 mg per 100 g versus 50.95 mg per 100 g), a significantly higher carbohydrate content (18.52 g per 100 g versus 0.24 g per 100 g), and a greater amount of dietary fiber (0.93 g per 100 g versus 0.00 g per 100 g). The analogue also had a much lower sodium content (592.59 mg per 100 g versus 1038.10 mg per 100 g) and a higher iron value (1.76 mg per 100 g versus 0.83 mg per 100 g) compared to the traditional meat product. Overall, the meat analogue product (Tofurky ham roast with glaze) was a lower-sodium option compared to the traditional meat product (Hormel Cure 81 Classic Boneless Ham). However, aside from the reduced sodium content, the meat analogue offers no significant nutritional advantages, especially when considering its high level of non-dietary fiber carbohydrates.

Chicken Nugget Products

Two chicken nugget products, one meat analogue product (Quorn brand chik'n nuggets) and one traditional product (Tyson fully cooked chicken nuggets), were used to form a comparison of nutrient specifications. The meat analogue product showed several differences in nutritional content compared to the traditional meat product. It had a lower energy value (203.39 kcal/100 g versus 300.00 kcal/100 g), a lower protein content (10.17 g/100 g versus 15.56 g/100 g), and a lower fat content (8.47 g/100 g versus 18.89 g/100 g). Additionally, the meat analogue contained less saturated fat (0.42 g/100 g versus 4.44 g/100 g) and less cholesterol (6.78 mg/100 g versus 44.44 mg/100 g). However, the meat analogue had a higher total carbohydrate value (24.58 g/100 g versus 16.67 g/100 g) and a significantly higher dietary fiber content (5.93 g/100 g versus 0.00 g/100 g). The sodium content was similar (449.15

mg/100 g versus 522.22 mg/100 g), as was the iron content (0.72 mg/100 g versus 0.91 mg/100 g). Overall, it can be concluded that the meat analogue product (Quorn brand chik'n nuggets) would likely be a healthier option compared to the traditional meat product (Tyson fully cooked chicken nuggets), due to its lower calorie and fat content and higher dietary fiber. However, the protein content of the meat analogue product was much lower, which should be taken into consideration when evaluating its nutritional value.

Summary of Nutritional Composition

In summary, the evaluation of the nutritional composition, in terms of macronutrient composition and other readily available minerals on the labels of food products, between modern meat analogue products and traditional meat products leads to inconclusiveness on which is generally healthier. It depends on the product and on which ingredients are used in both the meat analogue and traditional meat products. One consideration that should be clearer is that there is not a one size fits all approach when discussing the nutritional composition of meat analogue products versus traditional meat products, and the consumer should take the initiative to investigate the nutritional differences in the two types of products.

Conclusion

As previously mentioned, the meat analogue industry has made tremendous strides in the last several years and is predicted to reach a size of USD$21.23 billion globally by 2025. Predictions indicate that the marketplace will continue to mature and the products will improve in terms of their formulation and their nutritional application. While this chapter provided only a snapshot of the ingredients used to formulate modern meat analogue products and how the nutrition of these products compares with traditional meat products, the review does hopefully provide the framework for an unbiased assessment to be made. The overlying goal is that this chapter can be successfully used in present time by food manufacturers, dieticians, nutritionists, regulators, and most importantly consumers in efforts to make informed decisions regarding the claims of meat analogue products. Additionally, this chapter may be of use in the future to evaluate the growth and development of the meat analogue industry.

References

A. V. A., Resurreccion, Sensory aspects of consumer choices for meat and meat products, *Meat Sci.* 66 (2004) 11–20, http://dx.doi.org/10.1016/S0309-

Asioli, D., J. Aschemann-Witzel, V. Caputo, et al., Making sense of the "clean label" trends: a review of consumer food choice behavior and discussion of industry implications, *Food Res. Int.* 99 (2017) 58–71, http://dx.doi.org/10.1016/j.foodres.2017.07.022.

Biswas, A. K., V. Kumar, S. Bhosle, et al., Dietary fibers as functional ingredients in meat products and their role in human health, *Int. J. Livest. Prod.* 2 (2011) 45–54.

Bohrer, B. M., Nutrient density and nutritional value of meat products and non-meat foods high in protein, *Trends Food Sci. Tech.* 65 (2017) 103–112, http://dx.doi.org/10.1016/j.tifs.2017.04.016.

Damodaran, S., K. L. *Parkin, Fennema's Food Chemistry*, fifth edition, CRC press, 2017.

Elzerman, J. E., A. C. Hoek, M. J. van Boekel, et al., Appropriateness, acceptance and sensory preferences based on visual information: a web-based survey on meat substitutes in a meal context, *Food Qual. Prefer.* 42 (2015) 56–65, http://dx.doi.org/10.1016/j.foodqual.2010.10.006.

Field, C. J., L. Robinson, *Dietary fats, Adv. Nutr.* 10 (2019) 722–724, http://dx.doi.org/10.1093/advances/nmz052.

Foegeding, E. A., J. P. Davis, Food protein functionality: a comprehensive approach, *Food Hydrocolloid.* 25 (2011) 1853–1864, http://dx.doi.org/10.1016/j.foodhyd.2011.05.008.

Friedman, M., Nutritional value of proteins from different food sources. A review, *J. Agri. Food Chem.* 44 (1996) 6–29, http://dx.doi.org/10.1021/jf9400167.

Hall, K. D., A. Ayuketah, R. Brychta, et al., Ultra-processed diets cause excess calorie intake and weight gain: an inpatient randomized controlled trial of ad libitum food intake, *Cell Metab.* 30 (2019) 67–77, http://dx.doi.org/10.1016/j. cmet.2019.05.008.

Hu, F. B., Plant-based foods and prevention of cardiovascular disease: an overview, *Am. J. Clin. Nutr.* 78 (2003) 544S–551S, http://dx.doi.org/10.1093/ajcn/78.3.544S.

Huang, S., S. M. Vasquez Mejía, S. J. Murch, et al., Cooking loss, texture properties, and color of comminuted beef prepared with breadfruit (Artocarpus altilis) flour, *Meat Musc. Biol.* 3 (2019) 231–243, http://dx.doi. org/10.22175/mmb2018.11.0039.

Joye, I. J., Protein digestibility of cereal proteins, *Foods* 8 (2019) 199, http://dx.doi.org/10.3390/foods8060199.

Kyriakopoulou, K., B. Dekkers, A. J. van der Goot, Plant-based meat analogues, in: *Sustainable Meat Production and Processing,* Academic Press, 2019, pp. 103–126, http://dx.doi.org/10.1016/B978-0-12-814874-7.00006-7.

Liu, R. H., Health-promoting components of fruits and vegetables in the diet, *Adv. Nutr.* 4 (2013) 384S–392S, http://dx.doi.org/10.3945/an.112.003517.

Maki, K. C., M. L. Carson, M. P. Miller, et al., Hydroxypropylmethylcellulose and methylcellulose consumption reduce postprandial insulinemia in overweight and obese men and women, *J. Nutr.* 138 (2008) 292–296, http://dx.doi.org/10.1093/jn/138.2.292.

Malav, O. P., S. Talukder, P. Gokulakrishnan, et al., Meat analog: a review, *Crit. Rev. Food Sci. Nutr.* 55 (2015) 1241–1245, http://dx.doi.org/10.1080/.

Mancini, R. A., M. Hunt, Current research in meat color, *Meat Sci.* 71 (2005) 100–121, http://dx.doi.org/10.1016/j.meatsci.2005.03.003.

Monteiro, C. A., G. Cannon, J. C. Moubarac, et al., The UN Decade of Nutrition, the NOVA food classification and the trouble with ultra-processing, *Public Health Nutr.* 21 (2018) 5–17, http://dx.doi.org/10.1017/S1368980017000234.

Monteiro, C. A., G. Cannon, R. Levy, et al., NOVA. The star shines bright, *World Nutr.* 7 (2016) 28–38.

Mota, C., M. Santos, R. Mauro, et al., Protein content and amino acids profile of pseudocereals, *Food Chem.* 193 (2016) 55–61, http://dx.doi.org/10.1016/j.foodchem.2014.11.043.

Newswire, P. R., *Meat, Poultry & Seafood Market Size Worth $7.3 Trillion by 2025*, Accessed on June 23, 2019 at, Grand View Research, Inc., 2019 https://www.prnewswire.com/news-releases/meat-poultry–seafood-market-size-worth-73-trillion-by-2025-grand-view-research-inc-683844731.html.

Nosworthy, M. G., J. D. House, Factors influencing the quality of dietary proteins: implications for pulses, *Cereal Chem.* 94 (2017) 49–57, http://dx.doi.org/10.1094/cchem-04-16-0104-fi.

Pinstrup-Andersen, P., Nutrition-sensitive food systems: from rhetoric to action, *Lancet* 382 (2013) 375–376, http://dx.doi.org/10.1016/S0140.

Sadler, M. J., Meat alternatives – market developments and health benefits, *Trends Food Sci. Tech.* 15 (2004) 250–260, http://dx.doi.org/10.1016/j.tifs.

Topping, D. L., D. Oakenfull, R. P. Trimble, et al., A viscous fiber (methylcellulose) lowers blood glucose and plasma triacylglycerols and increases liver glycogen independently of volatile fatty acid production in the rat, *Brit. J. Nutr.* 59 (1988) 21–30, http://dx.doi.org/10.1079/BJN19880006.

Tuso, P. J., M. H. Ismail, B. P. Ha, et al., Nutritional update for physicians: plant-based diets, *Perm. J.* 17 (2013) 61–66, http://dx.doi.org/10.7812/TPP/.

Watson, D. B., Public health and carrageenan regulation: a review and analysis, in: *Nineteenth International Seaweed Symposium*, Springer, Dordrecht, 2007.

Chapter 8

Microalgae: A Promising Accessible Ingredient to Produce Meat Analogues

Abstract

Meat analogues have recently emerged as healthier and more sustainable alternatives to meet the increasing demand for meat. Microalgae, a novel protein source, hold significant potential as a key ingredient in these products. This chapter explores the use of microalgae in the production of meat analogues, focusing on four critical areas: their feasibility as an ingredient, industrial cultivation methods, production technologies, and the economic and technical challenges involved. With their high protein content and rich array of nutraceutical compounds, microalgae present substantial nutritional benefits. However, challenges remain, including underdeveloped texturing techniques, the need for effective odor and color removal, and the high costs associated with production. Addressing these issues will be essential for unlocking the full potential of microalgae in meat analogue products.

Keywords: microalgae, meat analogues, protein source, sustainability

Introduction

The United Nations statistics show that the total global population will reach 10 billion by 2050 (UN, 2019), which increases the concern for sustainability and food security. Also, the large population, the in- crease of people's wealth as well as urbanization will increase total food consumption and change diet choices. Food plays a crucial role in our lives, but current consumption patterns, particularly of animal-based foods, are not only inefficient but also pose significant risks to sustainability. This has led to increasing calls for reducing meat consumption. Many countries are exploring strategies to mitigate the negative impacts of meat production and consumption on both the environment and human health. In this context, meat analogues present a promising alternative, offering potential benefits in terms of environmental sustainability, health, and safety.

Studies have shown that livestock farming contributes approximately 14.5% of human-induced greenhouse gas (GHG) emissions (Gerber, FAO et al., 2013). For livestock production systems, carbon dioxide (CO2), methane (CH4), and nitrous oxide (N2O) are the three main GHGs emitted. The livestock industry accounts for 9% of global carbon dioxide emissions, 37% of global methane emissions, and 65% of global nitrous oxide emissions (Steinfeld, Gerber et al., 2006). Furthermore, the livestock industry requires an estimated 40% arable land and 56% agricultural freshwater consumption, which poses great pressures on limited arable land and fresh water resources (Mottet, de Haan et al., 2017; Weindl, Bodirsky et al., 2017). Coupling meat consumption with meat analogues would lead to significant environ- mental benefits. A plant-based diet is the most promising scenario to reduce environmental impacts. Eshel et al., estimated that replacing meat with plant-based alternatives could save 34%, 47%, and 38% of the national dietary land use, nitrogen fertilizer and GHG emissions in the United States of America (USA), respectively (Eshel, Stainier et al., 2019).

Studies have also associated total mortality and some diseases including obesity, type 2 diabetes, cardiovascular disease, stroke and colorectal cancer with diets containing red meat, while replacing animal proteins with that of plant was associated with lower mortality (Willett, Rockstrom et al., 2019). Guaschferre et al., (2019) suggested that substituting red meat with high-quality plant protein sources, as opposed to fish or carbohydrates, leads to more favorable changes in blood lipids and lipoproteins. This shift in diet from meat to plant-based foods is associated with lower risks of chronic diseases and overall mortality. Meat analogues present a promising alternative, offering significant nutritional advantages. These meat analogues contain 58% fewer calories, 96% less saturated fat, 33% less sodium, no cholesterol, and twice the amount of fiber compared to traditional meat burgers, making them a healthier and more contemporary dietary choice (Fresan, Mejia et al., 2019). Another important issue is the biosafety of meat. It can be threatened by many biological, chemical and physical hazards, where biological hazards are deemed to pose the highest risk of meat safety (e.g., zoonoses and microbiology contamination such as *Salmonella* and enterohaemorrhagic *E. coli* O157:H7). Meat analogues are generally considered less risky in terms of food safety issues because their raw materials primarily consist of plant-based proteins, which eliminates the risks associated with zoonotic diseases. Additionally, these plant-based ingredients undergo multiple processing steps, including high-temperature cooking, to achieve a meat-like texture, which significantly reduces the potential for microbial contamination. As a result, meat analogues offer a greener, healthier, safer, and more controllable food option.

In recent years, meat analogues have received more attentions. Current meat analogues are generally prepared using plant proteins. However, the yields and protein contents of these sources cannot satisfy the huge market demand for meat analogues. Arable land use is a crucial factor to consider, as the rate of soil erosion and degradation continues to increase globally. Consequently, land availability becomes a significant constraint on food production. To address this, there is a need to shift from traditional plant-based protein sources to more sustainable alternatives. Currently, people are exploring novel sources for producing meat analogues, including edible insects, fungi, and algae.

Microalgae are a promising protein resource, and have been studied for years as food ingredients. Up to now, microalgae have been applied in various industrial fields ranging from biodiesel production to functional foods and health care products. Recently, microalgae are regarded as a novel protein source for meat analogues. Microalgae contain a number of functional elements, such as peptides, carbohydrates, lipids, pigments, vitamins and minerals, which provide the consumer with a wide range of health benefits. Moreover, microalgae have a considerably high growth rate and can adapt to a wide range of growth conditions when compared with terrestrial crops. In general, microalgae are promising ingredients for meat analogues due to their rich nutritional value and high growth rates. This paper reviews the nutritional and production advantages of microalgae as ingredients for meat analogues compared to plant-based alternatives. It also summarizes the technologies used in the production of meat analogues, providing guidance for future research on microalgae-based meat analogues.

Microalgae as an Alternative Protein Resource

Protein is a critical component that influences both the texture and nutritional value of meat. Similarly, the quality and quantity of proteins in the raw ingredients play a significant role in determining the emulsification, gelation, viscosity, and texturization properties of meat analogues. This section reviews the composition and functions of microalgal proteins, with a focus on comparing them to traditional dietary protein sources. Additionally, nutraceutical compounds derived from microalgae are highlighted to emphasize the unique advantages of using microalgae as a primary ingredient in the production of meat analogues. Finally, a list of microalgae that have been approved for food applications is summarized at the end of this section.

Protein Content and Productivity

Microalgae are promising protein sources because their protein content can be up to 71% depending on species and cultivation conditions (Becker, 2007). *Arthrospira* sp. (marketed as spirulina) and *Chlorella* sp. are currently the most commonly sold microalgae in the market, their protein content is typically over 50% of dry weight, surpassing the typical 30–40% protein content in soybeans. The growth rate of microalgae is also superior comparing to conventional crop plants. It has been estimated that the yield of microalgae can reach 15–30 tons of dry biomass per unit area per year, whereas soybeans yield only 1.5–3.0 tons per unit area per year (Krimpen, Bikker et al., 2013). Therefore, microalgae are biologically a more efficient source for protein production.

Amino Acid Composition

Amino acid composition and the content of essential amino acids (EAAs) in dietary proteins are preliminary indices of protein nutritional quality. Although plant proteins are often deficient in one or more essential amino acids (EAAs), this issue can be addressed by combining proteins from alternative sources during the recipe development stage. The protein content in microalgae is notably superior to that of most other foods, and these microalgae often contain well-balanced amounts of EAAs. According to guidelines from the World Health Organization (WHO) and the Food and Agriculture Organization of the United Nations (FAO), as well as patterns from other protein sources, the levels of 7 essential amino acids in *Chlorella vulgaris* are comparable to those in beef. While its methionine and cysteine content is slightly lower, it still exceeds the levels found in the WHO/FAO reference standards.. The essential amino acid index (EAAI) scores for *Chlorella* sp. and Arthrospira platensis are 107.5 and 102.6 (the standard value is 100), which is comparable to casein, the main protein in milk; and are slightly higher than soybean meal (Kent, Welladsen et al., 2015; Zdunczyk, Minakowski et al., 1999). Moreover, the profile of EAAs vary substantially across microalgal species, hence balancing nutritional value with multiple microalgae could be an interesting area to explore.

Bioavailability

Bioavailability is a nutritional metric that evaluates the digestibility and absorption of proteins by human body, which also reflects protein quality. The following parameters - protein efficiency ratio (PER), biological value (BV), net protein utilization (NPU), and digestibility coefficient (DC, also referred to as true digestibility) - are used to compare microalgae with conventional foods. Casein, the main protein from milk is used as reference protein for scoring. The results show that the bioavailability of proteins from *Chlorella* sp. and *Arthrospira* sp. is overall lower than that of beef and casein, but is comparable to soy protein and better than wheat gluten.

Physicochemical Properties of Microalgal Protein

Solubility, emulsification, gelation, and foaming are key physicochemical properties of proteins that directly influence the preparation, processing, storage, and consumption of food (Kinsella, 1979). A thorough understanding of these properties is therefore crucial for the successful application of proteins in the production of meat analogues. In a 2019 review, Grossmann et al., examined several studies on microalgal proteins and concluded that the emulsifying and foaming properties of microalgal proteins were comparable to those of whey and soy proteins (Grossmann et al., 2019a).

In food terms, protein solubility commonly refers to the percentage of soluble proteins versus total protein content under given conditions. Protein solubility generally positively correlates to gelation, emulsification and foaming (Kinsella, 1979). Microalgal proteins display superior solubility than that of many protein sources. For example, glycinin, the main storage protein in soybeans, has $< 20\%$ solubility at pH 4.5–6.0; while that of a *Chlorella protothecoides* protein extract is $\geq 84.3\%$ at pH 2–12 (Grossmann et al., 2019b; Lakemond, de Jongh et al., 2000).

Stable protein emulsion is another important aspect to food processing. Emulsifying capacity (EC) and emulsion stability (ES) are key metrics to determine emulsifying ability of proteins. Ursu et al., reported an EC of 3740 ± 20 mL_{oil}/g protein and an ES of $72 \pm 1\%$ for proteins obtained from *C.*

vulgaris, which are higher compared to some commercial emulsifying proteins such as sodium caseinate (3250 ± 40 mL_{oil}/g protein, 62 ± 1%) and soy protein isolate (1780 ± 30 mLoil/g protein 61 ± 1%) (Ursu, Marcati et al., 2014). Bocker et al., utilized interfacial shear rheology to assess the emulsification potential of protein extracts from *A. platensis*. Their study demonstrated that both crude and water-soluble protein extracts could be efficiently emulsified. Furthermore, purified proteins were found to be effective in producing smaller emulsion droplets and creating stronger viscoelastic networks at the oil-water interface (Bocker, Bertsch et al., 2021).

Protein gelation is property often used in food preparation to bind ingredients. Several common methods can induce gelation, including salting, acidification, enzymatic catalysis and pressurization (Nicolai, 2019). Studies have demonstrated successful gelation of protein extracts from *A. platensis*, *Tetraselmis suecica* and *Chlorella sorokiniana* under different conditions (Grossmann et al., 2019a).

Foaming is the result of gas dispersed in liquid or solid materials, which can be a substantial impact on food texture (Foegeding & Davis, 2011). Due to increased viscosity with proteins and polysaccharides, microalgal protein extracts can produce stable foams. However, the foaming properties vary considerably among microalgal species. For example, Schwenzfeier et al., observed superior foam stability (FS) from the sample prepared with soluble proteins from Tetraselmis sp. at pH 5–7 compared with whey (Schwenzfeier, Lech et al., 2013). Proteins from *A. platensis* display high foam capacity (FC) (> 250%) at pH 10, but a very low foam stability (Benelhadj, Gharsallaoui et al., 2016). Proteins from *Chlorella pyrenoidosa* exhibit high foam stability (97% after 180 min), but the foam capacity (95%) is lower than soy proteins (235%) (Waghmare, Salve et al., 2016).

In summary, microalgae are great sources of proteins. Due to the diversity of microalgal species, the nutritional quality and physico- chemical properties of microalgal proteins can be adjusted to accommodate a wide range of applications. Currently, information of microalgal proteins is relatively sparse comparing to crop protein studies. However, microalgae can be a versatile ingredient in future food industry.

Nutraceutical Compounds from Microalgae

Microalgae are naturally abundant in pigments, including β-carotene, lutein, astaxanthin, fucoxanthin, and protein-pigment complexes like phycocyanin, which is a key metabolite in cyanobacteria. Both phycocyanin and these pigments have demonstrated significant health benefits for humans, including antioxidant and anti-cancer properties. As a result, they are widely used in health care products. Apart from their functional values, pigments in algal cells can be also served as the colorant when producing meat analogues in order to imitate the appearance of meat. Moreover, phycocyanobilin, the chromophore of phycocyanin, can be reduced to phycocyanorubin, a rubinoid pigment structurally similar to the plasma anti-oxidant bilirubin, which can potentially be exploited as a coloring ingredient for meat analogues (McCarty, Barroso-Aranda et al., 2009).

Microalgae are the main producers of long-chain polyunsaturated fatty acids (PUFA), e.g., eicosapentanoic acid (EPA) and docosahexaenoic acid (DHA) in nature. Studies have shown that EPA and DHA have great impact on neuronal, retinal and immune system development (Swanson, Block et al., 2012). Moreover, EPA and DHA can alleviate cardiovascular diseases in aspects of inflammation, peripheral artery disease, major coronary events, and anticoagulation (Swanson, Block et al., 2012). Marine fish is currently the primary dietary source of EPA and DHA, while microalgae-derived EPA and DHA are marketed as vegetarian supplements. It is likely that EPA and DHA from microalgae will be used in the production of fish analogues in the future.

Nearly all essential vitamins can be found in microalgae. The contents of vitamin A, B1 and B2 in A. platensis and *C. pyrenoidosa* are much higher than animal liver and spinach leaf (Becker, 2013). One vitamin that requires attention is vitamin B12, as meat is the primary source of this essential nutrient. However, it has been reported that the vitamin B12 level in soybeans is low or undetectable, while tempeh contains a higher amount, ranging from 0.7 to 8 μg/100 g (Nout & Rombouts, 1990). In contrast, the eukaryotic microalgae *Chlorella* sp. contains significantly higher levels of vitamin B12, ranging from 201.3 to 285.7 μg/100 g of dry biomass, and does so without the presence of inactive corrinoids (Kittaka-Katsura, Fujita et al., 2002). There are also clinical studies suggesting the serum level of vitamin B12 in vegans supplemented with *Chlorella* is twice than those who did not received the supplementation (Watanabe, 2007). Therefore, Chlorella sp. can be used as a functional ingredient for vitamin B12 in food industry. Unfortunately, despite *Arthrospira* sp. also contain vitamin B12, they also accumulate corrinoid compounds as pseudovitamin B12, which hampers the use of *Arthrospira* sp. (Watanabe, 2007).

The above findings highlight the diversity of microalgae as functional ingredients, it is important to recognize the strength and weakness of each microalga, in order to have knowledge-based food development process. Understanding the nutritional value of each ingredient plays a vital role for ingredients selection and rationing to produce analogous food products with comparable nutritional content.

Safety and Allergenicity of Microalgae

Food safety is the most important consideration for ingredients se- lection. Many common safety issues such as biological and chemical contaminations can be mitigated via tighter cultivation control. New species of microalgae are also being discovered over time, and many possess great potentials as food ingredients. An underlying safety risk is that some microalgal species from dinoflagellates, diatoms and cyano- bacteria, can produce toxins (Delia, Caruso et al., 2015). Thus, the biological toxicity of microalgae needs to be carefully investigated prior to food applications.

Allergenicity is another important factor in the food industry. Unlike soy and wheat, which are among the eight most common food allergens, the allergic effects of microalgal proteins are species-specific. Allergic reactions have been reported in individuals consuming A. platensis tablets, while proteins from *C. protothecoides* appear to have little or no allergic effects (Le, Knulst et al., 2014; Szabo, Matulka et al., 2013). It is worth noting that most individuals who reacted to A. platensis tablets are also prone to a broad range of other allergens. Currently, data on allergenicity potential of microalgae are still very limited, and with the available evidence, microalgae are not considered as the strong allergens.

Approved Microalgae for Foods

In China, *Crypthecodinium cohnii*, *Schizochytrium* sp. and *Ulkenia amoeboida* are only approved to be used as stock materials for food-grade DHA refinery, biomass consumption is prohibited. In USA, dietary microalgal species are categorized as GRAS (Generally Regarded as Safe) by U.S. Food and Drug Administration (FDA). In the EU, the license for food-grade microalgae is issued under Novel Food Regulation. This legislation stipulates that a safety assessment is required for foods or food ingredients that have not been used for human consumption to a significant degree in the EU before 1997. *Chlorella* sp. (*C. pyrenoidosa*, *C. vulgaris* and *C. luteoviridis*) and

Arthrospira sp. had been marketed and consumed as food ingredients before 1997, hence they are not subjected to the Novel Food Regulation (Champenois, Marfaing et al., 2015; Lafarga, Fernandez-Sevilla et al., 2020). Whereas *Aurantiochytrium limacinum*, *Euglena gracilis* and *Tetraselmis chui* were approved as recommended biological agents in 2019 with Qualified Presumption of Safety (QPS) status, and their applications require risk assessment from European Food Safety Authority (EFSA). The β-carotene derived from *Dunaliella salina* is authorized to be used as a food additive by the EU Scientific Committee for Food (SCF). Overall, *Arthrospira* sp., *Chlorella* sp., *Dunaliella* sp., *H. pluvialis*, and *Schizochytrium* sp. are the common species widely used as food ingredients across continents. *Arthrospira* and *Chlorella* have the largest market. More than 12 000 tons of Arthrospira biomass are produced per year, and the annual yield of Chlorella is estimated to be 5000 tons per year (Enzing, Ploeg et al., 2014).

Industrial Cultivation of Microalgae

Although legumes are often the main ingredient for meat analogues, their yields have no advantage over microalgae. Furthermore, the cultivation of these terrestrial plants is dependent on the environment and climate to a large extent. For example, soybean can only be successfully cultivated in regions with hot summers and optimum growing conditions at a mean temperature of 20–30°C. However, microalgae production has the potential to extend to areas and climates (e.g., deserts and seashores) where conventional agriculture is not feasible. Microalgae are not constrained by the natural environment and can be harvested year-round. They can be cultivated in various modes and systems, producing tailored biomass with high protein content and specific nutritional profiles. These characteristics make microalgae more competitive. In this section, the existing cultivation systems for microalgae are briefly reviewed. The aim is to evaluate the potentials of each system and outlook their future applications for mass production of algal biomass.

Cultivation System and Cultivation Mode

Cultivation System

The open raceway pond (ORP) is a classic type of photobioreactor and is widely used for industrial-scale algal cultivation. Its main advantages include simplicity in construction, customization, and operation, as well as being energy-efficient and scalable. However, the primary limitation of ORP systems is their susceptibility to biological and chemical contamination, a common issue with all open-loop systems. Contamination is particularly problematic in the production of food-grade materials, which restricts the use of ORPs primarily to the cultivation of non-dietary algal biomass.

ORP has been successfully applied for industrial production of *Arthrospira* sp. and *Chlorella* sp., the former is an alkaliphilic cyanobacterium, the high pH in the cultivation media is unsuitable to common biological contaminants to grow; Chlorella sp. however, relies on rapid growth rate to out-compete potential contaminants. Hence successful application of ORP for microalgae cultivation requires the presence of natural selective pressure, which renders many commercial strains and their applications incompatible. Another drawback of ORP is that its efficiency is highly dependent on environmental conditions, such as sunlight, temperature, humidity, and rainfall, which limits its use to specific geographic locations. To expand the use of ORP (oxidation-reduction potential) for microalgae cultivation, several strategies have been proposed to manage contamination risks. These strategies include: a) strain discovery to identify competitive microalgal species, b) optimization of cultivation conditions to promote microalgal growth while inhibiting contaminants, c) the use of antibiotics, prebiotics, and probiotics to protect microalgae, and d) the establishment of industrial standards for algal toxins to mitigate health risks associated with toxic algal contamination (Lam, Lee et al., 2018). Despite these measures, the use of ORP systems for microalgae production necessitates careful monitoring of both biological and chemical contaminants.

Closed-loop photobioreactors (PBRs) provide a more controlled environment, which is crucial for effective contamination management. Several types of PBRs have been designed, including tubular, plastic bag, flat-panel, and column reactors. The design of these PBRs is influenced by several key factors, including light distribution within the reactor, mixing efficiency, mass transfer efficiency, temperature control, and both capital and operating costs. These factors are essential for optimizing the performance and cost-effectiveness of PBRs in various applications (Huang, Jiang et al., 2017). With reduced risks of contaminations, microalgae cultivation in PBRs are often supplemented with organic carbon sources to improve cell density and growth rate. Due to the complexity of PBRs, closed-loop production of microalgae faces scaling issues. From an engineering perspective, enlarging the volume of cultivation chambers will alter light availability, which in turn will trigger knock-on effects on productivity (Jacob-Lopes, Queiroz Zepka et al., 2018). Scaling remains a key concern for the use of PBRs in food production, as food trade is typically measured in tons of biomass. Currently, the annual production of microalgal biomass is nearly negligible compared to the volume of trade in other food commodities.

Increasing number of concepts and designs are being proposed in recent years to improve the efficiency of PBR systems, some representative designs such as biofilm cultivation, membrane photobioreactor, and hybrid photobioreactors for microalgae production have showed a possibility for cheap and efficient microalgae cultivation and downstream processing (Jacob-Lopes, Queiroz Zepka et al., 2018). Conventional fermenters, which are a cheaper and easily scalable choice, can only be used in heterotrophic cultivation. There are also new types of fermenters (e.g., transparent fermenters with external and inner illumination) that allow microalgae ferment with light.

Cultivation Mode

Photoautotrophic cultivation is a process in which microalgae convert inorganic carbon into organic matter. Zhang et al., proposed a sophisticated coordination between photosynthesis and aerobic respiration in *Chromochloris zofingiensis* under mixotrophic cultivation, which enhanced energy and carbon metabolism. This coordination led to a higher biomass productivity compared to the combined productivity of photoautotrophic and heterotrophic cultivation (Zhang, Sun et al., 2017). Moreover, the CO_2 produced from aerobic respiration could potentially be channeled to photosynthesis at intracellular level under mixotrophic cultivation, which is potentially another mechanism to improve growth efficiency (Wang, Yang et al., 2014). However, during photoautotrophic cultivation, light energy becomes increasingly diluted among cells as the culture density increases, leading to a gradual reduction in photosynthesis. In mixotrophic cultivation, it becomes challenging to maintain a balance between photosynthesis and respiration.

Taken together, heterotrophic cultivation offers significant advantages for large-scale microalgae food production. The main benefits of this cultivation method include high biomass yield, precise and straightforward control of processing parameters, and relatively low technical barriers for large-scale production. These advantages are largely attributed to the extensive experience gained from traditional microbial fermentation in the food and pharmaceutical industries.

Tailored Biomass

By adjusting cultivation conditions, it is possible to stimulate the accumulation of specific metabolites, such as proteins, carbohydrates, or lipids. This ability to manipulate the chemical composition of the biomass allows for customization based on specific applications, providing greater flexibility in utilizing microalgal biomass as an ingredient for meat analogues. For instance, the abundance of nitrogen sources (such as $NaNO_3$, KNO_3, $(NH_4)_2CO_3$, CH_4N_2O and peptone) is closely correlated with protein yields in microalgae (Grossmann et al., 2019a). Guccione et al., reported that the protein content of *C. vulgaris* increased from 17.5% to 40.5% after switching the growth medium from a nitrogen-depleted to a nitrogen-rich one (Guccione, Biondi et al., 2014). The abundance of phosphorus is another factor closely linked to protein accumulations. Many microalgal species exhibits enhanced protein accumulation in phosphorus-rich conditions (Markou, 2012). Pigments content in microalgae can be a drawback for some applications, Caporgno et al., observed that *Auxenochlorella protothecoides* accumulated less chlorophyll in mixotrophic and heterotrophic cultivation modes. The emulsions properties of the respective biomass were also superior than those from phototrophic modes (Caporgno, Haberkorn et al., 2019).

The aforementioned examples highlight the potentials of configuring microalgal biomass for specific applications. In comparison with plants, the chemical compositions in microalgae can be altered dramatically, making microalgae a much more versatile ingredient for food processing.

Food Processing Technologies

Texturing Techniques

Extrusion is a continuous process that uses thermomechanical force to transform and restructure raw materials. It is the most commonly used technique to convert plant-based materials into fibrous structures. Traditional low moisture extrusion (< 35%) could produce meat analogues with sponge-like texture and the products requires rehydration before consumption (Guy, 2001). Plant-based materials processed by low moisture extrusion cannot form the fibrous texture that resembling meat. High moisture extrusion (>50%) is a technique that has been successfully utilized to create meat-like fibrous structures from soybeans and is currently the primary processing method for producing plant-based meat analogues. This process operates at a relatively low temperature (140-180°C), in contrast to low moisture extrusion, which occurs at around 180°C. The higher water content in high moisture extrusion reduces viscous dissipation, which in turn lowers the extrusion temperature (Cheftel, Kitagawa et al., 1992).

The high moisture extrusion process consists of three essential stages: mixing, cooking, and cooling. At the mixing stage, plant protein powders are mixed with water by a twin-screw extruder. Once the ingredients form a homogeneous state, the mixture is pushed through a cooking chamber with high pressure and high temperature. Finally, the cooked mixture passes through a long and narrow cooling die, where the proteins are aligned and texturized (Murillo, Osen et al., 2019). Several parameters, including water content, screw speed, and cooking and cooling temperatures, influence the texture of the product. Osen et al., (2015) concluded that the structural integrity of extrudates is primarily correlated with the formation of covalent disulfide bonds, while non-covalent interactions play a minor role. Although the process of forming the fibrous structure in the extruder remains somewhat unclear, several studies have attempted to explain the underlying mechanisms responsible for the development of fibrous texture. Murillo et al., suggested that the fibrous structure is caused by a phase separation process: a hot and homogeneous protein-water mixture is separated into water-rich and protein-rich domains upon cooling. Although still unable to predict the precise quantitative parameters such as speed, pressure and temperature, the hypothesis can explain the range of structures observed in extrusion processes (Murillo, Osen et al., 2019).

Currently, several plant-based protein sources, including soybeans, wheat gluten, and peas, are commonly used for high moisture extrusion. Additionally, other plant materials such as cottonseeds, canola/rapeseeds, peanuts, and sesame seeds have also been explored for producing specific textures (Riaz and Yada 2004). These ingredients could further expand the applications of extrusion technique in food processing industry.

Spinning is an alternative technique that can produce fibrous structure from plant materials. However, due to the demand for low pH, high salt concentrations and chemical additives, spinning is not yet a suitable process for sustainable production of meat analogues (Manski, van der Goot et al., 2007). Simple shear flow is an innovative technique based on the principle of flow-induced structuring. The Couette Cell, a shear cell device originally developed to study the behavior of dough under simple shear flow (Peighambardoust, van der Goot et al., 2004), has since been adapted for other applications. Manski et al., utilized the Couette Cell as a structuring technique for meat analogues by applying simple shear flow along with concurrent enzymatic crosslinking to induce an anisotropic fibrous structure in concentrated calcium caseinate (Manski, van der Goot et al., 2007). Building on this, Krintiras et al., improved the Couette Cell by enhancing heat management and materials handling, thereby successfully upscaling the production of plant-based meat analogues (Krintiras, Diaz et al., 2016). Despite the shear flow technique is still at the laboratory stage, recent advance in scaling is a sign of technological readiness for its industrial applications.

Information on using microalgal biomass to produce meat-like structures is sparse. There have been some attempts to substitute a portion of plant materials with microalgae to produce meat analogues via extrusion (Caporgno, Bocker et al., 2020; Grahl, Palanisamy et al., 2018; Palanisamy, Topfl et al., 2019). For example, Caporgno et al., showed that dry microalgal biomass combined with soy concentrates can form fibrillary textured extrudates via high moisture extrusion. Despite reduced texturing compared to full soy-based extrudates, the integration of microalgae improved the extrudate's nutritional profile by introducing vitamins B and E, with over 95% of the vitamins retained in the final product (Caporgno, Bocker et al., 2020). Similar studies have also shown that the addition of *Arthrospira* sp. to plant-based (lupin or soy) proteins can enhance the physicochemical and nutritional properties of the resulting meat analogues. (Grahl, Palanisamy et al., 2018; Palanisamy, Topfl et al., 2019). However, studies with a focus on using microalgal proteins as the main ingredient for processing is lacking. Martínez-Sanz et al., studied the nano- and micro- structure of extruded starch foams loaded with different concentrations of *Arthrospira* sp. biomass, as well as the effects of *Arthrospira* on the textural properties of the foams. They found that the foams with *Arthrospira* exhibited more densely packed and well-connected porous structures, and the foams were hardened with higher concentration of *Arthrospira* (Martínez-Sanz, Larsson et al., 2020). As unicellular organisms, the physical characteristics of microalgae can be very different to that of plant tissues. Therefore, targeted research on microalgal biomass processing is required to establish the knowledge framework for microalgae-based foods.

Flavor Adjustment Techniques

Control of Unpleasant Odors

Microalgae often carry unpleasant odors such as earthy/musty, grassy and fishy taste, which hampers their applications for foods. Studies have shown that volatile organic compounds (VOCs) in microalgae are the primary contributors to odors, with the composition of these VOCs varying depending on the species, growth phase, and cultivation conditions. For instance, the dried biomass of *Tetraselmis chui* is described as having a herbaceous flavor, while that of *Nannochloropsis gaditana* has a fishy taste (P´erez-Llore´ns, 2020). Hosoglu found that the VOCs responsible for the aroma characteristics are mainly hydrocarbons, aldehydes, alcohols, esters, ketones, lactones, short to medium-chain free fatty acids, phenolic and sulfur compounds (Hosoglu, 2018).

Various methods can be used to manage the odors in the final product. Milovanovi´c et al., observed that addition of nitrogen to growth medium can reduce production of VOCs (Milovanovi´c, Miˇsan et al., 2015). Hosoglu reported that the odor from Crypthecodinium cohnii shifts while cells moving through growth phases, implicating that flavor management could be achieved by altering harvesting time (Hosoglu, 2018). The cooking process is also a potential route to mitigate un- pleasant odors, as VOCs escape from the biomass.

Formulation of Meat-Like Flavor

Flavoring is crucial in the development of meat analogues. There are two main routes to achieve meat-like flavors: 1) the use of seasoning or spices, 2) chemical conversion of ingredients (amino acids and sugars) into flavoring through the Maillard reaction. Due to the complex physicochemical transformations that occur in raw materials and seasonings during the heating and pressurization processes within an extruder, achieving the desired flavor profile can be particularly challenging (Kyriakopoulou, Dekkers et al., 2019). As a result, optimizing the flavor of microalgae-based meat analogues may require extensive experimentation and iterative refinement.

Amino acids and simple sugars are crucial ingredients for flavoring, and sulfur plays an important role. Hofmann and Schieberle concluded that the sulfur-containing compounds generated by the Maillard reaction between cysteine/ribose are the main contributor to the roasty, meaty aroma (Hofmann & Schieberle, 1995). Similarly, sulfur-containing furans and thiophenes have been identified as key contributors to strong meat-like aromas (Evers,

Heinsohn et al., 1976). Additionally, the interaction of lipids within the Maillard reaction plays a crucial role in the development of characteristic meat-like flavors, further enhancing the sensory attributes of meat analogues (Farmer & Mottram, 1990).

Fat Imitation Techniques

In addition to the Maillard reaction, lipids play a crucial role in contributing to the flavor and texture of meat. The incorporation of fats or vegetable oils has been shown to enhance the juiciness, tenderness, and overall flavor profile of meat analogues (Egbert & Borders, 2006). Various lipid ingredients - such as canola oil, sunflower oil, coconut oil, cocoa butter, and corn oil - have been explored for use in the formulation of plant-based meat products (Bohrer, 2019). Despite this, current meat analogue products often contain relatively low levels of fat or oil, primarily because the commonly used plant protein ingredients are typically defatted during processing. Furthermore, while the addition of fats can improve sensory properties, it also reduces friction between ingredients during extrusion, which negatively affects shear force and, consequently, alters the desired texture development of the final product (Cheftel, Kitagawa et al., 1992). Gwiazda et al., found that an oil content above 15% would significantly reduce the formation of well-aligned fibrous structures (Gwiazda, Noguchi et al., 1987). A potential solution is to produce fat mimetics and fibrous structured components separately, and then combine them into meat analogues. The production of emulsion meat can be used as a reference for meat analogues development. First, chopped meat and myofibrillar proteins released from cutting are retrieved as the meat fraction. Salt, phosphates, and water are then added to the meat to create a viscous, water-based protein mass. Fat is finally incorporated to form an oil-in-water emulsion. Upon cooking, a gelation event occurs, resulting in a tender and meaty product (Kumar, 2019). Based on this principle, a microalgal version of meat analogue can be proposed, where animal proteins are replaced by microalgal proteins with strong gelling and emulsifying abilities, and fat is replaced by algal or plant oils. Since the characteristics of the viscous meat emulsion and cooked gel change with the functional properties of the proteins, extensive research is needed to develop fat mimetics using microalgal materials.

The Potential Technological Routes to Produce Microalgae-Based Meat Analogues

Both microalgal biomass and microalgal protein extracts can be tested for texturing. Direct use of microalgal biomass is an economical option, however, the presence of rigid cell walls and high oil content in some species can interfere texturing, and even reduce the bioavailability of nutrients. Therefore, protein extracts can be a complementing material for extrusion. After texturing, the fibrous products and fat mimetics can be assembled via emulsification.

The second aspect of producing meat analogues is flavor adjustment. Seasonings and colorants can be evaluated at various stages of food processing to assess their potential to induce desirable chemical reactions that contribute to flavor, appearance, and overall product quality. Furthermore, the diverse array of bioactive compounds found in microalgae presents a promising reservoir for novel chemical interactions. These interactions may lead to the development of unique flavors, but they also introduce the possibility of unforeseen chemical reactions that could pose new safety risks and require careful assessment. Hence microalgae as food ingredients possess many opportunities and risks, which require careful investigations in the near future.

Bottlenecks

Economic Constraints

Successful application of microalgal biomass in food manufacturing requires the production cost of microalgae to be comparable to conventional agricultural produce. Unfortunately, the cultivation of microalgae remains economically and energetically demanding. Currently, the viability of the microalgal biotechnology industry largely depends on the production of high-value nutraceutical products. Therefore, this section explores potential strategies for reducing production costs to enhance the economic feasibility of large-scale microalgae-based applications.

Scale up Microalgae Production

To scale up algal mass production, energy efficiency for biomass production needs to be addressed. At present, many cultivating conditions including temperature, carbon source, nitrogen source, light in- tensity and pH, and several strategies including fed-batch, semicontinuous and continuous have been optimized to achieve high biomass yields. To further expand the application of microalgae in industrial production, several strategies can be adopted. Current industrial production of microalgae is primarily conducted using open raceway ponds or closed photobioreactors (PBRs). However, these systems typically result in significantly lower biomass yields compared to heterotrophic fermenters. As a result, there is a growing need to further investigate and develop heterotrophic cultivation strategies for large-scale industrial applications. In parallel, screening for high-yield microalgal species represents another promising avenue. Despite the existence of tens of thousands of microalgal species, a vast majority remain unexplored. Identifying and utilizing more efficient strains could play a critical role in enhancing productivity and sustainability in microalgal biotechnology. The requirements of water, nutrients, light, CO_2, and appropriate temperature makes it expensive to scale up microalgae production. To reduce the use of water and nutrients, using wastewater as nutrients and recycling culture medium are also plausible. However, since the products are used as food, toxin, heavy metal and other ingredients that might cause adverse impacts should not be introduced. An extra sterilization step may also be needed for eliminating potential biological contaminants (Williams & Laurens, 2010). The limited global reserves of phosphorus make it challenging to produce microalgal biomass on a large scale. Anaerobic digestion offers a promising method for recycling the phosphorus in microalgal cells (Williams & Laurens, 2010). Since phosphorus is primarily distributed in lipids and nucleic acids, the residual material, after protein extraction and purification, can undergo anaerobic digestion for phosphorus recycling.

Due to the aquatic nature of microalgae, the industrial cultivation of microalgae requires large quantity of water. Water management is one of the greatest challenges for microalgae cultivation. Compared to open raceway ponds (ORPs), photobioreactors (PBRs) and fermenters offer more effective options for reducing water loss. Among these, PBRs are particularly energy-intensive and costly to operate. Furthermore, their closed-loop design presents challenges for efficient gas exchange, which can limit overall system performance. Pumping air, or CO_2-enriched air could generate substantial energy and economic expenses. Temperature control is a key issue, to benefit from natural light, PBRs are typically placed outdoor with transparent tubing and panels. Under bright sunlight, the temperature inside growth chambers rises rapidly. Another critical challenge in scaling up photobioreactors (PBRs) for industrial use is their high energy demand. The reliance on a stable and robust power supply can be a limiting factor, and more importantly, the net energy input required to produce microalgal biomass often exceeds the amount of energy stored within the biomass itself. This net energy loss undermines the sustainability and practical value of microalgae-based food production, making energy efficiency a key barrier to the broader adoption and scalability of microalgal systems. In response to this issue, a range of strategies and technological advancements have been developed in recent years to enhance the cultivation efficiency of microalgae, as comprehensively detailed in a recent review (Pierobon, Cheng et al., 2018).

Heterotrophic cultivation is often an underrated method to produce microalgae. It is a simple and reliable approach to growth biomass. However, organic carbon sources, which accounted for over 80% cost of the culture medium, are necessary in the heterotrophic mode. Since glucose is an expensive carbon source, industrial or agricultural waste such as molasses and cheese whey are better choices.

Cost of Biomass Processing

Despite microalgae are efficient photoautotrophs with higher growth rate than most plants, the industrial production of microalgae is currently inefficient. Due to the extremely small size of the cells, harvesting presents a significant challenge and accounts for approximately 20–30% of the total cost of the entire process. Chemical and biological harvesting methods are generally unsuitable for food applications due to safety and regulatory concerns. Gravity sedimentation, while inexpensive, is time-consuming and only effective for large, dense, and non-motile cells. In contrast, centrifugation provides high separation efficiency (greater than 90%) and broad applicability, making it a more effective and widely used method in cell harvesting. However, for large scale application, the energy consumption, maintenance and invest cost are all considerably high (Mata, Martins et al., 2010). Filtration is another common harvesting technique, which is apparently not suitable for species with small cells. The quality of the biomass harvested by filtration is better since the cells are less disrupted and chemical-free. Electro-flocculation technology offers several advantages over centrifugation, including broad species applicability and lower energy consumption. However, it also presents certain drawbacks, such as the need for regular electrode replacement and maintenance, which can introduce metal ion contamination and cause fluctuations in temperature and pH levels (Vandamme, Pontes et al., 2011). In industrial settings, combining multiple techniques can enhance overall recovery efficiency. For instance, sedimentation or electro-flocculation may be applied prior to centrifugation or filtration to reduce the processing volume, thereby improving energy efficiency. Additionally, cell disruption is a critical step in the preparation of microalgal ingredients, as the presence of intact cell walls significantly limits the bioavailability of nutrients for human consumption. Common industrial cell disruption techniques can be categorized as mechanical techniques (grinding, bead-milling, and high-pressure homogenization) and non-mechanical techniques (microwave assisted extraction, enzymatic hydrolysis, and acid/alkaline treatment). Due to low throughput and high cost, neither manual grinding nor cryogenic grinding is practical for industrial applications despite of the high efficiency for cell wall disruption. Compared to grinding, bead milling is a more effective method for breaking cells. However, the additional step of bead separation increases both the cost and energy consumption. High-pressure homogenization (HPH) yields better results than traditional cell disruption methods and is considered the most energy-efficient process based on protein release profiles (Grossmann et al., 2019a). Microwave-assisted extraction (MAE) is a simple and efficient technique with low energy consumption. Additionally, MAE demonstrates superior disruption ability for species with strong mechanical resistance (Barba, Grimi et al., 2015). Acid/alkaline treatment is not suitable for protein extraction as it causes protein degradation; whereas enzymatic treatments is difficult to use, due to the concern of side activities between the employed enzymes and the targeted proteins (Grossmann et al., 2019a). Overall, HPH and MAE are the top candidates to be selected as the disruption techniques of microalgae for food production. Extrusion is a novel choice; Wang et al., have applied this technique on disrupting the cell wall of pre-dewatering microalgae (Wang, Cheng et al., 2018).

Sustainability

Microalgae have always been regarded as a renewable, sustainable and environment-friendly resource due to the ability of converting solar energy and CO_2 to valuable molecules. Microalgae are capable of growing using flue and flaring gases, as well as polluted, saline, and brackish water, making them an attractive option for sustainable cultivation. However, despite the apparent simplicity of obtaining microalgal proteins in an environmentally friendly manner, the overall production process remains complex and, at present, not fully sustainable. Nevertheless, microalgae hold significant promise as a future sustainable protein source due to their high nutritional value and low land-use requirements. Several strategies are suggested to make it more sustainable.

Smetana et al., conducted a systematic life cycle assessment (LCA) on microalgae production for food applications, with two model species: *C. vulgaris* and *A. platensis*, and two cultivation scenarios: autotrophic and heterotrophic conditions. The study concluded that among the options, *C. vulgaris* biomass from the heterotrophic fermenter was the most efficient combination for protein production, and the conversion efficiency was comparable to animal proteins (Smetana, Sandmann et al., 2017). However, if fresh microalgal biomass, instead of dried biomass, was produced as the final product; the environmental impact of heterotrophic fermenter cultivation of *C. vulgaris* was nearly same as that of soybean meal. The authors also proposed several strategies to enhance the sustainability of microalgae production. These include the use of alternative carbon sources, modification of biomass drying methods, the adoption of renewable or alternative energy sources, and the selection of alternative microalgae species capable of utilizing food waste as a nutritional input. To minimize environmental impacts and promote a more sustainable production process, it is essential to optimize key factors such as the carbon sources used during cultivation and the processing methods employed at various stages of production. Besides, new technologies (method, devices, bioreactors, etc.) which exhibit higher utilization efficiencies on light, CO_2, water and nutrients need to be developed. This work highlighted the inefficient status of the current microalgae cultivation and processing systems, with emphasis on heat-drying cost. On the other hand, the conclusion had limitations, as the autotrophic growth scenario in the LCA study was based on a specific setting for the ORP cultivation system. Nonetheless, improving cultivation and processing efficiency should remain the key focus for microalgae production.

Technological Options to Produce Light-Colored Microalgal Ingredients

Visual impact is a prominent sensory trigger for appetites. In addition to the challenges on texturization and flavor adjustments, which are discussed in previous sections, color is another hurdle for microalgae-based meat analogues. Most microalgal species are phylogenetically closely related to plants, and their biomass typically exhibits a plant-like green coloration due to the presence of chlorophyll. To replicate the characteristic red-brown color of meat in meat analogues, a primary research focus has been the reduction or removal of these green pigments. This section summarizes the approaches explored to decrease chlorophyll content in microalgal biomass for improved color matching in meat alternatives.

Chemical Removal of Chlorophyll Pigments

An extra step to remove chlorophyll can be included in the pretreatment process of microalgae. Saponification is a common method to remove chlorophyll from microalgal biomass, and ethanol can assist the solubilization and removal of chlorophyll. Therefore, researchers have often combined ethanol with sodium hydroxide (NaOH) to effectively remove chlorophyll from microalgal biomass. For instance, Li et al., (2016) employed a solution of 100% ethanol and 1% NaOH in a 1:4 (v/v) ratio to treat dried microalgal biomass at 70 °C for one hour, achieving a 92.3% reduction in chlorophyll content. However, this method also resulted in a significant loss of total protein, with a reduction of 33.6%. Alternatively, Sathish et al., proposed a method that involves adjusting the pH to precipitate chlorophyll. Following alkali treatment, chlorophyll exists in its salt form and can be effectively precipitated by the subsequent addition of an acid solution (Sathish & Sims, 2012). However, this method resulted in a protein loss as well. Chen et al., applied bleaching earth to remove chlorophyll from crude biodiesel, achieving a chlorophyll removal efficiency of 99.1%. In addition to chlorophyll, bleaching earth can also remove carotene and other impurities (Chen, Liu et al., 2012). However, this method requires further investigation to confirm its feasibility for removing chlorophyll from algal biomass. While removing chlorophyll is not challenging, preserving the nutritional value remains a critical concern.

Reduce Chlorophyll Accumulation via Heterotrophic Cultivation

Heterotrophic cultivation is another option to reduce chlorophyll. Dark environments can inhibit photosynthetic pathways, hence result in a loss of chlorophyll-containing proteins. For example, *A. protothecoides* cultivated in heterotrophic modes accumulate fewer pigments and display yellow color under heterotrophic modes, while exposure to light result in a dark green biomass (Xiao, He et al., 2018).

Search for Chlorophyll-Deficient Microalgal Mutants

Mutation is the driving force of evolution; it naturally occurs in all species and enables adaptation to the dynamic environment. As unicellular organisms, microalgae are genetically diverse and prone to mutations. Finding an edible microalgal mutant that is deficient in chlorophyll biogenesis, or with reduced chlorophyll accumulation, could be a bypass to obtain lighter-colored biomass materials. It has been reported that a *C. vulgaris* mutant with reduced antenna size contain 50% less chlorophyll (Dallosto, Cazzaniga et al., 2019). Schüler et al., isolated yellow *C. vulgaris* mutants with an 80% reduction in chlorophyll content and also obtained a white mutant with a 99% reduction in chlorophyll. Interestingly, both mutants accumulated significantly more proteins, with the white mutant showing a 60% increase compared to the wild type (Schüler, Greque de Morais et al., 2020).

Targeted Gene Modifications

It can take substantial amount of work to screen algal species with reduced chlorophyll content from natural mutants, while it is easier to construct a low chlorophyll species by gene engineering technology based on the knowledge of the chlorophyll-related gene regulation mechanism. For example, a study on green microalga *Chlamydomonas reinhardtii* generated a chlorophyll-deficient mutant via disruption of phytoene synthase (PSY) (Mccarthy, Kobayashi et al., 2004). Targeted deletion of PSY homologs in edible microalgae presents a potential strategy for producing light-colored biomass materials. Notably, gene deletions that do not involve the introduction of foreign genetic elements are often regarded as natural modifications, which typically carry lower biosafety concerns. However, the use of genetic engineering techniques in food applications is subject to strict regulatory oversight. Therefore, it is essential to have a thorough understanding of relevant regional legislation before undertaking any practical work in this area.

The Development of Green and Efficient Protein Extraction Techniques

Although a collection of techniques has been used in microalgal protein extraction, the purity of current microalgal protein extracts is rather low (< 80%) compared to purified proteins from other sources (e.g., > 90% for whey protein/soy isolate). Therefore, novel techniques still need to be explored to develop greener and more efficient strategies for protein extraction and purification. Among the critical steps in this process are cell disruption and protein purification, which are essential for accessing and isolating proteins from within microalgal cells. Following cell disruption, the intracellular proteins are typically separated into two fractions based on their solubility in water: a soluble fraction and an insoluble fraction. In general, most proteins are extracted from the soluble fraction. After centrifugation, the supernatant is further purified. Isoelectric point precipitation is the most commonly used technique; however, this method can lead to water pollution due to the use of acid and alkaline. Filtration, on the other hand, is a gentler process that helps preserve the native state of the proteins. However, its efficiency is relatively low, likely due to protein aggregation and the presence of protein-polysaccharide complexes (Safi, Olivieri et al., 2017). Aqueous two-phase system (ATPS) is another technique for protein recovery, having the advantages of being mild, easily scalable, efficient and cost competitive. Nevertheless, its implementation is limited by the poor understanding of the partitioning mechanism and by the selection of the components (Garcia, Ruiz et al., 2018). It should be noted that procedures to extract food grade proteins from disrupted cells should not involve treatments with high temperatures and organic solvents. This may lead to protein denaturation and loss of protein functionality.

Conclusion

In this chapter, we review the potentials and challenges of using microalgae as ingredients to produce meat analogues. Protein content is the key element that defines the texture, flavor, color and nutritional value of meat, and it is the main focus for meat analogues development. Plant-based materials, with the exception of beans and peas, are generally poor protein sources. On the contrary, microalgae are plant-like photoautotrophs known for their high protein content, making them a promising alternative source of protein for the production of meat analogues. Emerging data suggests that both the quantity and quality of proteins derived from microalgae are suitable substitutes for those found in meat and soymeal. Moreover, microalgae are rich in a wide array of nutraceutical compounds, offering additional health benefits beyond basic nutrition. The chemical compositions in different algal species, and under different cultivation conditions, are diverse and dynamic. Hence, microalgal biomass represents a novel collection of ingredients that can be explored to make next generation meat analogues.

Despite microalgae's high efficiency in converting sunlight into biomass, the industrial cultivation of microalgae remains unsustainable. The main energy-consuming steps, including cell harvesting, dehydration, and disruption, need to be addressed. Heterotrophic cultivation is a promising mode for production of meat analogues from microalgae due to higher biomass density, more stable emulsification of protein, and lighter color than other modes. Moreover, heterotrophic cultivation also shows the lowest environmental impact, which presents a more sustainable choice.

References

Barba, F. J., Grimi, N., & Vorobiev, E. (2015). New approaches for the use of non-conventional cell disruption technologies to extract potential food additives and nutraceuticals from microalgae. *Food Engineering Reviews*, 7, 45–62.

Beach, E. F., Munks, B., & Robinson, A. (1943). The amino acid composition of animal tissue protein. *Journal of Biological Chemistry*, 148, 431–439.

Becker, E. W. (2007). Micro-algae as a source of protein. *Biotechnology Advances*, 25, 207–210.

Becker, E. W. (2013). Microalgae for human and animal Nutrition. In A. Richmond, & Q. Hu (Eds.), *Handbook of microalgal culture* (pp. 461–503). Oxford: Blackwell Publishing.

Benelhadj, S., Gharsallaoui, A., Degraeve, P., Attia, H., & Ghorbel, D. (2016). Effect of pH on the functional properties of Arthrospira (Spirulina) platensis protein isolate. *Food Chemistry*, 194, 1056–1063.

Biological toxins from marine and freshwater microalgae. In S. Parisi (Ed.), *Microbial toxins and related contamination in the food industry* (pp. 13–55). Heidelberg: Springer International Publishing AG Switzerland.

Bocker, L., Bertsch, P., Wenner, D., Teixeira, J., Bergfreund, S., Eder, S., Fischer, P., & Mathys, A. (2021). Effect of Arthrospira platensis microalgae protein purification on emulsification mechanism and efficiency. *Journal of Colloid and Interface Science*, 584, 344–353.

Bohrer, B. M. (2019). An investigation of the formulation and nutritional composition of modern meat analogue products. *Food Science Human Wellness*, 8, 320–329.

Caporgno, M. P., Bocker, L., Mussner, C., Stirnemann, E., Haberkorn, I., Adelmann, H., Handschin, S., Windhab, E. J., & Mathys, A. (2020). Extruded meat analogues based on yellow, heterotrophically cultivated Auxenochlorella protothecoides microalgae. *Innovative Food Science & Emerging Technologies*, 59, Article e102275.

Caporgno, M. P., Haberkorn, I., Bo¨cker, L., & Mathys, A. (2019). Cultivation of Chlorella protothecoides under different growth modes and its utilisation in oil/water emulsions. *Bioresource Technology*, 288, 121476.

Champenois, J., Marfaing, H., & Pierre, R. (2015). Review of the taxonomic revision of Chlorella and consequences for its food uses in Europe. *Journal of Applied Phycology*, 27, 1845–1851.

Cheftel, J. C., Kitagawa, M., & Queguiner, C. (1992). New protein texturization processes by extrusion cooking at high moisture levels. Food Reviews International, 8, 235–275. Chen, L., Liu, T., Zhang, W., Chen, X., & Wang, J. (2012). Biodiesel production from algae oil high in free fatty acids by two-step catalytic conversion. *Bioresource Technology*, 111, 208–214.

Dallosto, L., Cazzaniga, S., Guardini, Z., Barera, S., Benedetti, M., Mannino, G., Delia, A. S., Caruso, G., Melcarne, L., Caruso, G., Parisi, S., & Lagan`a, P. (2015).

EFSA. (2009). DHA and ARA and brain development - Scientific substantiation of a health claim related to docosahexaenoic acid (DHA) and arachidonic acid (ARA) and brain development pursuant to Article14 of Regulation (EC) No 1924/2006. *EFSA Journal*, 7, 1000.

EFSA. (2020). The 2019 updated list of QPS status recommended biological agents in support of EFSA risk assessments. *EFSA Journal,* 18, 5966.

Egbert, R., & Borders, C. K. (2006). Achieving success with meat analogs. *Food Technology,* 60, 28–34.

Enzing, C., Ploeg, M., Barbosa, M. J., & Sijtsma, L. (2014). *Microalgae-based products for the food and feed sector: An outlook for Europe.* Luxembourg: Publications Office of the European Union.

Eshel, G., Stainier, P., Shepon, A., & Swaminathan, A. (2019). Environmentally optimal, nutritionally sound, protein and energy conserving plant based alternatives to U.S. meat. *Scientific Reports,* 9, 1–11.

Evers, W. J., Heinsohn, H. H., Mayers, B. J., & Sanderson, A. (1976). Furans substituted at the three position with sulfur. In G. Charalambous, & I. Katz (Eds.), *Phenolic, sulfur, and nitrogen compounds in food flavors* (pp. 184–193). Washington, DC: American Chemical Society.

Farmer, L. J., & Mottram, D. S. (1990). Interaction of lipid in the maillard reaction between cysteine and ribose: The effect of a triglyceride and three phospholipids on the volatile products. *Journal of the Science of Food and Agriculture*, 53, 505–525.

Foegeding, E. A., & Davis, J. P. (2011). Food protein functionality: A comprehensive approach. *Food Hydrocolloids,* 25, 1853–1864.

Fresan, U., Mejia, M., Jaceldo-Siegl, K., Craig, W., & Sabate, J. (2019). Looking for a nutritive and sustainable source of protein? (P03-004-19). *Current Developments in Nutrition*, 3, 10.1093/cdn/nzz047.P03-004-19.

Garcia, E. S., Ruiz, C. A. S., Tilaye, T., Eppink, M. H., Wijffels, R. H., & van den Berg, C. (2018). Fractionation of proteins and carbohydrates from crude microalgae extracts using an ionic liquid based-aqueous two phase system. *Separation and Purification Technology,* 204, 56–65.

Garnett, T., Tilman, D., Declerck, F., & Wood, A. (2019). Food in the anthropocene: The EAT–lancet commission on healthy diets from sustainable food systems. *The Lancet,* 393, 447–492.

Gerber, P. J., FAO, R., Steinfeld, H., Henderson, B., Mottet, A., Opio, C., et al., (2013). *Mitigation Opportunities.* Rome: Food and Agriculture Organization of the United Nations (FAO).

Grahl, S., Palanisamy, M., Strack, M., Meierdinkel, L., Toepfl, S., & Morlein, D. (2018). Towards more sustainable meat alternatives: How technical parameters affect the sensory properties of extrusion products derived from soy and algae. *Journal of Cleaner Production,* 198, 962–971.

Grossmann, L., Hinrichs, J., & Weiss, J. (2019a). Cultivation and downstream processing of microalgae and cyanobacteria to generate protein-based technofunctional food ingredients. *Critical Reviews in Food Science Nutrition*, 45, 1–29.

Grossmann, L., Hinrichs, J., & Weiss, J. (2019b). Solubility and aggregation behavior of protein fractions from the heterotrophically cultivated microalga Chlorella protothecoides. *Food Research International,* 116, 283–290.

Guaschferre, M., Satija, A., Blondin, S. A., Janiszewski, M., Emlen, E., Oconnor, L. E., Campbell, W. W., Hu, F. B., Willett, W. C., & Stampfer, M. J. (2019). Meta-analysis of randomized controlled trials of red meat consumption in comparison with various comparison diets on cardiovascular risk factors. *Circulation,* 139, 1828–1845.

Guccione, A., Biondi, N., Sampietro, G., Rodolfi, L., Bassi, N., & Tredici, M. R. (2014). Chlorella for protein and biofuels: From strain selection to outdoor cultivation in a green wall panel photobioreactor. *Biotechnology for Biofuels,* 7, 84, 84.

Guy, R. (2001). *Extrusion cooking: Technology and applications.* Cambridge: Crc Press.

Gwiazda, S., Noguchi, A., & Saio, K. (1987). Microstructural studies of texturized vegetable protein products: Effects of oil addition and transformation of raw materials in various sections of a twin screw extruder. *Food Structure,* 6, 57–61.

Han, F., Huang, J., Li, Y., Wang, W., Wan, M., Shen, G., & Wang, J. (2013). Enhanced lipid productivity of Chlorella pyrenoidosa through the culture strategy of semi-continuous cultivation with nitrogen limitation and pH control by CO2. *Bioresource Technology,* 136, 418–424.

Hoffman, J. R., & Falvo, M. J. (2004). Protein - which is best? *Journal of Sports Science and Medicine,* 3, 118–130.

Hofmann, T., & Schieberle, P. (1995). Evaluation of the key odorants in a thermally treated solution of ribose and cysteine by aroma extract dilution techniques. *Journal of Agricultural and Food Chemistry,* 43, 2187–2194.

Hosoglu, M. I. (2018). Aroma characterization of five microalgae species using solid- phase microextraction and gas chromatography-mass spectrometry/olfactometry. *Food Chemistry,* 240, 1210–1218.

Huang, Q., Jiang, F., Wang, L., & Yang, C. (2017). Design of photobioreactors for mass cultivation of photosynthetic organisms. *Engineering,* 3, 318–329.

Jacob-Lopes, E., Queiroz Zepka, L., & Queiroz, M. I. (2018). *Energy from microalgae.* Springer.

Jorfi, R., Mustafa, S., Man, Y. B. C., Hashim, D. B. M., Sazili, A. Q., Farjam, A. S., Nateghi, L., & Kashiani, P. (2012). Differentiation of pork from beef, chicken, mutton and chevon according to their primary amino acids content for halal authentication. *African Journal of Biotechnology,* 11, 8160–8166.

Kent, M., Welladsen, H. M., Mangott, A., & Li, Y. (2015). Nutritional evaluation of Australian microalgae as potential human health supplements. *PloS One,* 10.

Kinsella, J. E. (1979). Functional properties of soy proteins. *Journal of the American Oil Chemists' Society,* 56, 242–258.

Kittaka-Katsura, H., Fujita, T., Watanabe, F., & Nakano, Y. (2002). Purification and characterization of a corrinoid compound from Chlorella tablets as an algal health food. *Journal of Agricultural and Food Chemistry,* 50, 4994–4997.

Krintiras, G. A., Diaz, J. G., Der Goot, A. J. V., Stankiewicz, A., & Stefanidis, G. (2016). On the use of the Couette Cell technology for large scale production of textured soy-based meat replacers. *Journal of Food Engineering,* 169, 205–213.

Kumar, Y. (2019). Development of low-fat/reduced-fat processed meat products using fat replacers and analogues. *Food Reviews International,* 1, 1–17.

Kyriakopoulou, K., Dekkers, B., & van der Goot, A. J. (2019). Plant-based meat analogues. In C. M. Galanakis (Ed.), *Sustainable meat production and processing* (pp. 103–126) (Academic Press).

Lafarga, T., Fern´andez-Sevilla, J. M., Gonz´alez-Lopez, C., & Aci´en-Fernandez, F. G. (2020). Spirulina for the food and functional food industries. *Food Research International,* 137, Article e109356.

Lakemond, C. M. M., de Jongh, H. H. J., Hessing, M., Gruppen, H., & Voragen, A. G. J. (2000). Soy glycinin: Influence of pH and ionic strength on solubility and molecular structure at ambient temperatures. *Journal of Agricultural and Food Chemistry,* 48, 1985–1990.

Lam, T. P., Lee, T. M., Chen, C. Y., & Chang, J. S. (2018). Strategies to control biological contaminants during microalgal cultivation in open ponds. *Bioresource Technology,* 252, 180–187.

Le, T.-M., Knulst, A. C., & Ro¨ckmann, H. (2014). Anaphylaxis to Spirulina confirmed by skin prick test with ingredients of Spirulina tablets. *Food and Chemical Toxicology,* 74, 309–310.

Li, T., Xu, J., Wu, H., Wang, G., Dai, S., Fan, J., He, H., & Xiang, W. (2016). A saponification method for chlorophyll removal from microalgae biomass as oil feedstock. *Marine Drugs,* 14, 162.

Li, X., Xu, H., & Wu, Q. (2007). Large-scale biodiesel production from microalga Chlorella protothecoides through heterotrophic cultivation in bioreactors. *Biotechnology and Bioengineering,* 98, 764–771.

Maffei, M., & Bassi, R. (2019). Combined resistance to oxidative stress and reduced antenna size enhance light-to-biomass conversion efficiency in Chlorella vulgaris cultures. *Biotechnology for Biofuels,* 12, 221. https://doi.org/10.1186/s13068-019-1566-9

Manski, Julita M, van der Goot, Atze J, Boom, Remko M, et al., (2007). Formation of fibrous materials from dense calcium caseinate dispersions. *Biomacromolecules*, 8(4), 1271–1279. https://doi.org/10.1021/bm061008p

Markou, G. (2012). Alteration of the biomass composition of Arthrospira (Spirulina) platensis under various amounts of limited phosphorus. *Bioresource Technology,* 116, 533–535.

Marquez, F. J., Sasaki, K., Kakizono, T., Nishio, N., & Nagai, S. (1993). Growth characteristics of Spirulina platensis in mixotrophic and heterotrophic conditions. *Journal of Fermentation and Bioengineering,* 76, 408–410.

Martínez-Sanz, M., Larsson, E., Filli, K. B., Loupiac, C., Assifaoui, A., Lo´pez-Rubio, A., & Lopez-Sanchez, P. (2020). Nano-/microstructure of extruded Spirulina/starch foams in relation to their textural properties. *Food Hydrocolloids,* 103, Article e105697.

Mata, T. M., Martins, A. A., & Caetano, N. S. (2010). Microalgae for biodiesel production and other applications: A review. *Renewable and Sustainable Energy Reviews*, 14, 217–232.

Mccarthy, S. S., Kobayashi, M. C., & Niyogi, K. K. (2004). White mutants of Chlamydomonas reinhardtii are defective in phytoene synthase. *Genetics,* 168, 1249–1257.

McCarty, M. F., Barroso-Aranda, J., & Contreras, F. (2009). Genistein and phycocyanobilin may prevent hepatic fibrosis by suppressing proliferation and activation of hepatic stellate cells. *Medical Hypotheses,* 72, 330–332.

Milovanovi´c, I., Miˇsan, A., Simeunovi´c, J., Kovaˇc, D., Jambrec, D., & Mandi´c, A. (2015). Determination of volatile organic compounds in selected strains of cyanobacteria. *Journal of Chemistry,* Article e969542, 2015.

Moser, C. R., & Herman, K. (2011). Method for the rapid determination of protein in meats using the CEM sprint protein analyzer: Collaborative study. *Journal of AOAC International,* 94, 1555–1561.

Mottet, A., de Haan, C., Falcucci, A., Tempio, G., Opio, C., & Gerber, P. (2017). Livestock: On our plates or eating at our table? A new analysis of the feed/food debate. *Global Food Security*, 14, 1–8.

Murillo, J. L. S., Osen, R., Hiermaier, S., & Ganzenmuller, G. (2019). Towards understanding the mechanism of fibrous texture formation during high-moisture extrusion of meat substitutes. *Journal of Food Engineering,* 242, 8–20.

National Health and Family Planning Commission of PRC. (2016). *The lists of novel food ingredients and common food ingredients.* http://www.nhc.gov.cn/sps/pztq/201612/712553a5f7554e0e9ec1dfdbcc91e99a.shtml.

Nicolai, T. (2019). Gelation of food protein-protein mixtures. *Advances in Colloid and Interface Science,* 270, 147–164.

Nout, M. J. R., & Rombouts, F. M. (1990). Recent developments in tempeh research. *Journal of Applied Bacteriology,* 69, 609–633.

Osen, R., Toelstede, S., Eisner, P., & Schweiggertweisz, U. (2015). Effect of high moisture extrusion cooking on protein–protein interactions of pea (Pisum sativum L.) protein isolates. *International Journal of Food Science and Technology*, 50, 1390–1396.

Palanisamy, M., Topfl, S., Berger, R. G., & Hertel, C. (2019). Physico-chemical and nutritional properties of meat analogues based on Spirulina/lupin protein mixtures. *European Food Research and Technology,* 245, 1889–1898.

Pe´rez-Llor´ens, J. L. (2020). Microalgae: From staple foodstuff to avant-garde cuisine. *International Journal of Gastronomy and Food Science*, 21, Article e100221.

Peighambardoust, S. H., van der Goot, A. J., Hamer, R. J., & Boom, R. M. (2004). A new method to study simple shear processing of wheat gluten-starch mixtures. *Cereal Chemistry*, 81, 714–721.

Pierobon, S. C., Cheng, X., Graham, P. J., Nguyen, B., Karakolis, E. G., & Sinton, D. (2018). Emerging microalgae technology: A review. *Sustainable Energy & Fuels,* 2, 13–38.

Riaz, M. N. (2004). Texturized soy protein as an ingredient. In R. Y. Yada (Ed.), *Proteins in food processing* (pp. 517–558) (Woodhead Publishing Limited).

Safi, C., Olivieri, G., Campos, R. P., Engelen-Smit, N., Mulder, W., Van den Broek, L., & Sijtsma, L. (2017). Biorefinery of microalgal soluble proteins by sequential processing and membrane filtration. *Bioresource Technology,* 225, 151–158.

Sarwar, G., Christensen, D. A., Finlayson, A. J., Friedman, M., Hackler, L. R., Mackenzie, S. L., Pellett, P. L., & Tkachuk, R. (1983). Inter- and intra-laboratory variation in amino acid analysis of food proteins. *Journal of Food Science Technology,* 48, 526–531.

Sarwar, G., Peace, R. W., Botting, H. G., & Brule, D. (1989). Digestibility of protein and amino-acids in selected foods as determined by a rat balance method. *Plant Foods for Human Nutrition,* 39, 23–32.

Sathish, A., & Sims, R. C. (2012). Biodiesel from mixed culture algae via a wet lipid extraction procedure. *Bioresource Technology,* 118, 643–647.

Schüler, L., Morais, G. E. D., Trov˜ao, M., Machado, A., Carvalho, B., Carneiro, M., Maia, I., Soares, M., Duarte, P., Barros, A., Pereira, H., Silva, J., & Varela, J. (2020). Isolation and characterization of novel Chlorella vulgaris mutants with low chlorophyll and improved protein contents for food applications. *Frontiers in Bioengineering and Biotechnology*, 8, Article e00469.

Schwenzfeier, A., Lech, F. J., Wierenga, P. A., Eppink, M. H. M., & Gruppen, H. (2013). Foam properties of algae soluble protein isolate: Effect of pH and ionic strength. *Food Hydrocolloids,* 33, 111–117.

Smetana, S., Sandmann, M., Rohn, S., Pleissner, D., & Heinz, V. (2017). Autotrophic and heterotrophic microalgae and cyanobacteria cultivation for food and feed: Life cycle assessment. *Bioresource Technology,* 245, 162–170.

Steinfeld, H., Gerber, P., Wassenaar, T., Castel, V., Rosales, M., Rosales, M., & de Haan, C. (2006). *Livestock's long shadow: Environmental issues and options*, FAO publications.

Swanson, D., Block, R., & Mousa, S. A. (2012). Omega-3 fatty acids EPA and DHA: Health benefits throughout life. *Advances in Nutrition*, 3, 1–7.

Szabo, N. J., Matulka, R. A., & Chan, T. (2013). Safety evaluation of whole algalin protein (WAP) from Chlorella protothecoides. *Food and Chemical Toxicology,* 59, 34–45.

Ursu, A. V., Marcati, A., Sayd, T., Sante-Lhoutellier, V., Djelveh, G., & Michaud, P. (2014). Extraction, fractionation and functional properties of proteins from the microalgae Chlorella vulgaris. *Bioresource Technology,* 157, 134–139.

Van Krimpen, M. M., Bikker, P., van de Meer, I. M., van de Peet-Schwering, C. M. C., & Vereijken, J. M. (2013). *Cultivation, processing and nutritional aspects for pigs and poultry of European protein sources as alternatives for imported soybean products.*

Van Mierlo, K., Rohmer, S., & Gerdessen, J. C. (2017). A model for composing meat replacers: Reducing the environmental impact of our food consumption pattern while retaining its nutritional value. *Journal of Cleaner Production,* 165, 930–950.

Vandamme, D., Pontes, S. C. V., Goiris, K., Foubert, I., Pinoy, L. J. J., & Muylaert, K. (2011). Evaluation of electro-coagulation–flocculation for harvesting marine and freshwater microalgae. *Biotechnology and Bioengineering*, 108, 2320–2329.

Waghmare, A., Salve, M. K., Leblanc, J. G., & Arya, S. S. (2016). Concentration and characterization of microalgae proteins from Chlorella pyrenoidosa. *Bioresources Bioprocessing*, 3, 16.

Wang, J., Yang, H., & Wang, F. (2014). Mixotrophic cultivation of microalgae for biodiesel production: Status and prospects. *Applied Biochemistry Biotechnology Advances,* 172, 3307–3329.

Wang, M., Cheng, H., Chen, S., Wen, S., Wu, X., Zhang, D., Yuan, Q., & Cong, W. (2018). Microalgal cell disruption via extrusion for the production of intracellular valuables. *Energy Policy,* 142, 339–345.

Watanabe, F. (2007). Vitamin B12 sources and bioavailability. *Experimental Biology and Medicine,* 232, 1266–1274.

Weindl, I., Bodirsky, B. L., Rolinski, S., Biewald, A., Lotze-Campen, H., Müller, C., Dietrich, J. P., Humpeno¨der, F., Stevanovi´c, M., Schaphoff, S., & Popp, A. (2017).

Livestock production and the water challenge of future food supply: Implications of agricultural management and dietary choices. *Global Environmental Change,* 47, 121–132.

WHOFAO. (1973). *Energy and protein requirements. WHO technical report series.*

Willett, W. C., Rockstrom, J., Loken, B., Springmann, M., Lang, T., Vermeulen, S. J.

About the Authors

Dr. Amin Abbasi (PhD)
Department of Food Science and Technology, Faculty of Nutrition and Food Sciences, Tabriz University of Medical Sciences, Tabriz, Iran
abasi@tbzmed.ac.ir; aminabasi.tbz.med.ac@gmail.com

Professor Hedayat Hosseini (D.V.M.; PhD)
Department of Food Science and Technology, National Nutrition and Food Technology Research Institute, Faculty of Nutrition Science and Food Technology, Shahid Beheshti University of Medical Sciences, Tehran, Iran
hedayat@sbmu.ac.ir

Amin Paydareh (B.S)
Department of clinical Nutrition and Dietetics, National Nutrition and Food Technology Research Institute, Faculty of Nutrition Science and Food Technology, Shahid Beheshti University of Medical Sciences, Tehran, Iran
amin.paidare80@gmail.com

Mahdis Nadian (B.S)
Department of Food Science and Technology, National Nutrition and Food Technology Research Institute, Faculty of Nutrition Science and Food Technology, Shahid Beheshti University of Medical Sciences, Tehran, Iran
mahdisnaadian@gmail.com

Melika Dashab (B.S)
Department of Food Science and Technology, National Nutrition and Food Technology Research Institute, Faculty of Nutrition Science and Food Technology, Shahid Beheshti University of Medical Sciences, Tehran, Iran
melikadashab1400@gmail.com

Irandokht Vakili (B.S)
Department of Food Science and Technology, National Nutrition and Food Technology Research Institute, Faculty of Nutrition Science and Food Technology, Shahid Beheshti University of Medical Sciences, Tehran, Iran
irandokht_vakili@yahoo.com

Fahimeh Ebrahimi (PhD)
Department of clinical Nutrition and Dietetics, National Nutrition and Food Technology Research Institute, Faculty of Nutrition Science and Food Technology, Shahid Beheshti University of Medical Sciences, Tehran, Iran
fahimeh.ebrahimi1373@gmail.com

Index

#

3D printing, xxii, 40, 42, 43, 46, 49, 51, 120, 124, 125, 126, 127, 128, 129, 131, 201, 237, 244, 265, 267, 277, 278, 279, 280, 281, 282, 283, 284, 285, 287, 288, 289, 290, 291, 292, 293, 294, 296, 297, 298, 299, 300, 301, 302, 303

A

adipocytes, 4, 12, 29, 30, 31, 35, 40, 42, 47, 57, 59, 60, 61, 62, 85, 86, 161
animal welfare, 3, 4, 14, 16, 17, 18, 21, 22, 61, 67, 95, 102, 114, 183, 196, 197, 218, 235
animal-friendly urban farms, 11
artificial meat synthesis, 29

B

beef patties, 13, 63, 92
bioavailability, 84, 109, 151, 156, 157, 159, 173, 222, 235, 236, 237, 238, 239, 241, 242, 244, 245, 246, 247, 248, 249, 250, 253, 254, 256, 258, 260, 270, 317, 329, 332, 342
biomaterial engineering, 21
biomechanical, 9
biophysical, 9, 33, 36
biopolymers, 32, 35, 296
biopsy, 6, 7, 10, 14, 29
bioreactor(s),xxi, 3, 10, 11, 12, 21, 22, 23, 43, 44, 45, 46, 67, 74, 76, 85, 97, 102, 103, 130, 170, 171, 172, 333, 340
bovine fetal calf serum, 10
brain development, 3, 337
business-to-consumer, 25

C

Campylobacter, 140, 183
cardio-metabolic health, 151, 153
carrageenan, 13, 43, 193, 201, 286, 295, 296, 299, 300, 302, 311
CCAAT/enhancer-binding protein, 59
cell attachment, 33, 41, 102
cell differentiation, 7, 51, 53, 69, 70, 101
cell engineering, 263
cellular agriculture, xix, 3, 4, 17, 19, 22, 24, 29, 30, 32, 46, 49, 53, 54, 65, 67, 69, 71, 89, 92, 95, 96, 101, 113
cellular meat, 3, 4, 5, 7, 8, 9, 10, 11, 12, 13, 15, 16, 17, 46
cellular meat production, 4, 7, 8, 10, 11, 13, 17
chemical crosslinking, 35, 50
chickpea protein, 215, 217, 257
clean meat, 4, 15, 47, 51, 89, 96
climate change, 53, 134, 136, 143, 147, 199, 206, 211, 225, 231
climate impact, 19
co-culture, 4, 14, 47, 76, 115, 271
commercialization, 3, 21, 24, 45, 65, 88, 95, 98, 102, 167, 168, 202, 227, 273, 287
computational fluid dynamics, 299
connective tissue cells, 4
consumer acceptance, xxi, 3, 4, 47, 55, 64, 73, 75, 87, 89, 92, 93, 97, 102, 113, 127,

133, 135, 145, 146, 147, 193, 197, 199, 204, 208, 211, 213, 223, 234
continuous cultivation, 339
control quality attributes, 84
conventional meat, xix, 3, 4, 13, 17, 62, 63, 64, 67, 75, 76, 77, 78, 79, 84, 85, 86, 87, 105, 172, 193, 202, 218, 228, 237, 245, 278
cost-effective, xxi, 3, 10, 11, 17, 25, 44, 45, 46, 50, 54, 64, 67, 70, 133, 145, 167, 186, 191, 202, 213, 220, 322
critical decision points, 5
cultured meat, xix, xxi, xxiii, 1, 3, 4, 15, 18, 19, 20, 21, 22, 23, 24, 25, 28, 29, 30, 31, 32, 41, 42, 44, 45, 46, 47, 48, 49, 50, 51, 52, 53, 67, 69, 70, 71, 73, 74, 75, 76, 77, 78, 79, 80, 81, 82, 83, 84, 85, 86, 87, 88, 89, 90, 91, 92, 93, 94, 95, 96, 97, 98, 100, 102, 104, 106, 108, 109, 110, 112, 113, 114, 115, 134, 138, 198, 210, 277, 278, 279, 290, 295, 297
cultured meat safety, 86, 95
cytotoxicity, 36, 48, 106, 107

D

decellularization, 36, 37, 47, 49, 50
dedifferentiated FAT, 61
dehydrothermal crosslinking, 34
dietary patterns, 19
differentiation, 6, 7, 8, 9, 10, 15, 22, 27, 29, 30, 31, 32, 33, 43, 46, 47, 48, 50, 54, 56, 57, 58, 59, 60, 61, 62, 67, 71, 85, 96, 99, 100, 103, 224, 290, 339
digestibility, 134, 148, 157, 170, 186, 220, 221, 222, 227, 230, 235, 236, 237, 238, 239, 241, 242, 243, 244, 245, 246, 247, 248, 249, 250, 251, 253, 254, 255, 256, 257, 258, 259, 260, 261, 270, 296, 301, 310, 317, 341

E

E. Coli, 183, 287, 314
electric field, 40, 245, 251, 254, 259, 260, 261
electrospinning, 39, 40, 43, 46, 47, 48, 238, 244, 267
electrostimulation, 9
embryo, 5, 29, 55
embryonic stem cells, 6, 7, 29, 64, 68, 98
environmental manipulation, 12
epithelial-mesenchymal transition, 57
ethical issues, 14, 54, 218
European Food Safety Authority (EFSA), 15, 104, 106, 107, 108, 109, 110, 321, 337, 338
European Union, 15, 25, 65, 85, 111, 338
extracellular matrix proteins, 6, 9, 63
extrusion-based 3D printing, 42, 279, 280

F

faba bean protein, 207, 256
fermentation techniques, 167
fetal bovine serum, 18, 23, 43, 64, 97, 100, 112, 114
fetus, 5, 14
fibro-adipogenic progenitors, 29
fibroblasts, 4, 12, 29, 30, 39, 48, 58, 71, 86
filamentous fungi, 119, 120, 121, 126, 127, 130, 140, 149, 167, 176, 187
film stacking, 39, 43, 45
flavor, 3, 12, 13, 25, 30, 31, 32, 44, 46, 59, 61, 67, 77, 79, 84, 86, 102, 122, 124, 140, 143, 145, 161, 162, 170, 181, 183, 187, 188, 189, 190, 191, 194, 195, 197, 202, 204, 209, 215, 224, 227, 228, 229, 233, 234, 237, 263, 264, 265, 268, 269, 270, 271, 272, 274, 275, 289, 293, 297, 327, 328, 329, 333, 336
FM Technologies, 10
Food and Agriculture Organization (FAO), 66, 96, 99, 103, 105, 136, 146, 147, 148, 182, 222, 231, 314, 316, 338, 342
food manufacturing, 122, 271, 277, 329
food safety, 12, 15, 16, 22, 45, 50, 55, 93, 95, 98, 99, 106, 110, 114, 141, 205, 209,

220, 255, 258, 274, 275, 287, 288, 291, 299, 314, 320, 321
food safety standards, 95
food security, xxi, 3, 17, 53, 67, 145, 147, 150, 182, 183, 218, 230, 313, 341
food systems, xix, xxii, 19, 20, 47, 89, 128, 145, 205, 207, 210, 211, 222, 223, 227, 261, 311, 338
foodborne pathogens, 83, 133, 135, 140, 183
formulation and processing, 197, 200
functional properties, xxi, 70, 189, 191, 192, 200, 213, 214, 215, 216, 222, 223, 227, 229, 231, 232, 233, 234, 240, 241, 242, 243, 244, 249, 250, 253, 254, 255, 256, 257, 258, 260, 261, 264, 271, 273, 274, 275, 282, 283, 284, 286, 298, 328, 337, 339, 342
fungal biomass, 119, 120, 121, 122, 123, 124, 125, 127, 129, 130, 170, 171, 173
fungi-based food products, 119, 122, 125, 126
fungi-based foods, 119, 126, 127, 133

G

gastrointestinal tract, 102, 140, 155, 235, 236, 237, 239
generally recognized as safe (GRAS), 65, 134, 150, 167, 174, 320
glycated hemoglobin, 153
green processing technologies, 235, 246, 254
greenhouse gas emissions, 18, 22, 88, 95, 120, 136, 148, 174, 198, 218, 253, 277
growth factors, 5, 6, 8, 9, 10, 11, 44, 45, 63, 64, 66, 100, 171, 290
growth media, 3, 7, 8, 9, 10, 11, 14, 17, 24, 67

H

hazard and critical control points (HACCP), 11, 12, 15, 16, 104, 111
hepatic progenitor cells, 43
high-density lipoprotein, 139, 153, 187
high-pressure process, 226, 235, 245, 246, 250, 254, 258, 265, 302
human corneal keratocytes, 43
human umbilical vein endothelial cells, 43
hydrophilic colloid, 285, 287

I

induced pluripotent stem cells, 6, 29, 99
industrial scale, 4, 10, 14, 45, 54
innovative technologies, xxii, 66, 277
insulin, 8, 47, 60, 100, 139, 153, 155, 156, 158, 160, 161, 162, 163, 164, 166, 175, 176
iron, 12, 74, 77, 82, 84, 87, 125, 173, 199, 204, 205, 268, 271, 277, 291, 306, 307, 308, 309

K

kinetics testing, 109

L

lab meat, 4
labeling design, 73, 75, 83, 84, 88
labelling, 11, 12, 15, 16, 65, 92
laser assisted methods, 41
laser-induced forward transfer, 41
lentil protein, 250
life cycle, 19, 20, 78, 85, 93, 96, 101, 102, 126, 127, 148, 206, 210, 333, 342
Listeria, 82, 149, 287
longevity, 5, 220
low field nuclear magnetic resonance (LF-NMR), 293, 294, 296, 299, 300
low-density lipoprotein, 139, 153, 162, 175, 187

M

Maillard reaction, 77, 204, 210, 265, 270, 275, 327, 328
market affordability, 138

meat alternatives, xix, xxi, xxii, 19, 24, 45, 50, 70, 74, 92, 93, 131, 133, 138, 144, 174, 183, 186, 188, 193, 196, 198, 199, 200, 202, 209, 210, 216, 218, 224, 225, 237, 287, 311, 333, 338
meat analogues, xxi, 45, 69, 145, 176, 181, 183, 184, 185, 186, 187, 189, 190, 191, 193, 194, 195, 198, 203, 208, 209, 210, 216, 235, 236, 237, 240, 241, 242, 243, 244, 245, 247, 249, 251, 254, 256, 257, 261, 263, 267, 270, 271, 273, 274, 289, 300, 305, 306, 310, 313, 314, 315, 317, 319, 321, 324, 325, 326, 327, 328, 329, 333, 336, 337, 339, 341
meat materials, 277, 278, 279, 280, 281, 285, 288, 290, 293, 294
mechanical stretch, 9, 63
mesenchymal stem cells, 29, 47, 50, 55, 57, 59, 99
mesenchymal stem/stromal cells, 57, 70
microalgae, 122, 129, 134, 138, 140, 150, 256, 313, 315, 316, 317, 318, 319, 320, 321, 322, 323, 324, 326, 327, 329, 330, 331, 332, 333, 334, 335, 336, 337, 338, 339, 340, 341, 342
microcarriers, 47, 113
micromolding technology, 35, 39
microwave heating, 249, 257, 260, 283, 286
modified atmosphere packaging, 79, 81, 91
mouthfeel, 12, 13, 26, 194, 237, 238
muscle cell proliferation, 6
muscle LIM protein, 6, 8
muscle satellite cells, 29, 47
muscle tissue engineering, 53, 70, 74, 299
mycoprotein, xix, xxi, xxiii, 87, 117, 119, 120, 121, 123, 126, 127, 128, 129, 130, 131, 133, 135, 136, 137, 139, 141, 142, 143, 144, 145, 146, 147, 148, 149, 150, 151, 152, 153, 154, 155, 156, 157, 158, 159, 160, 161, 162, 163, 164, 166, 167, 168, 169,170, 171, 172, 173, 174, 175, 176, 187, 188, 196
mycoprotein production, 127, 137, 144, 145, 149, 152, 167, 169, 171, 172
myoblast determination protein, 6, 8
myofibers, 8, 9, 19, 30, 103
myogenesis, 5, 9, 27, 28, 29, 32, 55, 68, 69
myogenic factor, 6, 8
myosatellite, 5, 6, 7, 10, 67, 100
myotubes, 8, 9, 27, 42, 93, 103

N

native ECM, 36, 49
nitrous oxide, 137, 182, 314
nutritional assessment, 167
nutritional benefits, 120, 121, 144, 145, 151, 184, 219, 313
nutritional composition, 25, 46, 65, 176, 305, 309, 337
nutritional value, 22, 119, 121, 127, 140, 149, 152, 159, 165, 181, 183, 197, 202, 203, 213, 217, 220, 221, 222, 227, 230, 254, 259, 265, 269, 278, 289, 291, 309, 310, 315, 317, 320, 332, 334, 336, 342

O

odor, 12, 191, 269, 270, 291, 313, 327
ohmic heating, 245, 251, 259

P

packaging technology, 73
palatability, 13, 30, 101, 181
pea protein, 215, 216, 217, 223, 230, 232, 234, 269, 270, 274, 275, 300
pesticides, 136, 143
physical crosslinking, 34, 35
physiologically-based toxicokinetic modeling, 110
plant-based alternatives, xxi, 3, 175, 181, 199, 253, 266, 314, 315
plant-based meat, xix, xxii, xxiii, 90, 91, 92, 93, 183, 197, 198, 199, 200, 201, 202, 204, 205, 206, 207, 208, 209, 210, 211, 226, 235, 236, 238, 241, 243, 244, 245, 253, 257, 261, 263, 264, 265, 266,

267, 268, 269, 270, 271, 272, 273, 274, 275, 305, 310, 325, 326, 328, 339
plant-based meat alternatives, xxii, 92, 197, 198, 199, 200, 201, 202, 204, 205, 209, 264, 273, 275
plant-based proteins, 152, 183, 185, 213, 215, 216, 219, 220, 221, 222, 223, 226, 227, 228, 230, 233, 234, 235, 238, 245, 247, 248, 252, 265, 272, 274, 305, 314
pluripotency, 29, 58, 68
pluripotent stem cells, 6, 29, 48, 55, 57, 67, 68, 71, 72, 99
polymerase chain reaction, 142
polyunsaturated fatty acids, 77, 173, 187, 319
post-mortem carcass, 13
preservative, 8
processed foods, 125, 305
proliferator-activated receptor gamma, 59
protein source, xxi, 3, 85, 87, 120, 126, 129, 131, 133, 134, 145, 150, 151, 152, 156, 157, 158, 159, 160, 163, 170, 175, 176, 184, 186, 198, 202, 208, 209, 215, 220, 221, 227, 229, 230, 234, 235, 237, 243, 244, 245, 253, 254, 255, 259, 265, 270, 313, 314, 315, 316, 317, 325, 336, 342
pulsed electric field, 235, 246, 252, 253, 254, 258, 261

R

reactive nitrogen species, 137
regulation, 11, 12, 15, 16, 19, 22, 27, 59, 65, 84, 85, 109, 111, 115, 139, 160, 162, 171, 209, 280, 311, 320, 335, 337
resident muscle stem cells, 29
retro-nasal cavity, 12
risk assessment, 15, 95, 98, 104, 105, 107, 108, 110, 112, 113, 321, 338

S

safe to eat, 32
safety, xix, xxi, xxii, 4, 15, 16, 24, 45, 50, 53, 54, 55, 65, 67, 73, 75, 78, 84, 85, 88, 90, 93, 95, 97, 101, 102, 104, 105, 106, 107, 109, 110, 113, 114, 135, 139, 140, 142, 147, 150, 166, 167, 174, 202, 205, 209, 226, 245,249, 255, 258, 273, 274, 275, 287, 289, 313, 314, 320, 329, 331, 342
Salmonella, 80, 82, 90, 140, 183, 314
scaffold manufacturing, 21, 46, 114
scaffold-network, 32
scaffolds, xxi, 7, 21, 22, 23, 24, 28, 31, 32, 33, 34, 36, 39, 40, 41, 42, 45, 46, 47, 48, 49, 50, 51, 52, 54, 61, 62, 64, 66, 68, 75, 76, 86, 102, 103, 113, 290, 295
selenium, 84, 270
self-assembly, 38, 50
self-renewing capacity, 5
serum-free medium, 51, 100, 113
short-chain fatty acids, 139, 154, 173
single-cell proteins, 134, 140, 141
skeletal muscle adaptation, 151, 156
sodium benzoate, 8
sodium dodecyl sulfate, 37, 196
solid-state fermentation (SSF), 167, 168, 169, 170, 171, 172, 175, 176, 271, 274
solvent-casting method, 33, 34
soy protein, 47, 68, 103, 113, 148, 164, 175, 181, 184, 185, 186, 188, 189, 191, 195, 201, 208, 209, 210, 214, 216, 223, 224, 227, 229, 230, 231, 232, 233, 234, 238, 240, 241, 242, 245, 251, 252, 253, 257, 266, 267, 273, 275, 282, 295, 298, 300, 317, 318, 339, 341
soy protein concentrates, 185, 227
stem cell(s), xxi, 4, 5, 6, 7, 8, 9, 10, 11, 18, 22, 29, 31, 48, 53, 54, 55, 56, 57, 58, 59, 60, 61, 63, 65, 66, 67, 68, 70, 71, 74, 75, 76, 91, 97, 98, 101, 111, 114, 115, 290
stromal vascular fraction, 60, 71
submerged fermentation, 128, 149, 167, 168, 169, 171
surface culture method, 147, 168, 172
sustainability, xxi, 3, 19, 24, 25, 52, 67, 71, 73, 75, 88, 90, 91, 92, 115, 127, 142,

145, 149, 151, 152, 197, 198, 202, 207, 209, 213, 218, 219, 228, 229, 244, 259, 274, 275, 278, 287, 292, 300, 313, 330, 331, 332, 333
sustainability in food production, 197
sustainable food production, 119
sustainable protein, 133, 145, 146, 152, 160, 163, 176, 181, 202, 260, 332
synthetic biology, xxii, 263, 264, 271, 272

T

taste, 12, 18, 22, 25, 32, 75, 86, 103, 113, 122, 135, 138, 185, 187, 191, 193, 194, 198, 202, 235, 264, 268, 270, 271, 278, 288, 291, 327
tenderness, 13, 25, 26, 31, 194, 293, 294, 328
texture, 3, 12, 13, 22, 26, 27, 30, 31, 32, 40, 41, 42, 46, 62, 63, 67, 73, 75, 77, 84, 86, 113, 121, 122, 123, 125, 126, 128, 135, 138, 143, 161, 162, 181, 183, 186, 187, 188, 191, 193, 194, 195, 197, 198, 200, 201, 202, 203, 204, 208, 223, 224, 235, 237, 238, 255, 263, 264, 266, 267, 268, 272, 280, 283, 286, 288, 289, 291, 297, 298, 299, 301, 310, 314, 315, 318, 325, 328, 336, 341
textured vegetable proteins, 184
texturization technology, 181
thyroid hormone, 8
tissue engineering, xix, 18, 20, 21, 22, 29, 32, 35, 36, 38, 39, 40, 41, 44, 45, 47, 49, 50, 51, 52, 54, 61, 62, 69, 70, 71, 74, 76, 88, 89, 95, 96, 112, 113, 257, 290, 299
tissue engineering technology, 32, 54, 74
tissue-engineered scaffold, 28, 33
transglutaminase, 238, 267, 274, 275, 282, 283, 286, 296, 297, 298, 300, 301, 303
Triton-X-100, 37

U

ultraviolet laser, 41
United States Department of Agriculture (USDA), 15, 85, 112, 115, 119, 131, 165, 192, 306
urbanization, 4, 151, 181, 313

V

vitamin B_{12}, 74, 84, 319, 342

W

waste, 8, 10, 11, 17, 32, 37, 45, 63, 87, 93, 125, 133, 135, 137, 142, 149, 168, 170, 175, 177, 203, 220, 277, 278, 292, 331, 333
water holding capacity, 207, 249, 293, 294, 302
wheat gluten, 101, 181, 185, 186, 189, 196, 201, 208, 215, 222, 223, 229, 273, 300, 317, 325, 341
willingness to pay, 64, 69, 84, 86

Z

zinc, 59, 84, 173, 199, 270, 277